Springer Tracts in Electrical and Electronics Engineering

Springer Tracts in Electrical and Electronics Engineering (STEEE) publishes the latest developments in Electrical and Electronics Engineering - quickly, informally and with high quality. The intent is to cover all the main branches of electrical and electronics engineering, both theoretical and applied, including:

- Signal, Speech and Image Processing
- Speech and Audio Processing
- Image Processing
- Human-Machine Interfaces
- Digital and Analog Signal Processing
- Microwaves, RF Engineering and Optical Communications
- Electronics and Microelectronics, Instrumentation
- Electronic Circuits and Systems
- Embedded Systems
- Electronics Design and Verification
- Cyber-Physical Systems
- Electrical Power Engineering
- Power Electronics
- Photovoltaics
- Energy Grids and Networks
- Electrical Machines
- Control, Robotics, Automation
- Robotic Engineering
- Mechatronics
- Control and Systems Theory
- Automation
- Communications Engineering, Networks
- Wireless and Mobile Communication
- Internet of Things
- Computer Networks

Within the scope of the series are monographs, professional books or graduate textbooks, edited volumes as well as outstanding PhD theses and books purposely devoted to support education in electrical and electronics engineering at graduate and post-graduate levels.

Review Process

The proposal for each volume is reviewed by the main editor and/or the advisory board. The books of this series are reviewed in a single blind peer review process.

Ethics Statement for this series can be found in the Springer standard guidelines here https://www.springer.com/us/authors-editors/journal-author/journal-author-helpdesk/before-you-start/before-you-start/1330#c14214

Sanjeev Kumar Raghuwanshi · Santosh Kumar ·
Ritesh Kumar

Geometric Feature-Based Fiber Optic Surface Plasmon Resonance Sensors

Springer

Sanjeev Kumar Raghuwanshi
Department of Electronics Engineering
Indian Institute of Technology (Indian School of Mines) Dhanbad
Dhanbad, Jharkhand, India

Ritesh Kumar
Department of Electronics and Communication Engineering
Shri Phanishwar Nath Renu Engineering College
Araria, Bihar, India

Santosh Kumar
Shandong Key Laboratory of Optical Communication Science and Technology
School of Physics Science and Information Technology
Liaocheng University
Liaocheng, Shandong, China

ISSN 2731-4200 ISSN 2731-4219 (electronic)
Springer Tracts in Electrical and Electronics Engineering
ISBN 978-981-99-7296-8 ISBN 978-981-99-7297-5 (eBook)
https://doi.org/10.1007/978-981-99-7297-5

This Springer imprint is published by the registered company Springer Nature Singapore Pte Ltd.
The registered company address is: 152 Beach Road, #21-01/04 Gateway East, Singapore 189721, Singapore

Preface

In the past three decades, a wide range of sensing applications based on optical fiber interferometric sensors have arisen. Wireless communication systems, satellite communication, advanced radar systems, healthcare gadgets, and electronic warfare are among the uses. In today's society, healthcare issues pose the greatest global challenge. Recent years have seen a dramatic increase in the need for speedy, dependable, accurate, and cheap detection methods and equipment in the biomedical industries, especially in less developed regions. Highly sensitive and accurate sensors for the detection of biological analytes have been realized using plasmonic-optical fiber sensors, which have shown outstanding capabilities in the field of health monitoring and diagnosis. The interferometric surface plasmon sensors offer label-free detection and real-time monitoring of molecular binding events. By manipulating the plasmon modes that such nanostructured metals support, a wide variety of intriguing optical features and functionality have been achieved by a variety of interferometric techniques, setups, and detecting schemes. New sensing methodologies and systems have been developed due to the sensitivity of plasmonic devices to variations in their local dielectric environment. Nanoscience and nanotechnology have largely been responsible for the extraordinary speed with which plasmonic nanostructures, thin films, and very sensitive optical characterization techniques have all been realized. Our book is based on the integration of surface plasmon resonance sensors with a variety of interferometric structures for the quantitative detection of chemical and biological substances in the biomedical domain. The most common methods for detection with plasmonic-optical fiber sensors and the many medical uses for this technology are detailed in this book. For increased sensitivity to biomolecules, the surface plasmon resonance sensor is presented in depth, including the geometrical optics theory correlating taper fiber sensor types, different shapes of the fiber, multi-channel sensing, and specialty fibers.

Organization of the Book

In the last three decades, surface plasmon resonance sensors have outperformed interferometric methods. Surface plasmon resonance (SPR) sensors provide chemical and biological detection benefits over traditional methods. New developments in plasmonic sensing techniques, detecting strategies, realization, setup, components used, and applications are discussed in this book. The plasmonic phenomenon and its application in optical fiber-based sensing are discussed in depth in the book's eight chapters.

In Chap. 1, different surface plasmon sensors and detection methods have been described. Sensitivity, range of the input signal, precision, highest resolution, accuracy, offset error, linear characterization, hysteresis loss, repeatability of the approach, and drift due to electronic components are essential parameters for sensing system performance analysis. Before discussing surface plasmon resonance structures and detection methods, Mach–Zehnder, Michelson, Sagnac loop, and Fabry–Perot interferometers are discussed. After explaining the operating principle, coupling setup, detecting technique, and design parameters, the surface plasmon resonance application is discussed.

Chapter 2 describes the geometrical optics theory of the taper fiber optic sensor. Correlating taper fiber sensor models with simulation results is essential for a thorough investigation. After highlighting the taper radii change, the taper angle and length were established. The model's propagation and tapering have been derived. The taper angle is crucial for sensing and transmission. Core radii and sensing length determine the taper angle. The taper angle is negligible, and taper ratio increases as taper angle decays and the taper radii decrease. Penetration depth and sensitivity increase as the incidence angle approaches the critical angle. Depth of penetration is crucial for sensing capabilities and rises with operating wavelength.

Chapter 3 discussed the U-type fiber optic sensor and how it can be used for SPR sensing. First, the basic ideas of U-type fiber optic sensors are introduced. U-type fiber optic SPR sensors have been the focus of current research and development, and this study examines their architecture and operating principles. U-type fiber optic sensor device design is also covered in this chapter. The research also explores the use of U-type fiber optic SPR sensors in other contexts, highlighting their adaptability and promise. At the end of each chapter is a brief overview that recaps the most important information presented there.

Chapter 4 deals with cascaded fiber optic surface plasmon resonance sensors. Multi-analyte/multi-channel sensing is possible with cascaded SPR sensors. This chapter covers a modified hetero-core structure fiber-based dual-channel SPR refractive index sensor. The typical hetero-core structure fiber may be shaped into a circular truncated cone with a varied polishing angle to change the resonance wavelength range. Using wavelength division multiplexing, two cascaded fiber optic surface plasmon resonance sensors may test two substances simultaneously. A cascaded multi-analyte SPR sensor is used to compare coagulant and anticoagulant concentrations for illness diagnosis.

The sensing concepts, device construction, and applications of symmetric and asymmetrically coated fiber optic SPR sensors are discussed in Chap. 5. The chapter starts out by talking about symmetric fiber optic SPR sensors and the advantages they have in terms of signal stability and reproducibility. In order to achieve dependable and precise sensing performance, it highlights the importance of using correct probe preparation techniques. The chapter continues with a discussion of asymmetrically coated (half-coated) fiber optic SPR sensors. The advantages and distinguishing sensing qualities of these sensors are discussed, including higher sensitivity and selectivity due to increased contact between the analyte and the coating. This chapter delves deeper into the coating strategies and functionalization processes for asymmetrically coated fiber optic SPR sensors in preparation for the sensing probe. It stresses the significance of achieving the desired sensor performance by optimizing probe production methods. In this chapter, we focus on one such application—the use of asymmetric fiber optic chemical detectors.

Analytical models from the finite element approach are presented in Chap. 6 on the D-type photonic crystal fiber performance used in SPR-based refractive index sensing. Sensitivity enhancements for SPR sensors are made possible by the coupling of SPPs made possible by hyperbolic metamaterial (HMM). HMM and 2D materials can be used to enhance the detection sensitivity of biosensors. A graphene monolayer with a D-shaped plastic optic fiber is a graphene monolayer with a D-shaped plastic optic fiber (a G/HMM/D-POF SPR sensor is shown). Based on a 3D $AuAl_2O_3$ HMM composite structure and a graphene sheet, an optical fiber SPR biosensor is presented.

In chapter seven, interferometric optical sensor concepts and applications are discussed. Interferometric sensors include Mach–Zehnder, Michelson, Sagnac loop, and Fabry–Perot types. These sensors measure temperature, pressure, stress, strain, etc. Interferometers use fibers' optical pathways to interfere with two beams. Interferometric configurations need beam splitting and beam combining. One optical route should be adjustable to external changes. Interferometric signal variations may reveal the target's wavelength, intensity, phase, frequency, and bandwidth. These modifications improve device sensitivity, accuracy, and dynamic range. Miniaturized fiber optic interferometers enable micro-scale applications. To replace bulky fiber optic components like combiners, beam splitters, and objective lenses, innovative fiber-scale devices are designed. SPR sensing accurately measures the refractive index of materials close to metal, unlike optical interferometers. Given the rising need to detect and analyze chemical and biological components in medicine, environmental monitoring, biotechnology, pharmaceuticals, and food monitoring, SPR sensor technology has great potential. This chapter's last part discusses interferometric SPR sensors' construction, content, performance, and application.

In the last chapter of this book, a review of the application of geometric-based SPR sensors is discussed. SPR sensors are used in chemical sensing, biomolecules, gas and liquid detection, medical diagnostics, chemical reaction rate monitoring, and biomolecular interaction. Due to their sensitivity and efficacy, PCF-SPR sensors have grown in popularity. These sensors outperform uncoated and prism-based optical sensors in sensitivity, selectivity, reaction time, recovery time, and repeatability.

PCF-SPR sensors with hollow cores, circular patterns, multiple sensing rings, quasi-periodic patterns, square arrays, D shapes, D-shape dual cores, and other designs can be used in a variety of chemical, biological, gas, and liquid detection applications. In medicine and chemistry, the PCF-SPR sensor may detect medicines, analytes, antigens, and antibodies. PCF-SPR sensors with varied structures may detect analyte refractive index variations. To increase sensitivity and range, change the air hole form, number, placement, and size. The authors thank Springer Nature, in particular the editor, who meticulously edited the book for publication.

Dhanbad, India — Sanjeev Kumar Raghuwanshi
Liaocheng, China — Santosh Kumar
Araria, India — Ritesh Kumar

Acknowledgments

The authors would like to express their heartfelt appreciation to the editorial board of Springer Nature for their meticulous efforts in editing and publishing this book, "Geometric Feature Based Fiber Optic Surface Plasmon Resonance Sensors." Their expertise and guidance have been invaluable in shaping the content and ensuring its quality.

In addition, the authors extend their deepest gratitude to the leaders of their institution for their unwavering support and encouragement throughout the writing and publication process. Their visionary leadership and commitment to fostering research and academic excellence have been instrumental in the realization of this work.

S. K. Raghuwanshi, currently serving as an Associate Editor of IEEE Sensors Journal, would like to acknowledge the support received from his wife, Indumati Raghuwanshi, and daughter, Navya Raghuwanshi. Their patience, understanding, and unwavering support during the writing of this book have been truly invaluable. He would also like to extend his heartfelt appreciation to his Ph.D. students, Vikash, Azhar, Tauseef, and Anas, for their invaluable assistance in the compilation of the book.

S. K. Raghuwanshi expresses his gratitude for the support provided by the project titled "Design and Development of Deployable Thin Film Based Evanescent Field Sensor to Check the Quality of Food from Adulteration." This project, funded by the Council of Scientific and Industrial Research (CSIR) India with Sanction No. 70(0077)/19/EMR-II and IIT (ISM) Project No. CSIR (32)/2019-2020/663/ECE, was sponsored by the Council of Scientific and Industrial Research (CSIR)-Central Scientific Instruments Organisation (CSIO) Chandigarh-160030. Furthermore, this work has also been carried out under the Research Grant with project reference no. SCP/2022/000271 and IIT (ISM) project no. DST(SERB)(356)/2022-2023/955/ ECE, funded by the Science and Engineering Research Board, Department of Science and Technology (SERB-DST), Government of India. The project titled "Design of a Webserver-Based Hybrid Physiological Sensor with Optical Cloth for Real-Time Health Specialist Care" aims to advance the field of biosensors and has provided valuable insights for this book.

Santosh Kumar would like to express his heartfelt thanks to his wife, Dr. Ragini Singh, and son, Ayaansh Singh, for their unwavering support and cooperation. Their belief in his work and constant motivation have been a source of inspiration throughout the journey. Additionally, the authors extend their deepest appreciation to their loving parents for their continuous encouragement and unwavering support. S. Kumar would like to extend gratitude to the Double-Hundred Talent Plan of Shandong Province, China, as well as Liaocheng University (Grant Number: 318052341). Additionally, the support received from the Science and Technology Support Plan for Youth Innovation of Colleges and Universities in Shandong Province, China (Grant Number: 2022KJ107) is greatly appreciated. S. Kumar expresses sincere gratitude to Prof. Zhang Bingyuan and Prof. Minghong Wang for their unwavering motivation and support throughout this endeavor.

Ritesh Kumar gratefully acknowledges the support from Shri Phanishwar Nath Renu Engineering College, Araria, Department of Science and Technology, Bihar. He expresses his deep sense of appreciation to his wife, Megha Kumari, and daughter, Riyanshi Gupta, for their constant support and cooperation. Ritesh Kumar is also grateful to his loving parents, Shri Ashok Kumar and Smt. Rita Devi, as well as his younger brother, Sumit Kumar, and all other family members, for their unwavering encouragement and support throughout the journey of writing this book.

Sanjeev Kumar Raghuwanshi
Santosh Kumar
Ritesh Kumar

Contents

1 Fiber Optic SPR Sensor—Past, Present, and Future 1
1.1 Introduction to Fiber Optic SPR Sensor 1
1.2 Exploring Some Concepts 2
1.2.1 Interference 4
1.2.2 Basic Terms 4
1.2.3 Interferometric Structures 6
1.3 Some Related Background Ideas 13
1.3.1 Coupling Configuration in SPR Sensing 13
1.3.2 Detection Scheme in SPR Sensing 17
1.3.3 Design Parameters 22
1.4 Applications 30
1.4.1 Temperature 31
1.4.2 Acoustics 31
1.4.3 Magnetic Field 33
1.4.4 Electric Field 33
1.5 About This Book 35
1.6 Summary 38
References 39

2 Taper Fiber-Based SPR Sensor 43
2.1 Introduction 43
2.2 Basics Concepts of Taper Fiber 44
2.2.1 Evanescent Field 44
2.2.2 Penetration Depth 44
2.2.3 Single-Mode Fibers (SMF-28) 45
2.3 Adiabaticity Criteria: Linear Taper Model 47
2.4 Key Components Required for Optical Fiber Communication 47
2.4.1 Optical Source (Laser) 48
2.4.2 Optical Spectrum Analyzer (OSA) 48
2.4.3 Fiber Couplers 48
2.5 Taper Fiber Sensor 49

2.6 Geometrical View of Tapered Fiber and Sensing Principle 49
2.6.1 Full-Section Geometrical View of Tapered Sensing Region 49
2.6.2 Half-Section Geometrical View of Tapered Sensing Region 50
2.6.3 Taper Angle 51
2.7 Taper Profile Technique 52
2.7.1 Variation of Angle Inside the Tapered Region 53
2.7.2 Number of Reflections Inside Tapered Region 55
2.7.3 Concept of Penetration Depth 56
2.8 Transfer Matrix Method and Transmission 58
2.9 Taper Fiber-Based Sensors 64
2.10 Summary 67
References 67

3 U-shape Fiber Optic-Based SPR Sensor 71
3.1 Introduction 71
3.1.1 Fiber Optic-Based SPR Sensors 72
3.1.2 Probe Design (Material and Geometry) 74
3.2 Basic Concepts of U-type Fiber Optic Sensor 76
3.3 U-type of Fiber Optic SPR Sensor Structure and Operation Principle 80
3.4 Recent Advances in U-type Fiber-Based Plasmon Sensors 86
3.5 Probe Design of U-type Fiber Optic Sensor 91
3.6 Application of a U-type Fiber Optic Surface Plasmon Resonance Sensor 95
3.7 Future Prospects of U-shaped SPR Sensor 99
3.8 Summary 102
References 103

4 Cascaded Fiber Optic SPR Sensor 107
4.1 Introduction 107
4.2 Basic Theory and Sensor Structure 111
4.3 Fabrication Process of the Cascaded Fiber Optic SPR Sensor 117
4.4 Principle and Method 120
4.5 SPR Enhancement with Cascaded Structure 122
4.6 Application of Cascaded Fiber Optic SPR Sensor 123
4.7 Future Prospects 128
4.8 Summary 130
References 130

5 Symmetric Versus Asymmetric Coated (Half Coated) Fiber Optic SPR Sensor 133
5.1 Symmetric Fiber Optics SPR Sensor 133
5.2 Asymmetric Coated (Half Coated) Fiber Optics SPR Sensor 138
5.3 Sensing Probe Preparation 143

5.4 Asymmetric Fiber Optic Chemical Detector ... 150
5.5 Summary ... 156
References ... 156

6 D-shape Fiber Structure-Based SPR Sensor ... 159
6.1 Introduction ... 159
6.2 SPR Sensor Based on Nanophotonics and Plasmonics ... 161
6.2.1 Nano Film D-shaped Plasmonic Fiber Sensor ... 163
6.2.2 Nano Film with a D-shaped PCF Plasmonic Sensor ... 165
6.2.3 Experimental Setup of D-shape Plasmonic Fiber Optic Sensor ... 167
6.3 Metamaterials-Based D-type of Fiber Optics SPR Sensor ... 167
6.3.1 A Metamaterial-Based Optical Fiber Sensor's Design and Implementation ... 171
6.3.2 Hyperbolic Metamaterial Used in Side-Polished Few-Mode Fiber Sensor ... 172
6.3.3 Design of a D-shaped Plastic Optical Fiber Sensor Based on Graphene and Hyperbolic Metamaterial ... 173
6.3.4 Metamaterial Based D-shape Plastic Optical Fiber Sensor for DNA Detection ... 174
6.3.5 Experimental Setup of Graphene and Hyperbolic Metamaterial-Based D-shape Plastic Optical Fiber Sensor ... 174
6.4 Graphene Injected into the D Type of Fiber Optics SPR Sensor ... 175
6.4.1 Geometrical Design Variations for Graphene-Based Sensors ... 176
6.4.2 Multilayer Sensing Mechanism of a PCF-Based SPR Sensor ... 177
6.4.3 Experimental Setup for Liquid Detection Using D-shaped SPR Sensor ... 178
6.4.4 Analysis of the Silver-Graphene-Coated PCF-SPR Sensor Numerically ... 180
6.5 Summary ... 182
References ... 183

7 Interferometric-Based SPR Sensors ... 189
7.1 Introduction ... 189
7.2 Interferometers with Various Fabrication Techniques and Their Detection Capability ... 191
7.2.1 Mach–Zehnder Interferometer ... 191
7.2.2 Michelson Interferometer ... 195
7.2.3 Sagnac Loop Interferometer ... 201
7.2.4 Fabry–Perot Interferometer ... 206
7.3 Interferometric SPR Sensors ... 212
7.4 Mach–Zehnder Interferometer SPR Sensors ... 212

7.4.1 Various Fibers in MZI Configuration 212
7.4.2 Tunable Coupling in MZI 214
7.4.3 Vertical Plasmonic MZI 215
7.4.4 Polarization Controlled MZI 218
7.5 Michelson Interferometer SPR Sensors 222
7.5.1 Prism-Based Pasmon Michelson Interferometer 222
7.5.2 Specialty Fiber-Based Plasmon Michelson Interferometer 223
7.5.3 Coherence Emission-Based Plasmon Michelson Interferometer 224
7.6 Sagnac Interferometer SPR Sensors 227
7.7 Fabry–Perot Interferometer SPR Sensors 231
7.8 Future Prospects 236
7.9 Summary 238
References 238

8 Application of Geometric-Based SPR Sensors 245
8.1 Introduction 245
8.2 SPR Based Sensor with a Different Geometrical Structure of PCF 247
8.3 Photonic Crystal Fiber of Various Geometries 248
8.3.1 Rectangular Hollow-Core PCF 248
8.3.2 Multiple Sensing Ring Photonic Crystal Fiber 249
8.3.3 Circular-Pattern Photonic Crystal Fiber 249
8.3.4 Quasi Photonic Crystal Fiber 251
8.4 Photonic Crystal Fiber of Different Geometries Based on SPR Sensors 252
8.4.1 Hollow-Core PCF-Based SPR Sensor 252
8.4.2 Square Array PCF-Based SPR Sensor 252
8.4.3 D-shape PCF-SPR Sensor 253
8.4.4 D-shaped PCF Polished on One Side SPR-Based Sensor 254
8.4.5 Surface Plasmon Resonance Sensor with D-shape Dual-Core PCF 255
8.4.6 PCF with a Dual Symmetrical Eccentric-Core Configuration-Based SPR Sensor 257
8.4.7 PCF-Based SPR Sensors Designed with Circular, Square, and Elliptical Air Holes 258
8.5 Experimental Set-up of Hollow Dual-Core PCF-SPR-Sensor 258
8.6 Numerical Investigation of D-shaped PCF-SPR Sensors with Various Material Coatings 261
8.7 Chemical Detection and Concentration Measurements PCF-Based SPR Sensors 263
8.7.1 Design and Analysis of Chemical Detection Sensor 265

8.7.2 Experimental Set-up for Chemical Substance Detection and Testing 265
8.8 Biological Detection/Biomedical Diagnosis Geometric-Based SPR Sensors 267
8.8.1 Design and Modeling of Biological Detection of Different Structures 270
8.9 Liquid and Gas Detection of Photonic Crystal-Based SPR Sensors 273
8.9.1 Design and Analysis of Liquid and Gas Sensor 275
8.10 Future Prospects 277
8.11 Summary 278
References 280

About the Authors

Sanjeev Kumar Raghuwanshi received his bachelor's degree in electronic and instrumentation engineering from S.G.S.I.T.S. Indore, Madhya Pradesh, India, and his master's degree in solid state technology from the Indian Institute of Technology, Kharagpur, in August 1999 and January 2002, respectively. Since July 2009, he has obtained a Ph.D. degree in the field of optics from the Department of Electrical Communication Engineering of the Indian Institute of Science, Bangalore, India. He is an associate professor in the Electronics Engineering Department of the Indian Institute of Technology (ISM) in Dhanbad, India. He was a Post-Doctoral Research Fellow from 2014 to 2015 at the Instrumentation and Sensor Division, School of Engineering and Mathematical Sciences, City University London, Northampton Square, London. He has published more than 100 peer-reviewed and indexed International SCI Journal papers in the last 10 years. He has published four books on the contemporary optical fiber domain. In the last 5 years, 4 Indian patents have been filed and published. He has been sanctioned and executed several R&D projects from different central government funding agencies, including the Department of Atomic Energy, the Government of India (GOI), the Indian Space Research Organization (ISRO), the Council of Scientific and Industrial Research (CSIR), etc. He served as a reviewer for journals like IEEE Transactions on Measurement and Instrumentation, IEEE Sensor Journal, IEEE Photonics Technology Letter, and IEEE Quantum Electronics, to mention a few. He is an editorial board member and reviewer for several Indian journals. He received the Erasmus Mundus Scholarship for his postdoctoral study. He is a Fellow of the Optical Society of India (OSI), a Life Member of IETE, a Senior Member of IEEE (USA), and a Life Member of the International Academy of Physics Sciences.

Santosh Kumar is a highly accomplished researcher and Professor with a Ph.D. degree from the Indian Institute of Technology (Indian School of Mines) in Dhanbad, India. Currently based at Liaocheng University, China, he specializes in optical fiber sensors, nano, and biophotonics, photonic and plasmonic devices, waveguides, interferometers and internet of things. Throughout his career, Dr. Kumar has successfully supervised twelve M.Tech. dissertations and mentored six Ph.D. candidates. His contributions to the field include the publication of over 325 research articles in prestigious SCI journals and conferences, with more than 5600 citations and an h-index of 45. His work has been featured in renowned journals such as Biosensors and Bioelectronics, Biosensors, Journal of Lightwave Technology, Optics Express, Optics Letters, Applied Physics Letters, ACS Applied NanoMaterials, and various IEEE Transactions. He has been recognized as one of the world's top 2% scientists by Stanford University during last three consecutive years. Driven by his commitment to advancing knowledge, he has presented his work at conferences held in China, India, Belgium, and the USA. Dr. Kumar is the author of there scholarly books, "2D Materials for Surface Plasmon Resonance-Based Sensors" (CRC Press, 2021), "Optical Fiber-Based Plasmonic Biosensors: Trends, Techniques, and Applications" (CRC Press, 2022), and Nanotechnology Advancement in Agro-Food Industry (Springer, 2023). His expertise is highly sought-after, reflected in his extensive reviewing contributions for over 1900 SCI journals published by renowned publishers such as IEEE, Elsevier, Springer, OPTICA, SPIE, Wiley, ACS, and Nature. Recognized for his achievements, Dr. Kumar is a Fellow of SPIE and a Senior Member of IEEE, SPIE, and OPTICA. He serves as an OPTICA Traveling Lecturer and holds the esteemed position of Chair of the Optica Optical Biosensors Technical Group. He has delivered numerous invited speeches and serves as a session chair for IEEE conferences. Additionally,

Dr. Kumar has made significant editorial contributions, serving as an Associate Editor for journals such as IEEE Sensors Journal, IEEE Internet of Things, IEEE Access, Frontiers of Physics, Biomedical Optics Express, and IEEE Internet of Things. Dr. Santosh Kumar's contributions extend beyond academia. He is currently a Shandong Provincial Distinguished Foreign Expert by the Department of Science and Technology of Shandong Province, China. With his vast expertise and substantial contributions to the field, Dr. Santosh Kumar continues to make remarkable strides in research, innovation, and scholarly leadership.

Ritesh Kumar received his M.Tech. and Ph.D. degrees from IIT (ISM) Dhanbad, Dhanbad, India. He was a postdoctoral fellow at IIT Bhubaneshwar. He served as an assistant professor (TEQIP faculty) for three years at Madan Mohan Malaviya University of Technology, Gorakhpur. He is currently an assistant professor with the Department of Electronics and Communication Engineering at Shri Phanishwar Nath Renu Engineering College, Araria, under the Department of Science and Technology, Govt. of Bihar, India. He has published more than 20 research articles in international SCI journals, conferences, and book chapters. He has one patent grant and two published patents. He is a Life Fellow Member of ISTE. He is capable of effective teaching and conducting high-quality research in the fields of microwave photonics, electronics, and communications engineering, as well as waveguides and interferometers. He has achieved two best paper awards at the Springer conference.

Abbreviations

AOTF	Acousto-Optic Tunable Filter
ASE	Amplified Spontaneous Emission
BIRI	Bend-Induced RI
BP	Balanced Photodetector
BPM	Beam Propagation Method
BSA	Bovine Serum Albumin
CCD	Charge Coupled Device
CEI	Complete Evaluation Indicator
CFU	Colony-Forming Units
CL	Collimation Lens
C-PML	Circular Perfectly Matched Layer
CPS	Coupled Surface Plasmons
CS	Chitosan
CVD	Chemical Vapor Deposition
DCF	Dispersion Compensation Fiber
DFF	Dispersion Flattened Fiber
DFPI	Dual-Cavity Fabry–Perot Interferometric
DLUWT	Double-Layer Uniform-Waist Tapered Fiber
DSF	Dispersion Shifted Fiber
DSF	Dispersion shifted Optical Fiber
DT	Dual Tapered
EA	Effective Area
EF-SPR	Electric-Field Assisted Surface Plasmon Resonance
EMT	Effective Medium Theory
ERI	Effective Refractive Index
EW	Evanescent Waves
EWA	Evanescent Wave Absorption
FBG	Fiber Bragg Grating
FDTD	Finite Difference Time Domain
FEM	Finite Element Method
FFT	Fast Fourier Transform

FMF	Few Mode Fiber
FOM	Figure of Merit
FPI	Fabry–Perot Interferometer
GNP	Gold Nanoparticles
HF	Hollow Fiber
HMM	Hyperbolic Metamaterial
HNDA	Hexagonal Nano Dome Arrays
HWP	Half-Wavelength Plate
IgG	Anti-Human Immunoglobulin G
IMOS	In-Fiber Multispectral Optical Sensing
IR	Iris Diaphragm
LbL	Layer by Layer
LC	Liquid Cell
LPFBG	Long-Period Fiber Gratings
MI	Michelson Interferometer
MIM	Metal Insulator Metal
MMF	Multimode Fiber
MSM	Multimode Single-Mode Multimode
MZI	Mach–Zehnder Interferometer
NA	Numerical Aperture
NCF	No Core Fiber
OSA	Optical Spectrum Analyzer
PAA	Polyacrylic Acid
PBS	Polarization Beam splitter
PC	Polarization Controller
PCA	Principal Component Analysis
PCF	Photonic Crystal Fiber
PCS	Plastic-Clad Silica
PCW	Photonic Crystal Waveguide
PD	Photodetector
PDMS	Polydimethylsiloxane
PE	Polyelectrolyte
PMF	Polarization Maintaining Fiber
PML	Perfectly Matched Layer
POF	Plastic Optical Fiber
PSP	Propagating Surface Plasmon Resonance
PZT	Piezo Transducer
RI	Refractive Index
RIE	Reactive Ion Etching
SCF	Solid Core Fiber
SC-PCF	Solid-Core Photonic Crystal Fiber
SH	Sulfhydryl
SI	Sagnac Interferometer
SI-NCF	Sagnac Interferometer with No-Core Fiber
SLD	Super Luminescent Diode

SLI	Sagnac Loop Interferometer
SMF	Single-Mode Fiber
SNR	Signal-to-Noise Ratio
SP	Surface Plasmon
SPP	Surface Plasmon Polaritons
SPR	Surface Plasmon Resonance
SPW	Surface Plasmon Wave
TCF	Thin Core Fiber
TE	Transverse Electric
TF	Thinned Fiber
TFBG	Tilted Fiber Bragg Gratings
TIR	Total Internal Reflection
TM	Transverse Magnetic
TNT	Trinitrotoluene
TPOF	Tapered Plastic Optical Fiber
VOC	Volatile Organic Compound
WGE	Whispering Gallery Effect
WS	Wavelength Sensitivity

Chapter 1
Fiber Optic SPR Sensor—Past, Present, and Future

1.1 Introduction to Fiber Optic SPR Sensor

In recent decades, a wide range of sensing applications based on optical fiber interferometric sensors have arisen. Wireless communication systems, satellite communication, advanced radar systems, healthcare gadgets, and electronic warfare are among the uses. Aside from low loss and high bandwidth, a fiber-based solution eliminates electromagnetic interference, corrosion, and information leakage via the information-carrying channel to the outside since there is no short-circuit potential, as opposed to transporting copper wire. This improves security and creates a secure transmission against eavesdropping on the information between the transmitter and receiver. Other military criteria include information on the surrounding temperature, radiation, severe vibration, shock, and mechanical stress, all of which are very helpful for survivability. A fiber optic sensor may be used to obtain this information. Interferometry, scattering, and fiber Bragg grating detections are the three primary categories of sensing. These strategies may be appropriate for a certain purpose.

Apart from conventional sensing approaches, surface plasmon resonance (SPR) sensors have promising advantages in the field of chemical and biochemical sensing. The first sensing application of SPR was reported by Leidberg et al. (1983). Thereafter, a lot of attention is given by researchers to the development of SPR-based sensors for chemical and biochemical sensing in industry and healthcare applications. Before going into the sensing capabilities of SPR, its basic principle should be defined. Surface plasmon resonance is a phenomenon that gives an oscillating charge density at the interface of metal and dielectric. The permittivity of both mediums is of opposite polarity. SPR gives the interrelationship of optical and biochemical domains, where optical signals are influenced by the material properties of biochemicals. The surface plasmon waves are excited by the matching condition of the electric field orientation of the incident optical signal with the charge density oscillation at the boundary. This principle of SPR is linked to the chemical–biomolecular interaction

S. K. Raghuwanshi et al., *Geometric Feature-Based Fiber Optic Surface Plasmon Resonance Sensors*, Springer Tracts in Electrical and Electronics Engineering,
https://doi.org/10.1007/978-981-99-7297-5_1

for sensing purposes. Further, the SPR-based system also incorporates optoelectronic components for electrical domain analysis of optical signals and their signal processing analysis.

Surface plasmon sensors have been reported in various configurations and detection techniques. Some of the most common configurations include prism-based, grating-based, and waveguide-based, as shown in Fig. 1.1. Prism-based configurations, based on Kretschmann geometry, are very efficient for sensing and are most widely used in SPR sensors. In this configuration, a major portion of the incident light is reflected from the metal–dielectric interface, and a fraction of the power of the incident waves transmits, which excites the SPR. In a grating-based structure, there are various discontinuities at the metal–dielectric interface, which result in the diffraction of the incident wave at various angles. These diffracted beams' momentum components at the interface vary from the incident wave by multiples of the grating wave vector. The waveguide-based configuration also works on Kretschmann geometry for exciting surface plasmon waves inside the waveguide. The waveguides may be optical fiber or integrated optical waveguides of small size and ruggedness. In this configuration, the optical waveguide can be controlled easily, and the properties of light can be efficiently controlled. There are various detection schemes for the interference signal, as depicted in Fig. 1.1. Some of these are optical heterodyne detection, polarimetry, and interferometry of the superimposed signals. Optical heterodyne detection is based on the optical mixing of two signals with different frequencies, which results in the summation or difference frequency of the mixing signal. The difference signal is named a "beat signal" with the same phase content as that of the input signals. Polarimetry is based on the polarization of an optical signal in two orthogonal polarization states, i.e., S-polarized and polarized components. In this technique, a polarizer is incorporated after the optical source, and the polarized wave contains the phase difference information for various interference patterns. The noise that appeared in the detected signal can be eliminated with the signal processing algorithm for large collections of data. The phase modulation of the signal contains a large number of harmonics, which contain the phase information. Interferometry is the interference of two signals coming from an optical source, and it falls under the category of optical instrumentation. In this technique, the resulting signal of high frequency in the optical domain is converted into an information signal of low frequency with a proper detection system. In this technique, there is a direct relationship between the intensity of the interfered signal and the phase induced. This chapter contains a detailed explanation of the working principles and applications of interferometry techniques.

1.2 Exploring Some Concepts

Instruments or devices that work on the principle of interference between signals are known as interferometers. The first interferometer was invented by American physicist Albert Michelson in 1890. It was an amplitude-splitting interferometer,

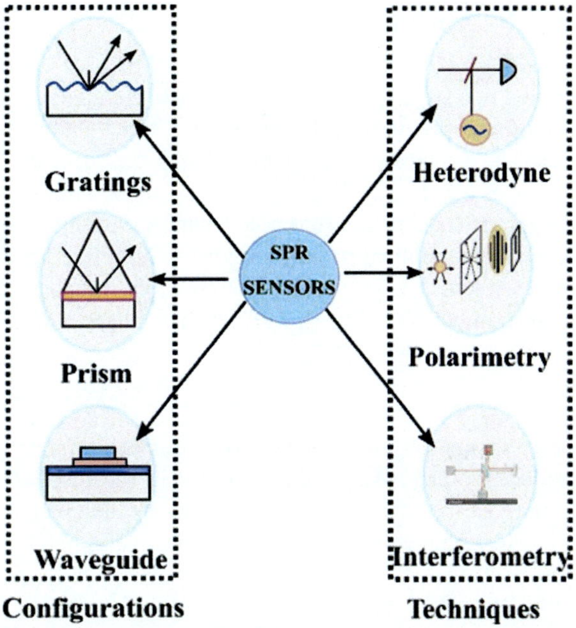

Fig. 1.1 Schematic diagram of various configurations and techniques for SPR-based sensing

and several experiments were demonstrated based on the interferometric concept (Lawson 2003). There may be intensity or phase modulation of optical signals inside the fiber. Intensity modulation characteristics can be used to identify the attenuation and scattering of light and grating plates, whereas interferometric interference, resonance, and polarization-dependent parameters can be identified by the phase modulation phenomena (Buck 2004).

The feasibility of its integration in a complex system may require a much more compact size, where conventional optical sensors fail to fulfill this requirement due to their bulk size. This can be accomplished by specialty fiber-based sensors, which reduce weight and occupied area, which is mostly desired in remote sensing applications. Some examples of intensity modulation-based sensors are microbend sensors, reflection-displacement sensors, and chemical-based sensors used in strain, temperature, pressure, and vibration sensing. The basic working principle of an interferometric structure is mature and can be understood as the combination of two signals with phase imbalance. An equally split part of the signal passes through two paths with different circumstances. One of the path signals is modulated by various physical parameters, which results in the phase difference of two signals. These signals are combined together, and information is collected in terms of phase or intensity.

1.2.1 Interference

When two signals with almost identical frequencies are combined together, a stable interference pattern can be observed. White light from two sources under interference leads to an interference pattern, where a specific color interferes with the same color of other light. The constructive and destructive patterns give bright and dark fringe patterns of interference depending on the amplitudes and phase conditions. Consider an optical path of length L with refractive index n, when an optical signal of wavelength λ is passed through it; the propagation constant β is a function of wavenumber $K = 2\pi/\lambda$; $\beta = nK$. The phase induced is given by

$$\varnothing = \beta L \tag{1.1}$$

For sensing purposes, one of the parameters, either β or L, should be changed to reflect the change in phase induced in the interference pattern. The number and nature of fringes can be used to identify the changes in the physical length of the propagating medium. Prediction of change in length by counting of fringes leads to better accuracy.

1.2.2 Basic Terms

Sensitivity: Sensitivity of sensors is defined as the change in output parameters with respect to the change in input parameters. In a basic sense, it can be defined as the smallest fractional change of the system or device output that can be measured. Suppose H is the transfer function, x is the input measuring parameter, and the sensitivity S for the input value of x_0 is given by

$$S = \frac{\partial H}{\partial x}|_{x=x_0} \tag{1.2}$$

The sensitivity of various sensors is defined as the change in input parameters necessary to achieve a standardized output change. Others describe it as a change in output voltage for a given change in input parameter. A common blood pressure transducer, for example, may have a sensitivity rating of 5 mV/V/mm Hg, which means that there will be a 10-mV output voltage for every volt of excitation potential and each mm Hg of applied pressure.

Range: It is the highest and/or lowest input parameter values that can be monitored. The dynamic range is defined as the sensor's whole range, from lowest to maximum. It gives the valid input range for which sensors can sense the information properly.

Precision: It is the degree to which a sensor measurement may be repeated when performing the same measurement; a highly exact sensor will always provide extremely identical results although this may not represent the "real" value, which is decided by the precision. Precision is a statistical measure of consistency, as a result, a measurement's ability to be consistently replicated. So, it is the degree of closeness that occurs among the set of measurements under stipulated conditions.

Resolution: Resolution defines the input parameter's lowest discernible incremental change that may be detected. Resolution may be stated as a percentage of the reading (or as the full-scale reading) or as an absolute value. The fundamental issue limiting a sensor's lowest feasible measurement is electrical noise in its output.

Accuracy: The largest discrepancy between the real value and the sensor output is defined as the accuracy of the sensor. It gives the degree of closeness between the measured value and the actual value. It can be defined in terms of absolute value or a percentage of relativeness. The ratio of the maximum deviation of a value indicated by the sensor to the ideal value is used to calculate accuracy. So, it is a measurement's capacity to match the actual value of a quantity being measured. Accuracy can be defined in terms of absolute error A_e and relative error R_e given in Eq. (1.3). In Eq. (1.3), M_v and T_v are measured value and true value, respectively:

$$\begin{aligned} A_e &= |M_v - T_v| \\ R_e &= \frac{|M_v - T_v|}{T_v} \end{aligned} \tag{1.3}$$

Offset: When the input is zero, the offset is the given output. Light sensors, for example, may still provide an output signal in a fully dark environment; this number is known as the offset.

Linearity: It is the degree to which the sensor's actual output curve deviates from the ideal curve, represented as a percentage of nonlinearity. The sensor's dynamic linearity refers to its capacity to respond to fast changes in input. Linearity error is defined as the greatest discrepancy between measured data and the data as estimated by the best straight-line equation of recorded data. It is often stated as a percentage of the whole scale. Full scale refers to the sensing device's complete range. The linear relation between input signal (x) and output of sensor (y) is given by

$$y = a + b.x \tag{1.4}$$

where a is the offset and b is the sensitivity or slope of the characteristics. The second term of Eq. (1.4) may be any power function, logarithmic, exponential, or similar nonlinear functions. The role of nonlinearity is important in the output characteristics. The nonlinearity error can be measured by taking the difference between points intercepted by two lines of the same slope in output characteristics. One point corresponds to the minimum value intercepted by one line, and another is the maximum intercepted by the second line on the output characteristics. The difference between the intercept value of the Y-axis and a parallel line passing through the largest deviation point is the total nonlinearity. The entire nonlinearity is estimated by estimating the biggest deviation at the X-axis's midway and observing the largest difference between the actual and computed values. Midpoint (X_{md}) on the X-axis for measuring parameters and the midpoint (Y_{md}) of the sensor output can be written as

$$X_{md} = X_{\min} + \frac{X_{\max} - X_{\min}}{2}$$
$$Y_{md} = Y_{\min} + \frac{Y_{\max} - Y_{\min}}{2} \tag{1.5}$$

The percentage of linearity can be approximated from the actual output (Y_a), given by

$$\%(\text{Linearity}) = \left| \frac{Y_{\min} + \frac{Y_{\max} - Y_{\min}}{2} - Y_a}{Y_{md} = Y_{\min} + \frac{Y_{\max} - Y_{\min}}{2}} \right| \times 100 \tag{1.6}$$

Hysteresis: It defines the reverse performance of the sensor for the same inputs. One can analyze the characteristics and behavior of the sensor when input varies in the reverse direction. The hysteresis characteristics of the sensor are useful in identifying the residual or offset value of the output parameter. It helps in determining hysteresis error, i.e., the biggest difference, expressed as a percentage of full scale, between the data points in a ramp from zero to full scale and back to zero.

Response Time: It defines how fast a sensor can respond to a change in input parameters, i.e., the time taken by the sensor to change its output when an input parameter changes from one point to another.

Repeatability: It is accurate under observational circumstances. Independent measurement results are produced in short intervals of time using the same procedure on identical measurement objects in the same test or measuring facility by the same operator using the same equipment (e.g., ability to stay stable during numerous heating and cooling cycles). When measuring minor changes, repeatability is often the most significant attribute.

Drift: Due to the presence of electronic components in the system, there is low-frequency fluctuation in the picture, which leads to low-frequency drift in the sensor's output. It depends on the age of the sensor and decreases with time.

1.2.3 Interferometric Structures

Interferometers are the most common type of phase-modulated sensor. These sensors can be Mach–Zehnder, Michelson interferometer, Sagnac loop, or resonator types, as depicted in Fig. 1.2. In the Mach–Zehnder-type structure of the interferometer, as shown in Fig. 1.2a, an optical signal from a light source is passed through a beam splitter, which splits the signal into two parts. The split power of these parts can be equal or can be distributed in unequal fractions with a specific coupler. The coupling coefficient of the coupler or reflectivity of the splitting medium can be adjusted to get the proper coupling ratio (Kumar et al. 2023a). One of the two paths is considered in the modulating environment, which takes advantage of the electro-optic interaction or changes in the optical properties of the arm due to physical parameter variation. This change in the optical signal can be in terms of phase or intensity, which is combined with the signal in another arm known as the reference signal. The reference signal is named so since it is passed through a medium without any external disturbance. These two signals are combined through a power combiner, and a change in the original source signal due to an external disturbance is observed on a detector, either in the optical domain or in the electrical domain. In Fig. 1.2b, a Michelson-type structure is depicted. In the case of a Mach–Zehnder modulator, two outputs are obtained through two arms; however, in the case of a Michelson interferometer, a signal output is obtained. Also, the latter can be used to probe a sample with light passing only a single time through it. The transmitted and reflected information are detected in the case of Mach–Zehnder and Michelson modulators, respectively. As shown in Fig. 1.2b, there are two reflecting surfaces through which split parts of coherent signals are reflected and interfere with each other. Except for reflection, the basic concept of modulation is similar to that of the Mach–Zehnder interferometer. One of the reflecting path lengths is fixed, and another reflecting path has a variable path length under the influence of some modulation scheme or external disturbance. The basic working principle can be seen in Fig. 1.3.

As shown in Fig. 1.3, the optical signal from a broadband light source is converted into monochromatic light with the help of a convex lens and split into two parts

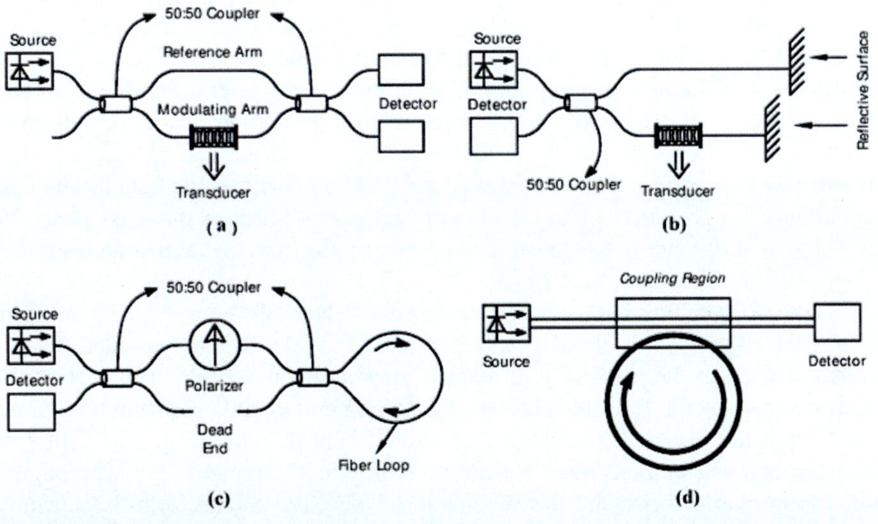

Fig. 1.2 Optical fiber sensor architectures: **a** Mach–Zehnder type; **b** Michelson type; **c** Sagnac loop; and **d** ring resonator type

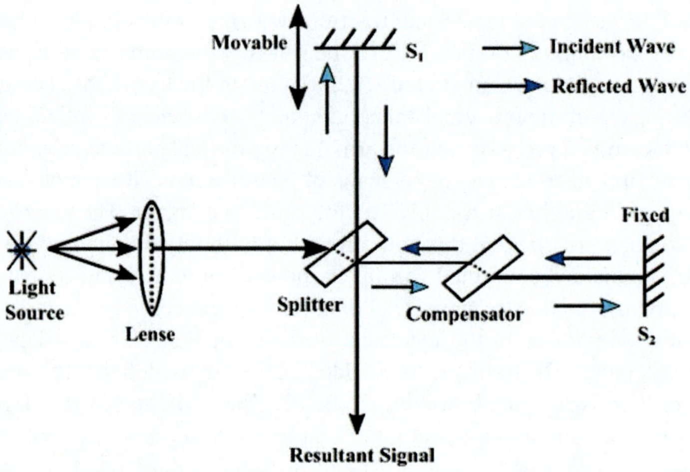

Fig. 1.3 Schematic for the working of the Michelson interferometer

through a splitter. Splitter is a half-silvered glass plate that is placed at a 45° inclination. At this splitter, reflection as well as transmission of the incident wave take place. Another glass plate is placed parallel to the splitter in the path of the transmitted wave. Each part is passed through different paths. One part has a fixed path length, and in another part, the path length is varied with a movable reflecting surface. The reflected waves from both parts are combined to get the resultant signal. The resultant signal has an observable interference pattern in terms of fringes due to the variable path length. If the path length difference between two arms is the sum of the integral multiple of wavelength and half wavelength, constructive interference takes place and there is the formation of a bright circular pattern.

However, if the effective path length difference is an integral multiple of wavelength, it leads to destructive interference and the formation of a dark pattern in circular form. The principle of the Michelson interferometer can be used to determine some basic parameters like the wavelength of monochromatic light through the spotting of fringe patterns, the thickness or refracting index of the glass plate, the wavelength difference between two close wavelengths from the laser source, and the temporal coherence of optical signals.

Sagnac loop defines the "optical rotation effect" due to the interference of incident and reflected waves in a circular path (Mock et al. 2021). Sagnac provided the first demonstration of the feasibility of an optical experiment capable of indicating the state of rotation of a frame of reference by making measurements within that frame. The schematic of the Sagnac loop interferometer is shown in Fig. 1.2c. The first Sagnac loop was demonstrated for the rotation effect (Sagnac 1913). The Sagnac loop is in closed form, and the schematic of the basic Sagnac loop is shown in Fig. 1.4. A high stability in the external disturbance is observed in the case of the Sagnac loop interferometer since it is free of phase control over two different paths.

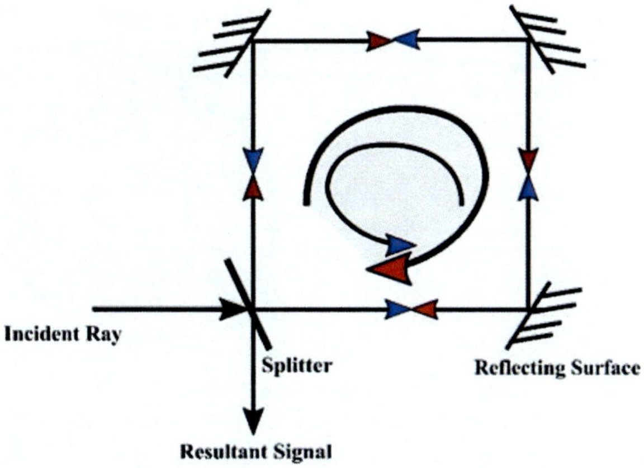

Fig. 1.4 Schematic diagram of the basic Sagnac loop

As shown in Fig. 1.4, incident rays are split into two parts through a splitter, and split rays travel in clockwise and anticlockwise directions guided through reflecting surfaces. The reflecting surface is made up of half-silver mirrors. These reflecting surfaces are placed at the corners of a rectangular, structured path. After a complete rotation, these two signals from both directions interfere, and the resultant signal gives some interference patterns. When the entire system is in motion with an angular rate Ω, there is a fringe shift ΔL in the interference pattern with respect to the interference pattern in the stationary case, and it is given by Post (1967),

$$\Delta L = \frac{4\Omega A\cos\theta}{\lambda_o c} \tag{1.7}$$

where A is the cross-sectional area of the closed optical path, λ_o is the wavelength of the source in vacuum, c is the velocity of light in free space, and θ is the angle between the axis of rotation and the normal of the area of the loop. The mirror locations must not alter under the influence of centrifugal force in order for the Sagnac experiment to be properly executed. Pattern alterations caused by interferometer distortion do not always result in a pure fringe shift and are thus identifiable from the predicted effect. Another criterion is that distortions be independent of rotational direction. On the rotating disc, the light source and fringe shift detection take place. Sagnac also demonstrated that the impact is unaffected by the form of the loop or the center of rotation.

The Sagnac loop can be placed in a ring structure (as shown in Fig. 1.2d), and its transfer function can be reconfigurable as a function of coupling coefficient and controlling the excess power loss (Vázquez et al. 2005) by changing the loop loss or electro-absorption (Djordjev et al. 2002; Armani et al. 2003). Apart from the sensing application, Sagnac interferometers are applicable in the implementation of logic gates (Menon et al. 2004a), positional switch incorporating fiber Bragg grating (Qu et al. 2000), ultrafine line laser (Shu et al. 2000), and optical filters (Menon et al. 2004b; Rabus et al. 2002). In Sagnac loop, signals are traveling in clockwise

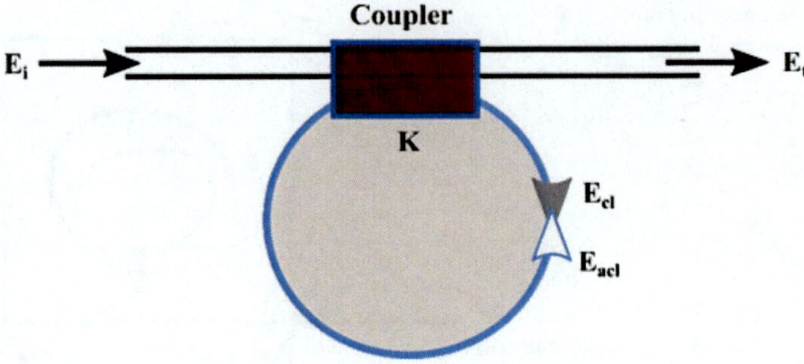

Fig. 1.5 Schematic of ring resonator structure

and anticlockwise directions; the formulation of transfer function can be performed based on coupling of ring structure with a straight propagation medium shown in Fig. 1.5. In Fig. 1.5, E_i and E_t are field strengths of incident and transmitted waves, respectively through the straightforward path. E_{cl} and E_{acl} are the field strengths of wave traveling in clockwise and anticlockwise directions, respectively through the ring resonator. Assume that K is the coupling coefficient and γ is the coupling loss; the input–output relation in the matrix form is given by following expression (Zhang and Lit 1994):

$$\begin{bmatrix} E_i \\ E_t \end{bmatrix} = \begin{bmatrix} \frac{1}{T} & -\frac{R}{T} \\ \frac{R}{T} & \frac{T^2-R^2}{T} \end{bmatrix} \begin{bmatrix} E_{cl} \\ E_{acl} \end{bmatrix} \tag{1.8}$$

where R and T are given by

$$\begin{aligned} T &= j\sqrt{K(1-\gamma)} \\ R &= \sqrt{(1-K)(1-\gamma)} \end{aligned} \tag{1.9}$$

A block diagram representation of the coupler can be seen as a more simple form, as indicated in Fig. 1.6a. Let K_1 and γ_1 be respective coupling coefficient and loss coefficient of the coupler, and the input–output relation in matrix form is given by

$$\begin{bmatrix} E_t \\ E_{cl} \end{bmatrix} = (1-\gamma_1) \begin{bmatrix} \sqrt{1-K_1} & j\sqrt{K_1} \\ j\sqrt{K_1} & \sqrt{1-K_1} \end{bmatrix} \begin{bmatrix} E_i \\ E_{ir} \end{bmatrix} \tag{1.10}$$

E_{ir} in Eq. (1.10) indicates the resultant of signals circulating inside the loop. In Sagnac loop, the interferometer part of the incident light is transmitted and the remaining part is reflected as shown in Fig. 1.6b. Fraction of light transmitted $\left(T_i = \frac{E_{2f}}{E_{1f}}\right)$ and fraction of light reflected $\left(R_i = \frac{E_{1b}}{E_{1f}}\right)$ through Sagnac loop is given by

$$T_i = (1-\gamma_1)(1-2K_1) \tag{1.11}$$

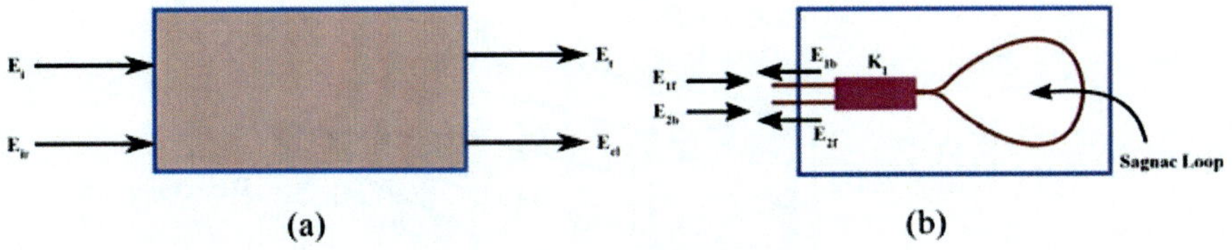

Fig. 1.6 **a** Block diagram representation of coupler for the transfer function in matrix form; **b** schematic of Sagnac loop for transfer function

$$R_i = j2(1-\gamma_1)\sqrt{K_1(1-K_1)} \tag{1.12}$$

In the above equations, attenuation of the propagating medium and signal retardation is not considered. The transfer function of Sagnac loop module box including coupler can be written as

$$TF = \begin{bmatrix} \frac{1}{R_i} & -\frac{T_i}{R_i} \\ \frac{T_i}{R_i} & \frac{R_i^2-T_i^2}{2} \end{bmatrix} \tag{1.13}$$

The overall transfer function of a cascade combination of two or more blocks incorporating couplers, Sagnac loops, and other fiber waveguides can be found by taking the product of the individual transfer functions. For example, consider a coupler section with a Sagnac loop in a cascade combination; the overall transfer function of the system is given by

$$TF_O = \begin{bmatrix} \frac{1}{T} & -\frac{R}{T} \\ \frac{R}{T} & \frac{T^2-R^2}{T} \end{bmatrix} \cdot \begin{bmatrix} \frac{1}{R_i} & -\frac{T_i}{R_i} \\ \frac{T_i}{R_i} & \frac{R_i^2-T_i^2}{2} \end{bmatrix} \tag{1.14}$$

The resonance wavelength is defined, where the denominator of the transfer function (TF_O) become zero. Phase dependency on the location of resonance wavelength can identified depending upon the real and imaginary behavior of R_i and T_i.

Another architecture used in optical sensing is the Fabry–Perot interferometer (FPI). This interferometer has multiple partial reflections between reflecting surfaces. The multiple reflected signals interfere with each other and create an interference pattern. Such an interferometer has high resolution. The working principle of a conventional Fabry–Perot interferometer is based on the diagram shown in Fig. 1.7. Basically, FPI structures are formed due to the cavity formation between two optical reflectors. There are two types of FPI structures, i.e., intrinsic and extrinsic FPIs. A schematic diagram of extrinsic FPI is shown in Fig. 1.8. In intrinsic FPI, a solid cavity exists between two inline reflecting surfaces. However, in the case of extrinsic FPI, there is a cleaved fiber tip and an external reflector placed inside a capillary tube separated by an air medium. This can be placed in a free medium instead of a capillary tube. In another type of extrinsic FPI, there may be film deposition after the cleaved facet of fiber, and the other end of the film and free space interface acts

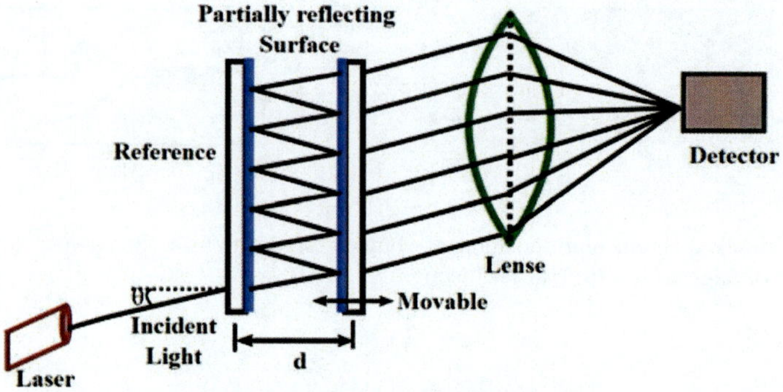

Fig. 1.7 Schematic diagram for the working of a conventional Fabry–Perot interferometer

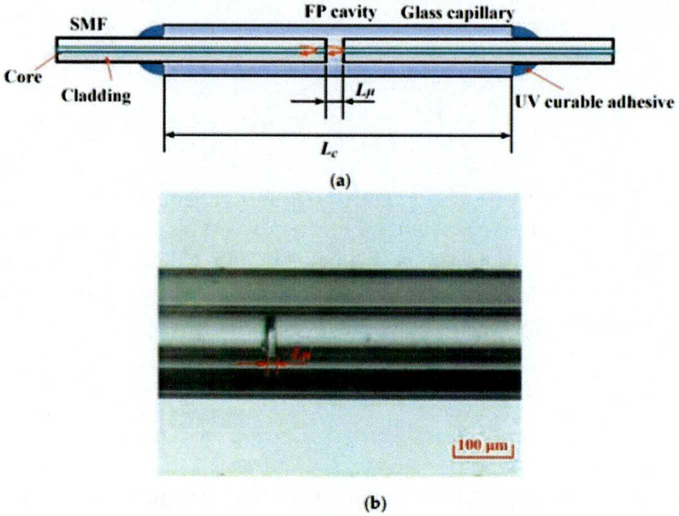

Fig. 1.8 Schematic of extrinsic Fabry–Perot interferometer **a** sensor and **b** microscopic view of the cavity. Adapted from Ma et al. (2019)

as a second reflecting surface, as depicted in Fig. 1.8. Extrinsic types of FPI have an easy fabrication process and provide a more stable interference pattern due to the air fiber interface.

In addition to the inherent properties of optical fiber, the above-mentioned architectures are numerously used in optical sensing applications. Applications of the discussed architectures in various sensing systems are described in the subsequent sections.

1.3 Some Related Background Ideas

In the present time, fast diagnosis is a key requirement in the fields of health care, environmental condition monitoring, and food safety. The technology used for this purpose is expected to have a low cost and very fast response time in real-time applications. Biosensing can be used in the detection of various biomolecules in healthcare diagnosis, bacterial disease, and pollutant particles in the environment (Anand et al. 2022). Surface plasmon resonance in the sensing application addresses the detection of the above-mentioned parameters, which is label-free sensing and gives real-time application of the events. Some diagnostic parameters like proteins, bacteria, nucleic acids, antibodies, and many more are considered in the development of SPR-based biosensors. Surface design plays a very important role in the development of efficient SPR sensors. There are various configurations, detection schemes, and design parameters used in SPR sensing, explained in the following subsections.

1.3.1 Coupling Configuration in SPR Sensing

Prism Coupler-based structure:

Most of the SPR devices work on the Kretschmann structure, which is based on the polarimetry of the incident signal. The conventional approach for the Kretschmann configuration is shown in Fig. 1.9. In this configuration, the optical signal is polarized through a polarizer and incident on the prism structure. The prism acts as a coupler, which was introduced by Kretschmann. The prism structure consists of two layers, namely glass and metal (usually gold). When incident light reflects at the dielectric–metal interface, part of the light interacts with the surface plasmon, and photon interaction with the surface plasmon results in surface plasmon polarization. The polariton generates an evanescent mode of electric field on either side of the metal with the same wavelength as that of the incident wave. At a particular angle of incidence and the proper value of the refractive index of the dielectric, polariton causes resonance to occur, known as surface plasmon resonance. Since resonance wavelength, i.e., peak of interference pattern, depends on the refractive index of the medium or various properties of the analyte like pH value, adsorption, chemical reaction, and electron affinity, the shift of wavelength can be used to identify the type of analyte. The resonance wavelength is also affected by the temperature, so temperature sensing can also be applied. As shown in Fig. 1.1, the sensing configuration may be prism-based, grating-based, or waveguide-based. In a surface plasmon resonance sensing system, real-time monitoring of changes in analytes can be observed.

The resonance occurs when the propagation constant of the surface plasmon matches that of the incident wave. In general, the propagation constant at the interface of the dielectric and metal is given by Homola (2008)

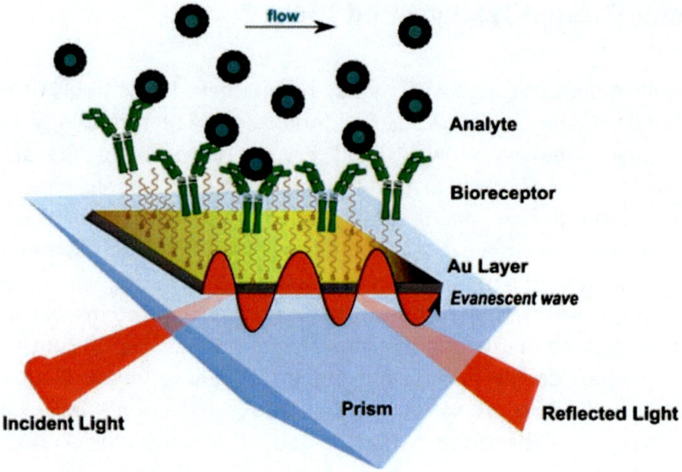

Fig. 1.9 Working principal of conventional SPR sensor based on Kretschmann structure. Adapted from Topor et al. (2023)

$$\beta_{sp} = \frac{2\pi}{\lambda}\sqrt{\frac{\epsilon_d \epsilon_m}{\epsilon_d + \epsilon_m}} \tag{1.15}$$

where ϵ_d and ϵ_m are dielectric constants of dielectric and metal, respectively, and λ is the wavelength of the optical signal. The propagation constant in Eq. (1.15) will stand for the guided mode when the real parts of the electric constants of metal and dielectric are equal in magnitude and opposite in polarity. The real part of the dielectric constant is positive for dielectric and negative for metal. Another important parameter in the analysis of surface plasmons is the propagation length (L), given by

$$L = \frac{\pi}{2\lambda}\frac{\epsilon_{m,i}}{\epsilon_m^2}\left(\frac{\epsilon_d \epsilon_m}{\epsilon_d + \epsilon_m}\right)^{\frac{3}{2}} \tag{1.16}$$

where $\epsilon_{m,i}$ represents an imaginary part of the dielectric constant for the metal layer. In Kretschmann configuration, surface plasmons are excited by transverse magnetic waves with the condition that the dielectric constant of the prism is higher than that of the sensing medium or analytes. Let ϵ_m be the dielectric constant of the analyte. Assume that K_x is the propagation constant of the incident wave, and the resonance condition is $K_x = \beta_{sp}$, expanded to

$$\frac{2\pi}{\lambda} \cdot n_p \cdot \sin\theta = \frac{2\pi}{\lambda}\sqrt{\frac{\epsilon_d \epsilon_m}{\epsilon_d + \epsilon_m}} \tag{1.17}$$

In Eq. (1.17), n_p is refractive index of the prism and θ is the angle of incidence and for surface plasmon resonance; it is given by

$$\theta = \sin^{-1}\left(\frac{1}{n_p}\sqrt{\frac{\epsilon_d\epsilon_m}{\epsilon_d + \epsilon_m}}\right) \tag{1.18}$$

Equation (1.18) represents that the incident polarized light, K_x can be coupled to the prism of refractive index, n_p, and incident at an angle of θ can be used to explore surface plasmon resonance.

Grating-based structure:

Figure 1.10 represents the grating-based structure for the incident light coupling in surface plasmon for SPR sensing. The first grating-based structure for SPR sensing was reported by Wood (1902). Grating-based angles work on the diffraction angle θ_d given by

$$\sin(\theta_d) = \sin\theta + d\frac{\lambda}{\Lambda} \tag{1.19}$$

where θ is the angle at which polarized wave is incident on the grating structure and Λ represents the grating period. Rayleigh wavelength in Eq. (1.19) is defined when $\sin(\theta_d) = \pm 1$. The relation between incident wave and grating configuration at resonance condition can be written as (Phillips 2008)

$$n_d \sin(\theta_d) + d\frac{\lambda}{\Lambda} = \pm\sqrt{\frac{\epsilon_d\epsilon_m}{\epsilon_d + \epsilon_m}} + \Delta\beta\frac{\lambda}{2\pi} \tag{1.20}$$

In Eq. (1.20), n_d is the refractive index of the sensing medium, and $\Delta\beta$ represents change in propagation constant for grating structures. It shows that resonance wavelength is a function of the refractive index of the medium, and SPR exists when a polarized wave incident at θ fulfills the above condition. The grating-based structures have attracted great attention from the researchers due to their simple fabrication and easy optical-based disk method, laser interference lithography, and

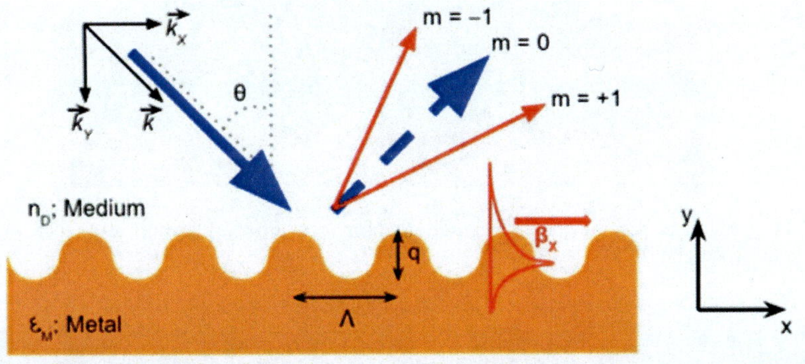

Fig. 1.10 Schematic diagram of grating-based structure for surface plasmon resonance. Adapted from Prabowo et al. (2018)

wet etching methods. The diffracted beams' momentum components at the interface vary from the incident wave by multiples of the grating wave vector. The optical wave may couple to the SPW if the total component of momentum at the interface of a diffracted order is equal to that of the SPW. However, the main difficulty associated with this structure is the complex theoretical modeling, which makes it difficult to analyze grating-based structures as compared to prism couplers.

Waveguide-Based Structure:

Embedded optical waveguide SPR sensors look promising for the creation of multiplex sensing devices on a single chip, allowing for effective reference and multi-component sensor analysis of complicated data. When a metal–dielectric interface is periodically warped, the incoming optical wave is diffracted, resulting in a sequence of beams angled away from the surface. The waveguide structure may consist of a slab, single-mode integrated optical waveguides, and a channel. Several methods for modifying the sensor's working range have been investigated, including the use of waveguides made of low refractive index glass, a buffer layer, a high refractive index overlayer, and more sophisticated multilayer structures. Figure 1.11 represents the waveguide coupling-based structure for SPR sensing. In this structure, the optical signal is guided into the metal thin layer region deposited on the substrate. When the guided wave and surface plasmon wave fulfill the phase match condition, the surface plasmon is excited in the region above the metal film layer. The waveguide-based structure is effective in the design of an on-chip SPR sensor; however, it is difficult to analyze angular modulation in this scheme.

A comparative study of the above three configurations, prism coupler-based, grating structure, and waveguide coupling, is given in Table 1.1 in terms of their advantages and limitations.

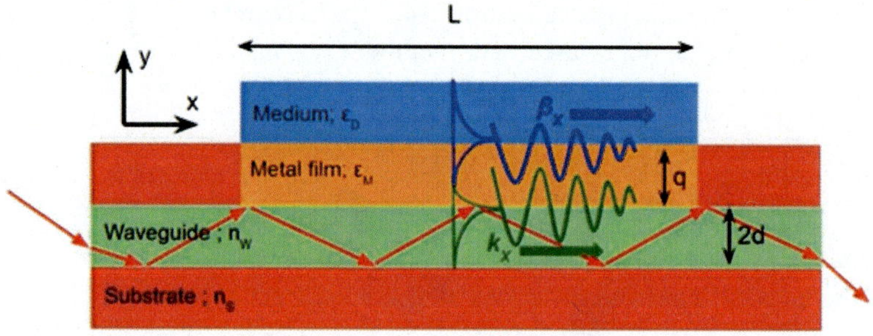

Fig. 1.11 Schematic diagram for waveguide coupling for surface plasmon resonance. Adapted from Prabowo et al. (2018)

Table 1.1 Comparative study of SPR coupling configuration

References	Coupling configurations	Advantages	Limitations
Huang et al. (2006), Ruffato et al. (2012), Rossi et al. (2018), Gupta et al. (2016)	Prism coupler configuration	• High sensitivity • Easy handling	• Complex compatibility with opto-mechatronics • Difficulty in miniaturization
Rossi et al. (2018), Homola et al. (1999), Couture et al. (2013)	Grating configuration	• Highly integrable • Miniaturization on-chip applications	• Low sensitivity • Complicated manufacturing
Prabowo et al. (2018), Huang et al. (2006), Kashyap and Nemova (2009)	Waveguide-based configuration	• Miniaturized • Highly sensitive	• Angular modulation not applicable • SPR with polychromatic light only

1.3.2 Detection Scheme in SPR Sensing

Analysis of the sensing parameters can be done in terms of angular change, amplitude change, and wavelength shift. Detection schemes are based on the following parameters:

- Angular
- Wavelength
- Intensity
- Phase.

The angular or angle of incidence method is applicable when the system uses a monochromatic light source. However, the wavelength interrogation method is applicable when the system incorporates a broadband source of light or polychromatic light. In amplitude or intensity detection techniques, the intensity of the reflectivity is measured at a particular value of wavelength or angle of incidence. This technique is useful in SPR imaging, in which a contrast image can be pictured for the sensing region. Figure 1.12 depicts these three detections in the reflection spectrum. As indicated in Fig. 1.12, there are two different reflection spectra at different dip positions for the refractive indices n_1 and n_2. This shift can be interpreted as the change in wavelength or change in phase inferred from the change in refractive index. Also, intensity variation can be observed for these spectra, and intensity modulation occurs due to changes in their intensity values.

Wavelength modulation and changes in refractive index may be detected by scanning the wavelength of the incoming light or by using a spectrometer with a broadband light source. Wavelength modulation-based systems do not involve mechanical setup, which results in relatively stable systems compared with others. The wavelength modulation uses wavelength scanning of white light followed by optical filtering. The optical filters used in these techniques are acoustic–optical tunable

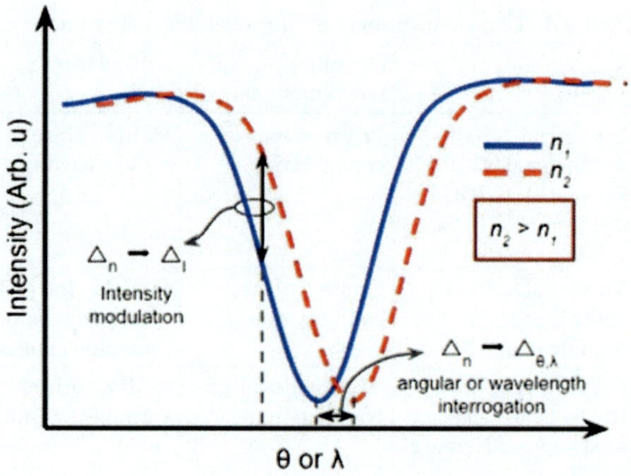

Fig. 1.12 Angular, wavelength, and intensity detection of surface plasmon resonance. Adapted, 2018, MDPI (Prabowo et al. 2018)

filters. The acousto-optic tunable filter (AOTF) is a kind of electro-optical device that acts as an electronically adjustable excitation filter, modulating the intensity and wavelength of numerous laser lines from one or more sources at the same time. The schematic diagram for the wavelength modulation-based detection system is shown in Fig. 1.13. A wavelength-swept laser source of 50 nm wavelength is used to have an optical signal in the range of 770–820 nm with a maximum power of 20 mW and a linewidth of 0.02 nm. The optical signal is passed through single-mode fiber (SMF), collimation lens (CL), and iris diaphragm (IR) and fed to the optical filter. A detailed analysis of optical filters can be found in Lee et al. (2016). The filtered signal is then analyzed on a DAQ system. A top and side view of the wavelengths during the filtering process is shown in the inset of Fig. 1.13.

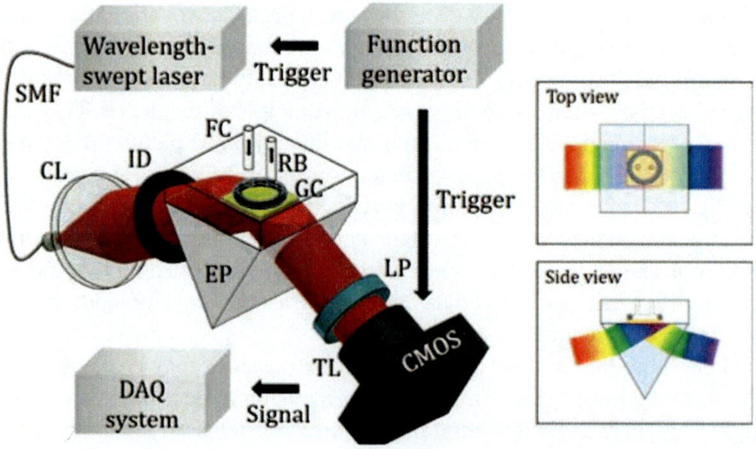

Fig. 1.13 Schematic diagram demonstrating a wavelength modulation-based system. Reprinted with permission from Optics Letters. Copyright, 2022, Optica (Bak et al. 2018)

Refractive change is considered in wavelength modulation-based detection; the refractive index change is a function of the operating wavelength (λ), which is given by the Sellmeier equation given by Ghatak and Thyagarajan (1998)

$$n(\lambda) = \sqrt{1 + \frac{C_1\lambda^2}{\lambda^2 - D_1^2} + \frac{C_2\lambda^2}{\lambda^2 - D_2^2} + \frac{C_3\lambda^2}{\lambda^2 - D_3^2}} \quad (1.21)$$

In Eq. (1.21), C_1, C_2, and C_3 and D_1, D_2, and D_3 represent Sellmeier coefficients. The dielectric dependence of the plasma wavelength (λ_p) and collision wavelength (λ_m) of the metal is given by Sharma and Gupta (2007a)

$$\epsilon_m(\lambda) = \epsilon_{m,r} + j\epsilon_{m,i} = \frac{\lambda^2\lambda_c(\lambda_c - j\lambda)}{\lambda_p^2(\lambda_c^2 + \lambda^2)} \quad (1.22)$$

where $\epsilon_{m,r}$ and $\epsilon_{m,i}$ represent real and imaginary parts of the metal's dielectric constant. The normalized transmitted power available at one end of the fiber due to the broadband optical signal at the other end of surface plasmon resonance is given by Singh and Gupta (2010)

$$P_{trans,norm} = \frac{\int_{\theta_{cr}}^{\frac{\pi}{2}} R_P^{N_{ref}(\theta)} \frac{n_1^2 \sin\theta\cos\theta}{(1-n_1^2\cos^2\theta)^2}}{\int_{\theta_{cr}}^{\frac{\pi}{2}} \frac{n_1^2 \sin\theta\cos\theta}{(1-n_1^2\cos^2\theta)^2}} d\theta \quad (1.23)$$

where θ_{cr} is the critical angle for total internal reflection, θ is the angle at which the optical signal is launched inside the fiber core, and n_1 is the refractive index of the fiber core. R_p represents reflection coefficient. Number of reflections inside the fiber core of sensing length L and core diameter D is $N_{ref}(\theta) = L\tan\theta/D$. The reflectivity in multi-layered structures can be analyzed using the N-matrix method given by the characteristic's matrix M (Hansen 1968):

$$M = \begin{bmatrix} M_{11} & M_{12} \\ M_{21} & M_{22} \end{bmatrix} = \begin{bmatrix} \cos\beta_k & \frac{-j\sin\beta_k}{q_k} \\ -jq_k\sin\beta_k & \cos\beta_k \end{bmatrix} \quad (1.24)$$

where q_k and β_k are given by

$$q_k = \frac{(\epsilon_k - n_1^2\sin^2\theta_1)^{\frac{1}{2}}}{\epsilon_k}$$
$$\beta_k = \frac{2\pi d_k}{\lambda}(\epsilon_k - n_1^2\sin^2\theta_1)^{\frac{1}{2}} \quad (1.25)$$

where d_k and ϵ_k are thickness and dielectric constant of kth layer. The overall reflectivity over the N-layer structure is given by

$$r_p = \frac{(M_{11} + M_{12}q_N)q_1 - (M_{21} + M_{22}q_N)}{(M_{11} + M_{12}q_N)q_1 + (M_{21} + M_{22}q_N)} \quad (1.26)$$

Resonance condition for the excitation of surface plasmon is given by Sharma and Gupta (2007a)

$$\frac{2\pi}{\lambda} n_1 \sin\theta = \left[\frac{2\pi}{\lambda} \left(\frac{\epsilon_m n_s^2}{\epsilon_m + n_s^2} \right)^{\frac{1}{2}} \right] \tag{1.27}$$

where ϵ_m is dielectric constant of metal layer and n_s is refractive index of the sensing layer of the structure. In multilayer configurations, the electromagnetic field's coupling to the surface plasmon waves' propagation constant at the outer metal–insulator (analyte) interface is strongly influenced by the outer metal's plasma frequency and damping constant, as well as the coupling of the electromagnetic field to the mobile charges of the various metal layers.

In contrast, intensity modulation is generally utilized for SPR imaging and is based on detecting variations in reflectance intensity at a given incident angle or wavelength (Sun et al. 2010; Wong et al. 2013; Howe et al. 2019). The intensity-based detection can be enabled with the transformation of the change in polarization state to be affected during the excitation of the surface plasmon (Sun et al. 2010). In this scheme, a parallel light beam passes through a polarizer with its light transmission axis at an acute angle to the vertical direction, and the emergent light is elliptical polarized in a fixed state after the excitation of surface plasmons with a fixed refractive index of the sample at the surface of the gold film, as shown in Fig. 1.14. If a quarter wave plate is put behind the prism exit surface with its fast axis pointing in the same direction as the elliptical polarization axis, the emerging light will become linearly polarized, which may be stopped by a crossed polarizer placed after the wave plate. The charge-coupled device (CCD) at the end of the light path then displays a dark patch known as the extinction point, and the matching refractive index is known as the extinction refractive index (Homola and Yee 1998).

The electric field output ($E_{o/p}$) when reflected from a quarter wave plate and polarizer is defined by John matrix and given by

$$E_{o/p} = \frac{1}{\sqrt{2}} \begin{bmatrix} \cos^2\theta_{pol} & \frac{1}{2}\sin 2\theta_{pol} \\ \frac{1}{2}\sin 2\theta_{pol} & \sin^2\theta_{pol} \end{bmatrix} \begin{bmatrix} 1 - j\cos 2\theta_{QWP} & -j\sin 2\theta_{QWP} \\ -j\sin 2\theta_{QWP} & 1 + j\cos 2\theta_{QWP} \end{bmatrix} E_{ref} \tag{1.28}$$

where θ_{pol} and θ_{QWP} are the angles introduced by polarizer and wave plate, respectively in the electric field component of the reflected signal (E_{ref}) from CCD. The electric field output has two polarization states, i.e., surface polarized (E_{o/p_s}) and p-polarized (E_{o/p_p}). The light intensity can be written as

$$I_{o/p} = \left| E_{o/p_s} \right|^2 + \left| E_{\frac{o}{p_p}} \right|^2 \tag{1.29}$$

The phase difference between s-polarization and p-polarization is shown in Fig. 1.15, where α' is the orientation angle of the ellipse formed due to the different

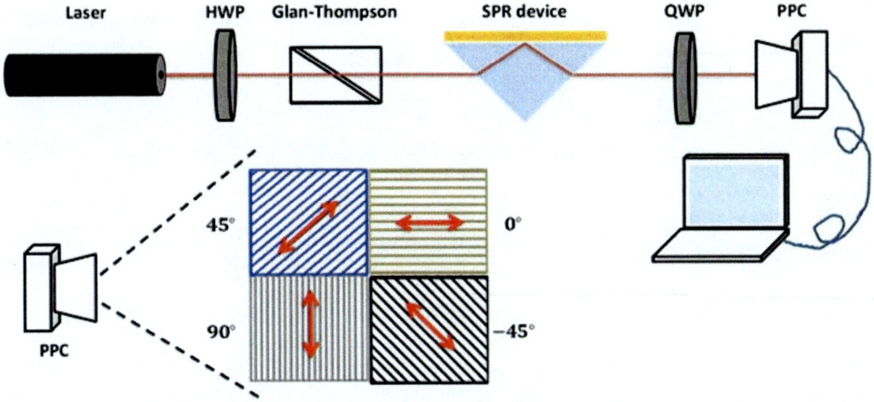

Fig. 1.14 Schematic diagram for the phase-sensitive SPR based on simultaneous polarization measurement with common-path interferometry. QWP: quarter-wave plate; HWP: half-wave plate; PPC: pixelated polarization camera. Adapted from Li et al. (2021a)

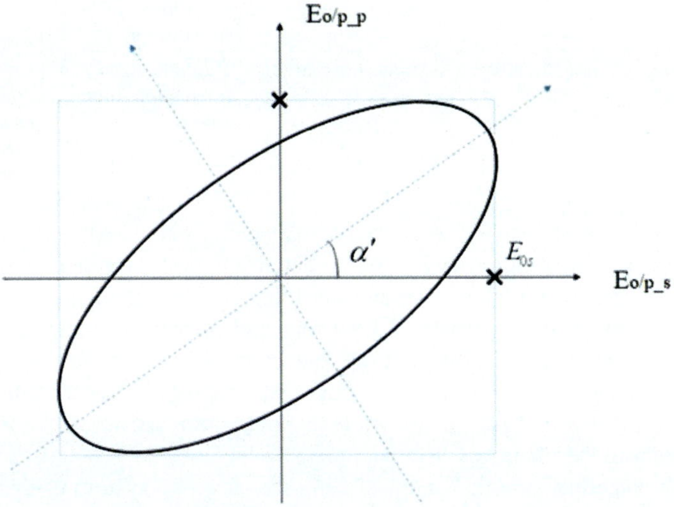

Fig. 1.15 Phase difference between S-polarization and P-polarization states

magnitudes of the polarization states. To get the most performance out of optical parts and systems, polarization behavior must be precisely controlled. Various polarizations will have various effects on characteristics including reflectance, insertion loss, and beam splitter ratios. Because it may be used to send messages and take accurate measurements, polarization is particularly crucial. The polarization state of an optical beam may transmit important information even when the light intensity is constant (Table 1.2).

Table 1.2 Comparative study of various modulation schemes in terms of advantages, field of application, and limitations

References	Modulation techniques	Advantages	Detecting parameters	Limitations
Kashif et al. (2014), Puiu and Bala (2016), Aparna and Tetala (2023)	Angle modulation	Noise elimination, high resolution	Health care, protein screening	Low stability and scan rate due to mechanical angle sensor
Kashif et al. (2014), Puiu and Bala (2016), Aparna and Tetala (2023)	Wavelength modulation	Relatively high stable due to non-mechanical set up	Antigen–antibody interaction	Expensive components
Zeng et al. (2007)	Intensity-based detection scheme	Simple configuration, effective real-time monitoring	Biomolecular interaction, food safety, environmental monitoring	Low sensitivity
Li and Zhang (2007)	Phase modulation	Highest sensitivity	Biomolecular interaction, food safety	Complex integration system, high precision instruments

The beam's polarization may be used to determine how various material interactions (such as magnetic, chemical, and mechanical) have changed it. Such polarization shifts may be tailored to run sensors and measuring tools. Utilizing the reflection, absorption, and transmission characteristics of the materials employed in these components will allow for this polarization control. P-polarization denotes the component within the plane, while S-polarization denotes the component perpendicular to the plane.

A phase detection scheme is applicable in the case of a system using a coherent monochromatic source of light. This scheme has a less complex hardware structure compared to the previous schemes due to the involvement of phase extracting equipment and the requirement for a locked-in amplifier. The phase detection measurement system for SPR is shown in Fig. 1.16.

1.3.3 Design Parameters

There are various factors like sensitivity, operating range, type of sensing environment, reproducibility, resolution, precision, and many more that must be kept in

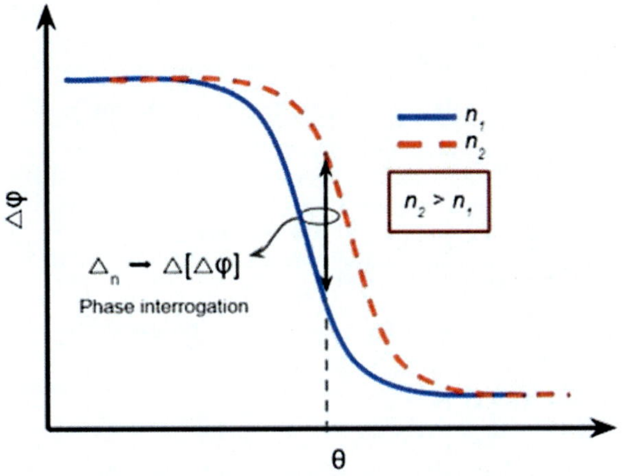

Fig. 1.16 Phase detection scheme employed in surface plasmon resonance. Adapted from Prabowo et al. (2018)

mind during the probe fabrication process of SPR sensors. The optimal sensor is one that produces repeatable results and has excellent detection accuracy, sensitivity, and operating range. Numerous changes have been made to the fiber optic SPR probe's design to accomplish this.

Selection of metal: The choice of metal depends on its ability to explore surface plasmon characteristics. Oxidation and corrosion are important parameters that should be considered during the selection of metal for stable SPR sensing. Some of the common metals are gold (Au), silver (Ag), copper (Cu), and aluminum (Al). Al and Cu are easily affected by oxidation and corrosion, which demands external protection. It is found that Au and Ag are generally used metals in SPR sensing; however, there are different evaluation factors in the performance analysis. Ag used in SPR sensing gives a narrow spectral width with good accuracy, but it is unstable and chemically reactive, which leads to careful handling in a chemical environment. In a liquid or gaseous environment, Ag gets oxidized easily. On the other hand, Au-based SPR sensors are sensitive, stable, and immune to oxidation, but lower accuracy is observed. A combination of Au and Ag in an SPR sensor can result in better sensitivity, good accuracy, and chemical stability. The combination of two different methods is popularly used in the design of SPR sensors. For example, Zynio et al. (2002) used two different metals, Au and Ag, in the plasmonic layer to enhance the sensitivity of the SPR sensor. In this structure, Au is used on the outer surface of the metallic layer.

The response of surface plasmon resonance to various refractive indices of analytes for gold, silver, and bimetallic structures is shown in Fig. 1.17. Ag provides narrower spectral characteristics, which give a high signal-to-noise ratio (SNR), and also provides immunity against oxidation. Silver layers were sandwiched between gold layers and a glass prism in the case of bimetallic layers. While the bimetallic films gave a shift that was almost equal to that of the pure gold layers, the changes in the resonance angle recorded on gold layers were 10% larger than on silver layers. Al

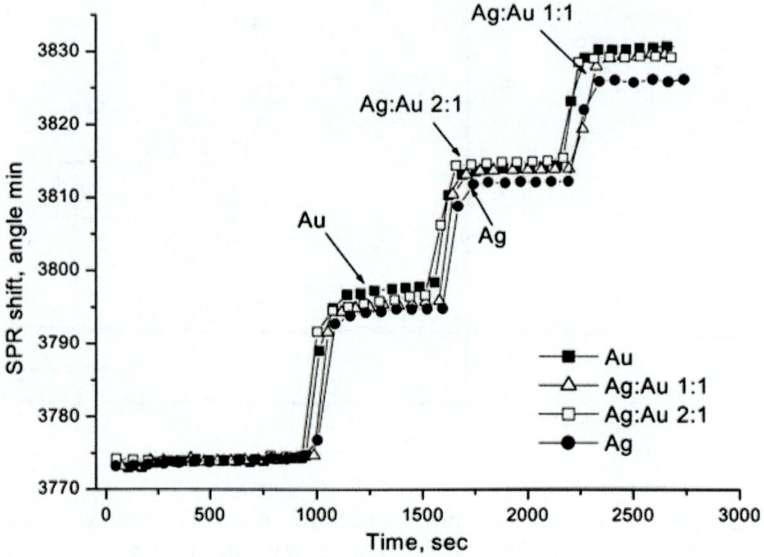

Fig. 1.17 Response of surface plasmon resonance in various refractive index of analytes for gold, silver, and bimetallic structure. Adapted from Zynio et al. (2002)

has a broad range of reflectivity, but it has lower sensitivity. There are other metals, Cu and Al, that have been investigated for surface plasmon resonance (Mitsushio et al. 2006). Spectral response and reflectivity have been shown for Au, Cu, Ag, and Al in Fig. 1.18a, and normalized intensity for various refractive indices is shown in Fig. 1.18b. A spectral response has been shown for a 20% methanol solution of benzyl alcohol.

In Sharma and Gupta (2007a), the performance analysis of SPR sensing, various metal combinations as well as metallic layer thickness are investigated. Figure 1.19 shows sensitivity and SNR variation in various combinations of metals for the inner thickness fraction. As discussed above, Au has the highest sensitivity, and Al has the lowest sensitivity. As indicated in Fig. 1.19a, the least sensitive combination for any ratio of their respective thickness values is Cu–Al. Combinations of Ag–Au and Cu–Au also provide high sensitivity. However, if the proportion of inner gold layer thickness is considered to be greater than 0.3, the Au–Al scenario offers better sensitivity than any other combination. It suggests that a thick Au layer with a thin, 2–4 nm-thick Al layer covering may provide a considerable amount of sensitivity. In the analysis of SNR, for a single metal, Au shows the lowest SNR, but Ag provides the highest SNR. In the case of bimetallic structures, as indicated in Fig. 1.19b, Cu–Al is clearly superior to all others. The Cu–Al example is marginally superior to the Ag–Al combination.

Geometry of the fiber: In order to improve the sensitivity of the SPR sensor, various geometries of optical fiber have been reported in the past. Tapering is one of the commonly used techniques for changing the geometry of fiber. It changes the core

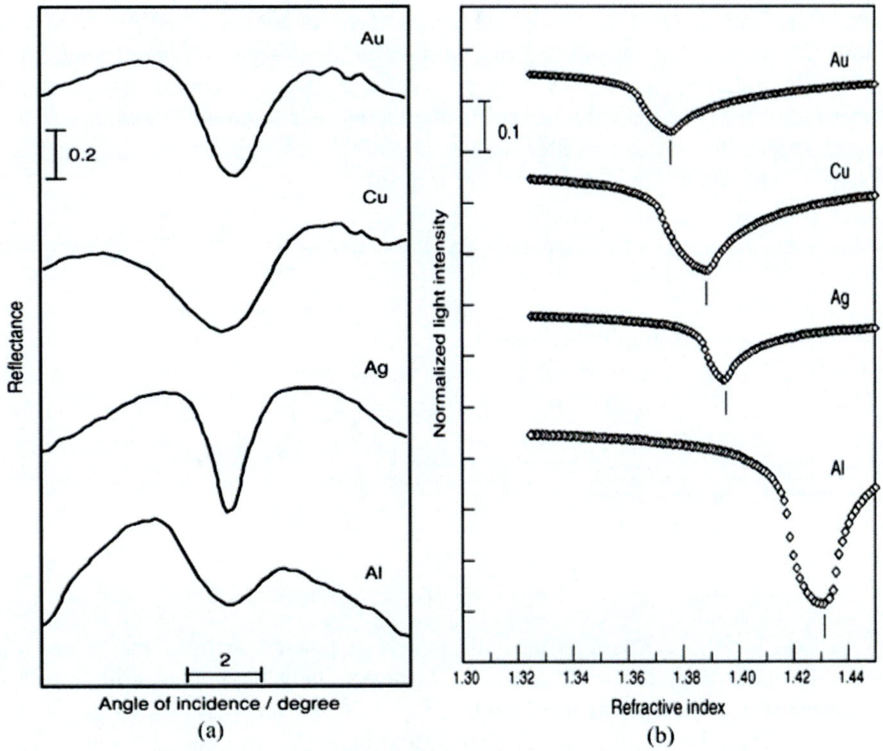

Fig. 1.18 **a** Reflectance of Au, Cu, Ag, and Al deposited over a silica substrate (Mitsushio et al. 2006); **b** normalized light intensity from the response of metal-deposited SPR. Reprinted with permission from Sensors and Actuators A: Physical. Copyright, 2006, Elsevier (Mitsushio et al. 2006)

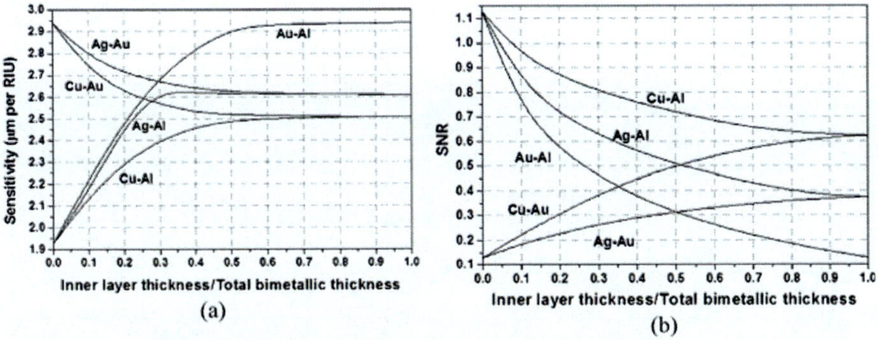

Fig. 1.19 **a** Sensitivity and **b** signal-to-noise ratio of various bimetallic structures as a function of inner layer thickness fraction. Reprinted with permission from Journal of Applied Physics. Copyright, 2007, AIP Publishing (Sharma and Gupta 2007a)

radius, and various tapering profiles are incorporated for enhanced sensitivity. Verma et al. (2008) reported various tapering profiles in the design of tapered fiber optic sensors to enhance the sensitivity in physical and chemical environments. These tapering profiles are, namely, linear, parabolic, and exponential. The radius of the tapered region for various profiles is a function of the propagation distance (z) along the axis of the fiber. The radius for a linear profile is

$$\rho_l(z) = \rho_i - \frac{z}{L}(\rho_i - \rho_o) \tag{1.30}$$

For exponential–linear tapering region,

$$\rho_{el}(z) = (\rho_i - \rho_o)\left(e^{-\frac{z}{L}} - \frac{z}{L}e^{-1}\right) + \rho_o \tag{1.31}$$

For parabolic tapering region,

$$\rho_p(z) = \left(\rho_i^2 - \frac{z}{L}\left(\rho_i^2 - \rho_o^2\right)\right)^{\frac{1}{2}} \tag{1.32}$$

In Eqs. (1.30)–(1.32), L is the total length of tapered section, and ρ_i and ρ_o are radii of the core at $Z = 0$ and $Z = L$, respectively. Various tapering profiles and their effects on tapering are shown in Fig. 1.20. Figure 1.20a shows the radius variation of the linear, exponential–linear, and parabolic profiles as a function of distance along the axis of the fiber. Figure 1.20b shows the change in sensitivity of various tapered profiles for different values of the taper ratio between 1 and 3. The sensitivity is enhanced with the increased value of the taper ratio, and a fast increment in sensitivity occurs for an exponential–linear profile.

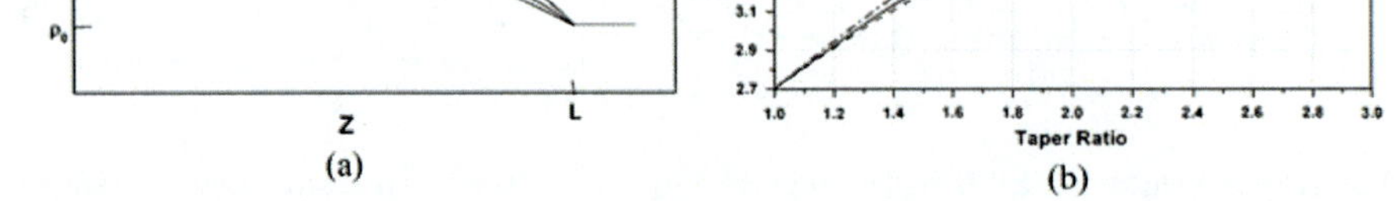

Fig. 1.20 **a** Radius variation as function of distance along the axis of fiber core for linear, exponential–linear, and parabolic taper profiles (Verma et al. 2008); **b** variation in sensitivity with taper ratio. Reprinted with permission from Optics Communications. Copyright, 2008, Elsevier (Verma et al. 2008)

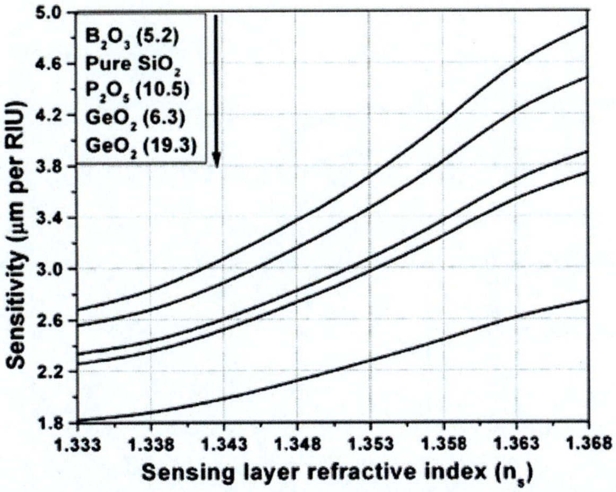

Fig. 1.21 Effect of doping on the sensitivity of SPR due to a change in the refractive index of the sensing layer. Reprinted with permission from Optics Communications. Copyright, 2007, Elsevier (Sharma and Gupta 2007b)

Dopants for sensitivity enhancement: Surface plasmon resonance is affected by the fiber core refractive index, which can be taken into account for the sensitivity change through the various dopings inside the core of the fiber. Silica is a commonly used material for fiber core. Sharma et al. (2007b) investigated the effects of doping on sensitivity and signal-to-noise ratio enhancement. In this work, pure silica was doped with several oxides, including germanium (GeO_2), boron (B_2O_3), and phosphorus (P_2O_5). Figure 1.21 shows the effect of doping on the sensitivity of SPR due to a change in the refractive index of the sensing layer. Between B_2O_3 (5.2) and GeO_2 (19.3) dopants, the simulation forecasts a sensitivity increase of almost 50%. In addition, the sensor's sensitivity drops dramatically when the doping concentration of GeO_2 is raised from 6.3 to 19.3 mol%. In addition, it has been observed that the use of a single metal layer or a bimetallic arrangement makes no difference to the influence of dopants on the sensitivity of the fiber optic SPR sensor.

Multi-metallic coating: Silver and gold are both employed for metallic coatings on the base of the prism or the core of the fiber. Gold's chemical stability and greater sensitivity of its resonance parameter to changes in the sensor layer's refractive index make it an attractive material. In contrast, silver's SPR curve is more convex, meaning that it has a better signal-to-noise ratio (SNR) and hence greater detection accuracy. The imaginary component of the metal's dielectric constant determines the resonance curve's sharpness. Compared to gold, silver's SPR curve has a shorter breadth due to its bigger value of the imaginary component of the dielectric constant, leading to a better SNR, or detection accuracy. However, the true component of the metal's dielectric constant determines the resonance curve shift. Gold exhibits a larger shift in the resonance parameter in response to a change in the refractive index of the sensing layer than silver does because the real portion of the dielectric constant is larger in the case of gold than silver. Silver's oxidation makes it chemically unstable. To address the above issue, a thin surface layer along with an Au coating is required to enhance the sensitivity. In this context, a novel resonant metal film structure was disclosed

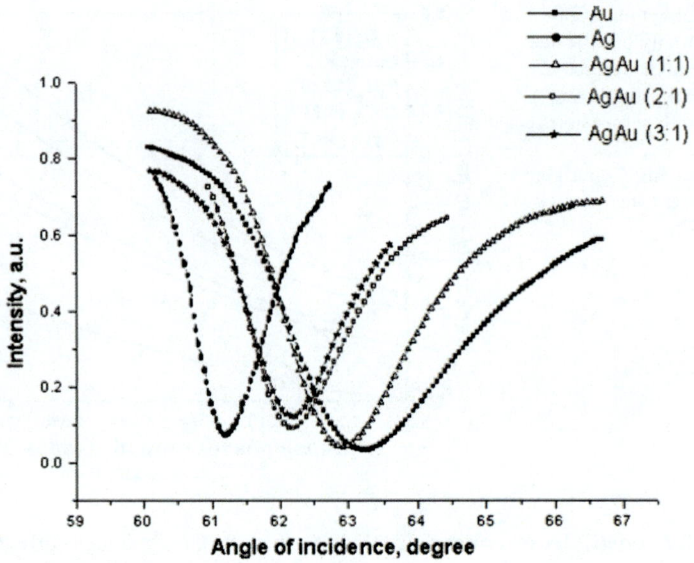

Fig. 1.22 Resonance angle shift due to single metal coating and bimetallic coating. Adapted, 2002, MDPI (Zynio et al. 2002)

(Zynio et al. 2002), one that uses angular interrogation to probe bimetallic layers (with gold as the outer layer) on a prism basis. Because of this, Au is chemically stable and exhibits a greater shift of the resonance angle in response to changes in the ambient refraction index. Despite its low chemical durability, the signal-to-noise ratio of SPR chemical sensors is improved by Ag due to its smaller resonance curve. Figure 1.22 shows variations in resonant angle shift due to single metal coatings and combinations of bimetallic layers. There is a significant decrease in the resonance half width, which results in better accuracy of SPR sensing.

Fiber shape for proper incidence: In order to get the ray's angle of incidence with the normal to the core–cladding contact near the critical angle, a U-shaped probe is used. As illustrated in Fig. 1.20 (Verma and Gupta 2008), A bidimensional model was used to analyze an SPR-based fiber optic sensor with a homogeneous semimetal-coated U-shaped probe. In this structure, all electric vectors and rays of p-polarized light entering the fiber are supposed to stay inside the U-shaped probe's bending plane. As the bending radius decreases, the sensitivity of SPR increases. The U-shaped probe is made by cutting off a section of the fiber's cladding from its midsection. By applying heat, the central, unclad portion of the fiber is shaped into a U. As can be seen in Fig. 1.23, the U-shaped SPR probe that will undergo theoretical analysis comprises a bended portion that is unclad (layer 1) and a length of its bottom region that is covered with a metallic layer (layer 2).

The critical angle of a guided ray in the sensing area (bent region) is the angle by which the beam is deviated from the normal to the core–cladding contact. Assume

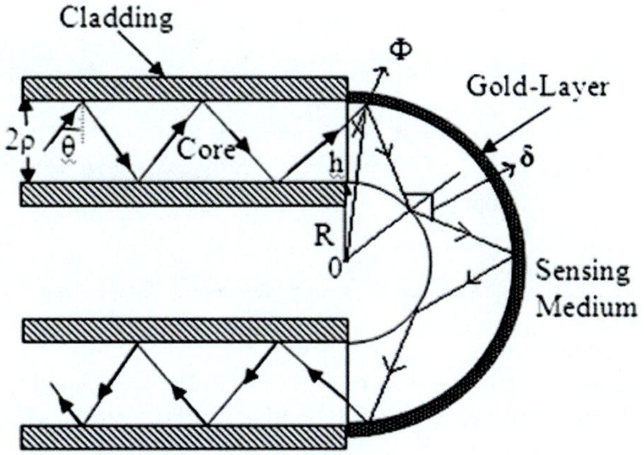

Fig. 1.23 Schematic diagram of U-shaped fiber for surface plasmon resonance. Adapted from Gupta and Verma (2009)

that θ represents the angle of incidence at the core–cladding interface of the fiber; the angle transformed at the outer and inner regions of the bent section is given by

$$\varnothing = \sin^{-1}\left[\frac{R+h}{R+2\rho}\sin\theta\right] \tag{1.33}$$

$$\delta = \sin^{-1}\left[\frac{R+h}{R}\sin\theta\right] \tag{1.34}$$

All parameters of Eqs. (1.33) and (1.34) are indicated in Fig. 1.20.

Side polishing: Fiber-optic SPR probes have also been made using single-mode fiber. Side-polished single-mode optical fiber with a thin metal coating for a surface plasmon resonance sensor. The surface plasmon wave between the metal and the sensing medium is excited by the guided mode traveling in the fiber. If the two modes are in phase with one another, resonance will occur. When compared to SPR sensors that use multimode optical fibers, single-mode optical fiber-based sensors have more sensitivity and precision. In contrast to multimode fiber optics, however, their production requires a higher level of complexity and sophistication. The refractive index may be measured with just a small quantity of sample when using a side-polished half-block SPR sensor, which is a major benefit. The guided mode of a mono-mode optical fiber and a surface plasmon wave supported by a thin metal overlayer are shown to interact in resonance, resulting in a unique optical fiber sensor that authors have described in their work (Homola and Slavik 1996).

Light from a polychromatic source is shone into the optical fiber at one end if the spectral interrogation technique is utilized, as illustrated in Fig. 1.24. For rays traveling at an angle between the critical angle (which varies with the numerical aperture of the fiber and the light wavelength) and about 90°, complete internal reflection occurs. Because of this, surface plasmons at the fiber core–metal layer contact are excited by the evanescent field. The evanescent field's connection with

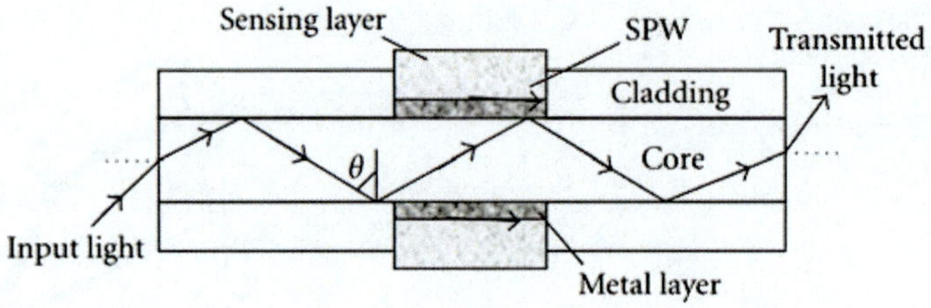

Fig. 1.24 Side polished fiber for SPR sensing. Adapted from Gupta and Verma (2009)

surface plasmons is very sensitive to the wavelength of the light, the parameters of the fiber, the geometry of the fiber, and the characteristics of the metal layer.

1.4 Applications

There are a wide variety of uses for fiber optic sensors based on surface plasmon resonance for the quantitative detection of chemical and biological substances. The quality of food, medical diagnosis, and ecological research are all examples. To detect these, the medium around the metallic coating is altered in refractive index. The refractive index of the medium is altered either immediately or indirectly by the measurand. There are various structures of interferometers, as discussed in Sect. 1.2 of this chapter, and the applications of these structures are depicted in Fig. 1.25.

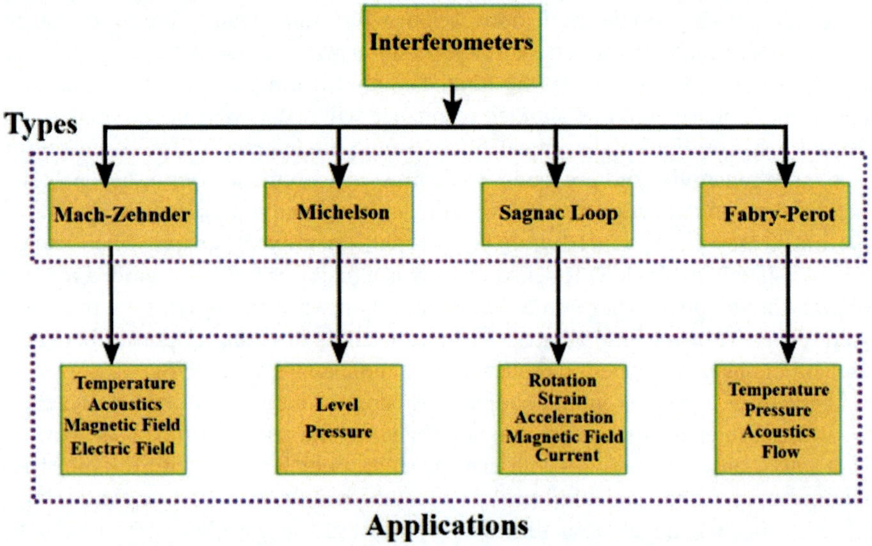

Fig. 1.25 Types of interferometers and their applications

The applications of surface plasmon resonance are described in brief as follows.

1.4.1 Temperature

The refractive index of analytes changes with temperature, which can be utilized in SPR sensing. Temperature sensing in SPR is reported for prism-based structures and angular modulation techniques (Turhan-Sayan 2003). Titanium dioxide and silicon acrylate, which have a significant thermo-optic coefficient, were proposed as sensing layers for use with the SPR method for temperature monitoring. Any physical factor that can cause variations in the refractive indices and thicknesses of materials forming the SPR configuration will generate significant changes in the plasmon dispersion relations and, thus, in the SPR spectrum, as the extreme sensitivity of the SPR phenomenon relies on the collective oscillation of free electrons at the metal surfaces. The sensitivity of SPR sensors has been found to degrade as a function of temperature, which is one of these parameters. Furthermore, by adopting an appropriate thickness for the two layers, surface plasmon wave penetration may be limited to only the metal and sensing layers. Further, the remote sensing of temperature based on SPR sensing is reported (Sharma and Gupta 2006), which theoretically analyzes the system in terms of angle of incidence, metallic layer, and sensing region. A temperature effect is observed with changes in the geometric structure and thickness of the metallic layer. Several methods have been proposed to counteract the detrimental effects of temperature on SPR sensors. However, this feature of the SPR phenomenon may be used for temperature sensing if the SPR shape is chosen appropriately. In thin-film sensing systems that rely on SPR, temperature-dependent readings are still underutilized.

1.4.2 Acoustics

Subcutaneous microvasculature imaging, tumor detection in early stages, and measuring blood oxygen levels are just a few examples of how this technology has been put to use during the last several decades in the study of physiopathology. Frequency domain analysis of photoacoustic signals allows for quantitative assessment of microscopic properties, such as size, shape, orientation, and density, of biological particles randomly dispersed in tissues. The ultrasonic detector's acoustic frequency response must be adequate for this purpose. Because of the physical nature of the piezoelectric material, the piezoelectric transducers frequently utilized in the majority of current photoacoustic imaging setups often function across a very limited bandwidth (centered at their resonance frequency) (Zhou et al. 2011; Xu et al. 2012). An emerging contender for photoacoustic sensing is optical surface wave sensing technology, such as a surface plasmon resonance sensor, due to its ultrafast temporal

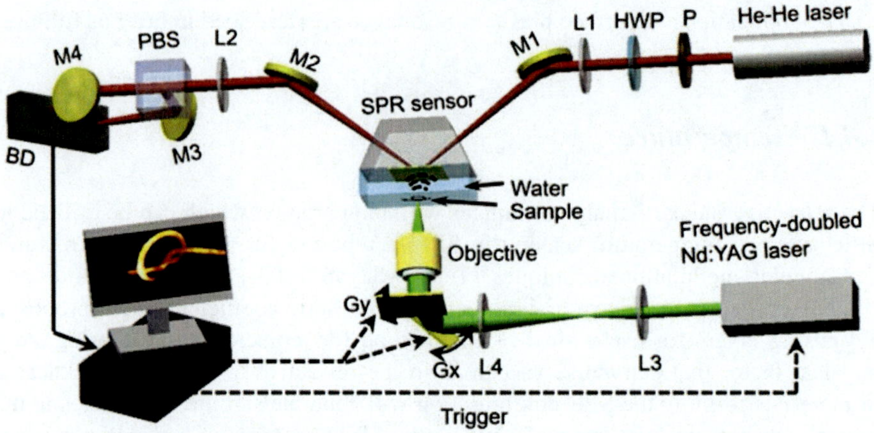

Fig. 1.26 Schematic diagram for the SPR-based high-speed spectroscopic optical-resolution photoacoustic microscopy system. Reprinted with permission from Photoacoustics. Copyright, 2021, Elsevier (Yang et al. 2021)

response and tightly localized evanescent field at the interface of the sensing material and coupling medium (Yang et al. 2018; Wang et al. 2015). Using a broadband SPR sensor kept at rest for wide directivity photoacoustic signal detection and a galvanometer for high-speed optical raster scanning, a fast spectroscopic optical-resolution photoacoustic microscopy system was developed for dynamic imaging and spectroscopic analysis of the moving objects (Yang et al. 2021) as shown in Fig. 1.26.

To sum up, a polarizer (P) and a half-wavelength plate (HWP) are used to modify the linear polarization of 632.8-nm interrogation light from a He–Ne laser. The interrogation light is poorly focused by an achromatic lens (L1) before being incident on the prism–water interface at a 71.9° incidence angle. Since an Au film is resistant to oxidation in air and water, it provides high stability for real-world biomedical photoacoustic applications, making it the material of choice for designing the SPR sensor for photoacoustic imaging. The surface of the Au film is stimulated by an SP field with an elliptical shape. In order to cut down on background noise, polarization-differential light detection is used. This involves a polarization beam splitter (PBS) to divide the linearly polarized reflected beam into p- and s-polarized components and a balanced photodetector (BP) to assess the light intensity. A galvanometer (Gx and Gy) is used for the raster scanning in high-speed photoacoustic imaging, keeping the SPR sensor and sample in one place so that the photoacoustic signal may be acquired without any movement. After being collimated by lenses L3 and L4, the galvanometer refocuses the photoacoustic illumination beam produced by a frequency-doubled Nd:YAG laser.

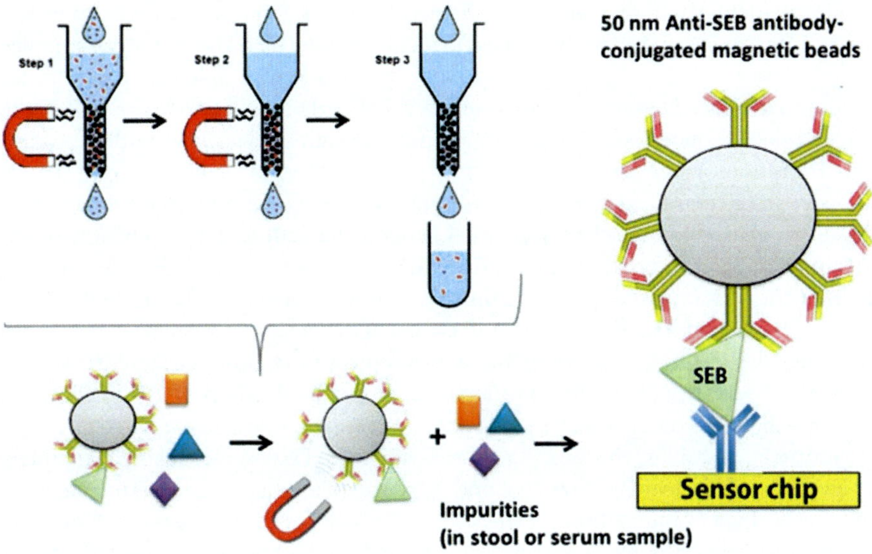

Fig. 1.27 Processing steps of samples with colloidal antibody-coupled nanoparticles and their representation. Adapted from Nguyen et al. (2015)

1.4.3 Magnetic Field

While magnetic nanoparticles have been used for bioseparation for decades, their use in bioassays is more recent. In part because of these benefits—faster binding rates, better miscibility, more specificity, and a larger surface-to-volume ratio—these materials have garnered a lot of interest. The price of magnetic nanoparticles is lower than that of other plasmonic nanoparticles. Figure 1.27 shows the processing steps of samples with colloidal antibody-coupled nanoparticles and their representation. Here, Staphylococcal enterotoxin B was isolated and concentrated from genuine complicated matrices using the beads' magnetic properties before being fed into their small SPR device.

1.4.4 Electric Field

Graphene's 2D structure makes it very sensitive to adsorbed gaseous molecules since each carbon atom essentially functions as a surface atom. Due to its high electron mobility at ambient temperature, it is a promising material for use in ultra-fast sensing. To effectively build and develop gas sensors based on surface plasmon resonance, graphene's aforementioned benefits may be used. Many scientists have long been interested in the negative impacts of heavy metal ions on ecosystems and human health. A quick and easy sensing device for screening pollution in water bodies, particularly in distant regions, is necessary for a viable detection approach. Many different methods have been used to identify heavy metal ions in water samples, including inductively coupled plasma mass spectroscopy, ion-selective electrodes, anodic stripping voltammetry, electrochemical impedance spectroscopy, and quartz crystal microbalance spectroscopy. Applying an external electrical field to produce ion adsorption on the sensor surface is a novel method of detection presented by Kyaw et al. (2015). The term "Electric-Field-Assisted Surface Plasmon Resonance" (EF-SPR) is used to describe the method being discussed. A small mechanism built within an SPR chip provides the optical functionality. As fan-in/fan-out coupling devices, the SPR chip integrates the diffractive optical elements. The reflected beam is photographed in a motionless state to establish the detection system. A black strip denoting an SPR angle appears in the reflected beam at resonance. The SPR surface (gold film) served as the working electrode in this application of EF-SPR, which included the addition of a single electrode to the top of a flow channel.

Similarly, Localized Surface Plasmon Resonance (LSPR) biosensing principle is a powerful technique that utilizes the optical properties of nanomaterials to detect and analyze biological interactions at the nanoscale. LSPR biosensors typically consist of metallic nanoparticles, such as gold and silver, which exhibit unique optical properties due to the collective oscillation of their free electrons when excited by light. The LSPR biosensing principle operates on the basis of changes in the local refractive index surrounding the metallic nanoparticles. When biomolecules, such as proteins and DNA, bind to the nanoparticle surface, they induce a shift in the resonance frequency of the plasmons, resulting in a measurable change in the optical response, such as a shift in the wavelength of the absorbed or scattered light. This shift can be quantified and correlated with the concentration or activity of the target analyte, providing valuable information about the presence or behavior of specific biomolecules (Agrawal et al. 2020a, b; Kumar et al. 2021). Nanomaterials play a crucial role in LSPR biosensors by providing enhanced sensitivity and selectivity. Their unique optical properties, large surface-to-volume ratio, and tunable surface chemistry make them ideal candidates for capturing and detecting biomolecules with high specificity. Nanomaterials can be functionalized with ligands or receptors that selectively bind to the target analyte, enabling precise detection and quantification (Kumar et al. 2023b; Li et al. 2021b; Raghuwanshi et al. 2021). LSPR biosensors have significant clinical importance in various areas of health care. They have been widely applied in medical diagnostics, drug discovery, and personalized medicine. These sensors offer rapid and label-free detection of biomarkers, enabling early diagnosis of diseases, monitoring of therapeutic treatments, and assessment of disease progression. They have the potential to revolutionize clinical diagnostics by providing accurate, sensitive, and real-time analysis of biological samples, such as blood, urine, and saliva, with minimal sample preparation. Moreover, LSPR biosensors offer advantages such as portability, miniaturization, and compatibility with multiplexed analysis, allowing simultaneous detection of multiple analytes. This capability is particularly valuable in point-of-care testing, where quick and reliable diagnostic results are required for timely decision-making and patient management

(Singh and Kumar 2022; Singh et al. 2020). In summary, the LSPR biosensing principle, coupled with the use of nanomaterials, holds great promise for advancing clinical diagnostics. These sensors offer sensitive, selective, and real-time detection of biomolecules, enabling early disease detection, monitoring of therapeutic responses, and improving patient care. The integration of LSPR biosensors into clinical practice has the potential to enhance disease management, facilitate personalized medicine, and contribute to improved healthcare outcomes (Agrawal et al. 2020c, d; Kumar et al. 2022; Singh et al. 2022).

1.5 About This Book

In the last three decades, surface plasmon resonance sensors have outperformed interferometric methods. Surface plasmon resonance (SPR) sensors provide chemical and biological detection benefits over traditional methods. In the first chapter, different surface plasmon sensors and detection methods have been described. Sensitivity, range of the input signal, precision, highest resolution, accuracy, offset error, linear characterization, hysteresis loss, repeatability of the approach, and drift due to electronic components are essential parameters for sensing system performance analysis. These characteristics are briefly given to help novices comprehend sophisticated sensing systems. Before discussing surface plasmon resonance structures and detection methods, Mach–Zehnder, Michelson, Sagnac loop, and Fabry–Perot interferometers are discussed, presenting SPR coupling and detection technologies. Prism, grating, and waveguide SPR couplings are most prevalent. Comparing various coupling setups' pros and cons Angular, intensity, wavelength, and phase change detection systems analyze sensor parameters. The advantages, applications, and drawbacks of different modulation methods are also compared. SPR sensor probe manufacture must consider sensitivity, operating range, sensing environment, repeatability, resolution, accuracy, and more. Sensor fabrication involves metal choice, geometry configuration, dopant effect, multi-metallic coating, fiber shape for light coupling inside the core, and side polishing. After explaining the operating principle, coupling setup, detecting technique, and design parameters, surface plasmon resonance is used for sensing. This chapter's last part briefly discusses the book's substance.

Geometrical optics theory describes the taper fiber optic sensor in Chap. 2. Correlating taper fiber sensor models with simulation results is essential for a thorough investigation. After highlighting the taper radii change, taper angle and length were established. The model's propagation and tapering have been derived. The taper angle is crucial for sensing and transmission. Core radii and sensing length determine the taper angle. The taper angle is negligible. Taper ratio increases as taper angle decays and taper radii decrease. Penetration depth and sensitivity increase as the incidence angle approaches the critical angle. Depth of penetration is crucial for sensing capabilities and rises with operating wavelength. Incident angle diminishes it. The critical angle has a maximum propagation constant of zero at ninety degrees. It lowers more than the critical angle. The propagation constant decreases with wavelength between the critical angle and 90°.

In Chap. 3, we will talk about the U-type fiber optic sensor and how it can be used for SPR sensing. First, the basic ideas of U-type fiber optic sensors are introduced. U-type fiber optic SPR sensors have been the focus of current research and development, and this study examines their architecture and operating principles. U-type fiber optic sensor device design is also covered in this chapter. The research also explores the use of U-type fiber optic SPR sensors in other contexts, highlighting their adaptability and promise. At the end of each chapter is a brief overview that recaps the most important information presented there.

Chapter 4 describes cascaded fiber optic surface plasmon resonance sensors. Multi-analyte/multi-channel sensing is possible with cascaded SPR sensors. This chapter covers a modified hetero-core structure fiber-based dual-channel SPR refractive index sensor. The typical hetero-core structure fiber may be shaped into a circular truncated cone with a varied polishing angle to change the resonance wavelength range. Using wavelength division multiplexing, two cascaded fiber optic surface plasmon resonance sensors may test two substances simultaneously. A cascaded multi-analyte SPR sensor is used to compare coagulant and anticoagulant concentrations for illness diagnosis.

The sensing concepts, device construction, and applications of symmetric and asymmetrically coated fiber optic SPR sensors are discussed in Chap. 5. The chapter starts out by talking about symmetric fiber optic SPR sensors and the advantages they have in terms of signal stability and reproducibility. In order to achieve dependable and precise sensing performance, it highlights the importance of using correct probe preparation techniques. The chapter continues with a discussion of asymmetrically coated (half-coated) fiber optic SPR sensors. Advantages and distinguishing sensing qualities of these sensors are discussed, including higher sensitivity and selectivity due to increased contact between the analyte and the coating. This chapter delves deeper into the coating strategies and functionalization processes for asymmetrically coated fiber optic SPR sensors in preparation for the sensing probe. It stresses the significance of achieving the desired sensor performance by optimizing probe production methods. In this chapter, we focus on one such application—the use of asymmetric fiber optic chemical detectors. Potential applications in fields as diverse as pharmaceutical research, food safety analysis, and environmental monitoring are discussed, with a focus on the great sensitivity and specificity of these techniques. Symmetric and asymmetrically coated fiber optic SPR sensors are summed together to round up the chapter. Their rudiments, methods of preparation, and potential uses are all covered. This chapter explains how SPR sensors work, how they may be optimized, and the various sensing tasks that could benefit from using them.

Analytical models from the finite element approach are presented in Chap. 6 on the D-type photonic crystal fiber performance used in SPR-based refractive index sensing. Sensitivity enhancements for SPR sensors are made possible by the coupling of SPPs made possible by hyperbolic metamaterial (HMM). HMM and 2D materials can be used to enhance the detection sensitivity of biosensors. A graphene monolayer with a D-shaped plastic optic fiber is a graphene monolayer with a D-shaped plastic optic fiber (a G/HMM/D-POF SPR sensor is shown).

Based on a 3D $AuAl_2O_3$ HMM composite structure and a graphene sheet, an optical fiber SPR biosensor, were presented. According to a detailed theoretical and experimental investigation, in the visible to near-infrared range, the SPR resonance peak position may be tunable by changing in the HMM system the number of Au layers. A PCF-based plasmonic sensor with a graphene-coated D shape has been proposed. The sensor's many structural and material parameters have been fine-tuned. To detect biological species such as cells, proteins, and DNA, plasmonic fiber optic biosensors combine the flexibility and compactness of optical fibers with the high sensitivity of nanomaterials. Plasmonic fiber optic biosensors have the potential to revolutionize clinical diagnostics, drug discovery, food process control, illness, and environmental monitoring because of their small size, precision, low cost, and ability to be remotely and distributedly sensed. The stacking interaction of graphene aids in the adsorption of biomolecules, but it also inhibits metals like silver from oxidizing. To model a silver–graphene PCF-SPR refractive index sensor, researchers used the full vector finite element method (FEM). The finite element method is used to do numerical simulations of a surface plasmon resonance (SPR) sensor based on a D-shaped photonic crystal fiber (PCF) to detect changes in the refractive index of liquid analytes (FEM). The performance of the SPR-PCF sensor coated with a graphene layer can be improved by using silver as the plasmonic metal. In order to prevent active plasma material from oxidizing, dielectric materials (such as graphene) are employed (silver). There are several advantages to using the D-shaped PCF-SPR sensor shown here over other types of PCF-SPR sensors, including its ease of manufacture, high sensitivity, low cost, and ability to reuse the sensor.

In Chap. 7, interferometric optical sensor concepts and applications are discussed. Interferometric sensors include Mach–Zehnder, Michelson, Sagnac loop, and Fabry–Perot types. These sensors measure temperature, pressure, stress, strain, etc. Interferometers use fibers' optical pathways to interfere with two beams. Interferometric configurations need beam splitting and beam combining. One optical route should be adjustable to external changes. Interferometric signal variations may reveal the target's wavelength, intensity, phase, frequency, and bandwidth. These modifications improve device sensitivity, accuracy, and dynamic range. Miniaturized fiber optic interferometers enable micro-scale applications. To replace bulky fiber optic components like combiners, beam splitters, and objective lenses, innovative fiber-scale devices are designed. SPR sensing accurately measures the refractive index of materials close to metal, unlike optical interferometers. Given the rising need to detect and analyze chemical and biological components in medicine, environmental monitoring, biotechnology, pharmaceuticals, and food monitoring, SPR sensor technology has great potential. This chapter's last part discusses interferometric SPR sensors' construction, content, performance, and application.

In the last chapter of this book, a review of the application of geometric-based SPR sensors is discussed. SPR sensors are used in chemical sensing, biomolecules, gas and liquid detection, medical diagnostics, chemical reaction rate monitoring, and biomolecular interaction. Due to their sensitivity and efficacy, PCF-SPR sensors have grown in popularity. These sensors outperform uncoated and prism-based optical sensors in sensitivity, selectivity, reaction time, recovery time, and repeatability. PCF-SPR sensors with hollow cores, circular patterns, multiple sensing rings, quasi-periodic patterns, square arrays, D shapes, D-shape dual cores, and other designs can be used in a variety of chemical, biological, gas, and liquid detection applications. In medicine and chemistry, the PCF-SPR sensor may detect medicines, analytes, antigens, and antibodies. PCF-SPR sensors with varied structures may detect analyte refractive index variations. To increase sensitivity and range, change the air hole form, number, placement, and size.

SPR sensing is popular because of its great sensitivity. SPR-based sensors may detect biological and chemical analytes, antibody–antigen interactions, and medical diagnostics. SPR is a powerful optical detection tool for real-time label-free biomolecular interaction analysis in biomedical applications due to its label-free sensing, rapid responsiveness, and high sensitivity. This commonly used method is very sensitive to molecule-induced refractive indices. Fiber optics innovations like PCF are exciting. SPR-based PCF sensors are a promising use of PCF because of its design flexibility and ease of optical modification. Changing the analyte refractive index improves sensing. SPR-microstructured chemical sensor. The refractive index (RI) sensing optical fiber was created for visible to near-infrared wavelengths and gas–liquid pollution detection to achieve coupling, manufacturing convenience, inert gold with plasmonic properties, and analyte. The sensing layer covers MOF outside. Computers examine sensors. The wavelength, amplitude, and full-vector FEM approaches define sensor qualities. FEM builds and analyzes an SPR-based MOF-RIS sensor. Precious gold's chemical stability in water made it a plasmonic material. It is plasmonic because of its flexibility. MOF-SPR sensors provide excellent sensing and industrial compatibility. Ozone monitoring, CO_2 detection, and other uses Environmentalists test water quality. The refractive index range is vast, sensitivity is strong, linear correlation is closer to one, and linear correlation is better. Thus, it has prospective biological and chemical sensing implications. Gold nanowire-based PCF sensors can detect SPR variations using the FEM. For real-time detection, the sensor analyte was put outside the optical fiber. The air hole and gold wire sizes were compared to sensor detection. The sensor was designed for liquids with a refractive index of 1.33–1.36. SPR sensors monitor refractive index. Finite element analysis examined sensor construction and performance. System layers and border scattering characteristics absorb energy. Optimizing structural parameters for higher sensitivity, confinement loss was evaluated for air hole radius, gold line radius, and analyte refractive index.

1.6 Summary

In this chapter, the advancement of surface plasmon resonance sensors over the conventional interferometric approach are explored over the last three decades. Apart from conventional sensing approaches, SPR sensors have promising advantages in the field of chemical and biochemical sensing. Surface plasmon sensors have been reported in various configurations and detection techniques. In the performance analysis of a sensing system, it is important to know about the basic parameters, like sensitivity, range of input signal, precision, highest resolution, accuracy, offset error if any, linear characterization, hysteresis loss, repeatability of the approach, and drift due to electronic components. All these parameters are described in brief, which will help beginners understand a complex sensing system. Before going to the various structures and detection techniques involved in surface plasmon resonance, the working principles of conventional interferometric-based structures like Mach–Zehnder interferometers, Michelson interferometers, Sagnac loop interferometers, and Fabry–Perot interferometers are discussed. Further, various coupling methods and detection schemes involved in SPR are presented. The most commonly used coupling configurations are prism-coupled, grating-coupled, and waveguide-coupled SPR. All these coupling configurations are compared in terms of their advantages and limitations. An analysis of the sensing parameters is demonstrated in terms of angular, intensity, wavelength, and phase change detection schemes. A comparative study of various modulation schemes in terms of advantages, field of application, and limitations is also given. There are various factors like sensitivity, operating range, type of sensing environment, reproducibility, resolution, precision, and many more that must be kept in mind during the probe fabrication process of SPR sensors. Important design elements, including metal choice, geometry configuration, dopant effect, multi-metallic coating, fiber form for light coupling within the core, and side polishing, are discussed as they pertain to sensor production. After covering the basic details of the working principle, coupling configuration, detection scheme, and various design parameters, the application of surface plasmon resonance is discussed in various aspects of sensing. In the last section of this chapter, the content of the book is briefly discussed.

References

Agrawal N et al (2020a) Detection of L-cysteine using silver nanoparticles and graphene oxide immobilized tapered SMS optical fiber structure. IEEE Sens J 20(19):11372–11379

Agrawal N et al (2020b) Development of uric acid sensor using copper oxide and silver nanoparticles immobilized SMSMS fiber structure-based probe. IEEE Trans Instrum Meas 69(11):9097–9104

Agrawal N et al (2020c) Development of dopamine sensor using silver nanoparticles and PEG-functionalized tapered optical fiber structure. IEEE Trans Biomed Eng 67(6):1542–1547

Agrawal N et al (2020d) Ultra-sensitive cholesterol sensor using gold and zinc-oxide nanoparticles immobilized core mismatch MPM/SPS probe. J Lightwave Technol 38(8):2523–2529

Anand U et al (2022) Recent advances in the potential applications of luminescence-based, SPR-based, and carbon-based biosensors. Appl Microbiol Biotechnol 106(8):2827–2853. https://doi.org/10.1007/S00253-022-11901-6

Aparna GM, Tetala KK (2023) Recent progress in development and application of DNA, protein, peptide, glycan, antibody, and aptamer microarrays. Biomolecules 13(4):602

Armani D, Kippenberg T, Spillane S (2003) Ultra-high-Q toroid microcavity on a chip. Nature 421(6926):925–928

Bak S, Kim GH, Jang H, Kim J, Lee J, Kim CS (2018) Real-time SPR imaging based on a large area beam from a wavelength-swept laser. Opt Lett 43(21):5476–5479

Buck J (2004) Fundamentals of optical fibers, vol 50. Wiley

Couture M, Zhao SS, Masson JF (2013) Modern surface plasmon resonance for bioanalytics and biophysics. Phys Chem Chem Phys 15(27):11190–11216

Djordjev K, Choi S et al (2002) Vertically coupled InP microdisk switching devices with electroabsorptive active regions. IEEE Photonics Technol Lett 14(8):1115–1117

Ghatak A, Thyagarajan K (1998) An introduction to fiber optics. Cambridge University Press

Gupta BD, Verma RK (2009) Surface plasmon resonance-based fiber optic sensors: principle, probe designs, and some applications. J Sens

Gupta BD, Shrivastav AM, Usha SP (2016) Surface plasmon resonance-based fiber optic sensors utilizing molecular imprinting. Sensors 16(9):1381

Hansen WN (1968) Electric fields produced by the propagation of plane coherent electromagnetic radiation in a stratified medium. JOSA 58(3):380–390

Homola J, Slavik R (1996) Fibre-optic sensor based on surface plasmon resonance. Electron Lett 32(5):480–482

Homola J, Yee SS (1998) Novel polarization control scheme for spectral surface plasmon resonance sensors. Sens Actuators B Chem 51(1–3):331–339

Homola J, Yee SS, Gauglitz G (1999) Surface plasmon resonance sensors. Sens Actuators B Chem 54(1–2):3–15

Howe CL, Webb KF, Abayzeed SA, Anderson DJ, Denning C, Russell NA (2019) Surface plasmon resonance imaging of excitable cells. J Phys D Appl Phys 52(10):104001

Huang JG, Lee CL, Lin HM, Chuang TL, Wang WS, Juang RH, Lin CW et al (2006) A miniaturized germanium-doped silicon dioxide-based surface plasmon resonance waveguide sensor for immunoassay detection. Biosens Bioelectron 22(4):519–525

Homola J (2008) Surface plasmon resonance sensors for detection of chemical and biological species. ACS Publ 108(2):462–493. https://doi.org/10.1021/cr068107d

Kashif M, Ashrif A, Bakar A, Arsad N, Shaari S, Michelson M-Z (2014) Development of phase detection schemes based on surface plasmon resonance using interferometry. mdpi.com 14:15914–15938. https://doi.org/10.3390/s140915914

Kashyap R, Nemova G (2009) Surface plasmon resonance-based fiber and planar waveguide sensors. J Sens

Kumar S et al (2021) MoS_2 functionalized multicore fiber probes for selective detection of shigella bacteria based on localized plasmon. J Lightwave Technol 39(12):4069–4081

Kumar S et al (2022) Optical fiber-based plasmonic biosensors: trends, techniques, and applications. CRC Press

Kumar R, Singh Y, Raghuwanshi SK, Chandra S, Nadeem D (2023a) Delay and dispersion investigation of optical components for microwave photonic filter. Lect Notes Electr Eng 877:693–703. https://doi.org/10.1007/978-981-19-0312-0_69

Kumar S et al (2023b) Optically active nanomaterials and its biosensing applications—a review. Biosensors 13(1):85

Kyaw HH, Boonruang S, Mohammed WS, Dutta J (2015) Design of electric-field assisted surface plasmon resonance system for the detection of heavy metal ions in water. AIP Adv 5(10):107226

Lawson PR (2003) Optical interferometry comes of age. Sky Telesc 105(5):30

Lee HS, Cho SW, Kim GH, Jeong MY, Won YJ, Kim CS (2016) Parallel imaging of 3d surface profile with space-division multiplexing. Sensors 16(1):129

Li Z, Zhang Y (2007) The methods and applications of SPR phase detection. Optoelectron Technol 27(4):220

Li MC, Chen KR, Kuo CC, Lin YX, Su LC (2021a) A simple phase-sensitive surface plasmon resonance sensor based on simultaneous polarization measurement strategy. Sensors 21(22):7615

Li M et al (2021b) Gold/zinc oxide nanoparticles functionalized tapered SMF-MMF-SMF-based sensor probe for uric acid detection. SPIE/COS Photonics Asia 11895. SPIE

Liedberg B, Nylander C (1983) Surface plasmon resonance for gas detection and biosensing. Sens Actuators 4:299–304 (Elsevier)

Ma Z, Cheng S, Kou W, Chen H, Wang W, Zhang X, Guo T (2019) Sensitivity-enhanced extrinsic Fabry-Perot interferometric fiber-optic microcavity strain sensor. Sensors 19(19):4097

Menon V, Tong W, Xia F, Li C (2004a) Nonreciprocity of counter propagating signals in a monolithically integrated Sagnac interferometer. Opt Lett 29(5):513–515

Menon V, Tong W, Forrest SR (2004b) Control of quality factor and critical coupling in microring resonators through integration of a semiconductor optical amplifier. IEEE Photonics Technol Lett 16(5):1343–1345

Mitsushio M, Miyashita K, Higo M (2006) Sensor properties and surface characterization of the metal-deposited SPR optical fiber sensors with Au, Ag, Cu, and Al. Sens Actuators A 125(2):296–303

Mock J, Medlin J et al (2021) Demonstrating the Sagnac effect using tabletop optics on a rotary platform. aip.scitation.org 31(1):100008. https://doi.org/10.1063/10.0006346

Nguyen HH, Park J, Kang S, Kim M (2015) Surface plasmon resonance: a versatile technique for biosensor applications. Sensors 15(5):10481–10510

Phillips KS (2008) Homola J (ed) Surface plasmon resonance-based sensors. Anal Bioanal Chem 390(5):1221–1222. https://doi.org/10.1007/S00216-007-1821-Y

Post EJ (1967) Sagnac effect. Rev Mod Phys 39(2):475–493. https://doi.org/10.1103/REVMODPHYS.39.475

Prabowo BA, Purwidyantri A, Liu KC (2018) Surface plasmon resonance optical sensor: a review on light source technology. Biosensors 8(3):80

Puiu M, Bala C (2016) SPR and SPR imaging: recent trends in developing nanodevices for detection and real-time monitoring of biomolecular events. Sensors 16(6):870

Qu R et al (2000) Configurable wavelength-selective switch based on fiber grating and fiber loop mirror. IEEE Photonics Technol Lett 12(10):1343–1345. https://doi.org/10.1109/68.883824

Rabus DG, Hamacher M, Troppenz U, Heidrich H (2002) High-Q channel-dropping filters using ring resonators with integrated SOAs. IEEE Photonics Technol Lett 14(10):1442–1444

Raghuwanshi SK, Kumar S, Singh Y (2021) 2D materials for surface plasmon resonance-based sensors. CRC Press

Rossi S, Gazzola E, Capaldo P, Borile G, Romanato F (2018) Grating-coupled surface plasmon resonance (GC-SPR) optimization for phase-interrogation biosensing in a microfluidic chamber. Sensors 18(5):1621

Ruffato G, Zacco G, Romanato F (2012) Innovative exploitation of grating-coupled surface plasmon resonance for sensing. Plasmon Princ Appl 2012:419–444

Sagnac G (1913) The luminiferous ether is detected as a wind effect relative to the ether using a uniformly rotating interferometer

Sharma AK, Gupta BD (2006) Theoretical model of a fiber optic remote sensor based on surface plasmon resonance for temperature detection. Opt Fiber Technol 12(1):87–100

Sharma AK, Gupta BD (2007a) On the performance of different bimetallic combinations in surface plasmon resonance based fiber optic sensors. J Appl Phys 101(9):093111

Sharma AK, Gupta BD (2007b) Influence of dopants on the performance of a fiber optic surface plasmon resonance sensor. Opt Commun 274(2):320–326

Shu X, Jiang S, Huang D (2000) Fiber grating Sagnac loop and its multiwavelength-laser application. IEEE Photonics Technol Lett 12(8):980–982

Singh S, Gupta BD (2010) Simulation of a surface plasmon resonance-based fiber-optic sensor for gas sensing in visible range using films of nanocomposites. Meas Sci Technol 21(11):115202

Singh R, Kumar S (2022) Cancer targeting and diagnosis: recent trends with carbon nanotubes. Nanomaterials 12(13):2283

Singh R et al (2020) Etched multicore fiber sensor using copper oxide and gold nanoparticles decorated graphene oxide structure for cancer cells detection. Biosens Bioelectron 2020(168):112557

Singh R et al (2022) Selective colorimetric detection of cancer cells based on iron/copper nanocatalyst peroxidase activity. IEEE Sens J 22(11):10492–10499

Sun B, Wang X, Huang Z (2010) Study on intensity-modulated surface plasmon resonance array sensor based on polarization control. In: 2010 3rd international conference on biomedical engineering and informatics, vol 4. IEEE, pp 1599–1602

Topor CV, Puiu M, Bala C (2023) Strategies for surface design in surface plasmon resonance (SPR) sensing. Biosensors 13(4):465

Turhan-Sayan G (2003) Temperature effects on surface plasmon resonance: design considerations for an optical temperature sensor. J Lightwave Technol 21(3):805

Vázquez C, Vargas SE, Manuel Sánchez Pena J (2005) Sagnac loop in ring resonators for tunable optical filters. J Light Technol 23(8). https://doi.org/10.1109/JLT.2005.850793

Verma RK, Gupta BD (2008) Theoretical modelling of a bi-dimensional U-shaped surface plasmon resonance based fibre optic sensor for sensitivity enhancement. J Phys D Appl Phys 41(9):095106

Verma RK, Sharma AK, Gupta BD (2008) Surface plasmon resonance based tapered fiber optic sensor with different taper profiles. Opt Commun 281(6):1486–1491

Wang T, Cao R, Ning B, Dixon AJ, Hossack JA, Klibanov AL, Hu S et al (2015) All-optical photoacoustic microscopy based on plasmonic detection of broadband ultrasound. Appl Phys Lett 107(15):153702

Wong CL, Chen GCK, Li X, Ng BK, Shum P, Chen P, Olivo M et al (2013) Colorimetric surface plasmon resonance imaging (SPRI) biosensor array based on polarization orientation. Biosens Bioelectron 47:545–552

Wood RW (1902) XLII. On a remarkable case of uneven distribution of light in a diffraction grating spectrum. Dublin Philos Mag J Sci (London, Edinburgh) 4(21):396–402. https://doi.org/10.1080/14786440209462857

Xu G, Dar IA, Tao C, Liu X, Deng CX, Wang X (2012) Photoacoustic spectrum analysis for microstructure characterization in biological tissue: a feasibility study. Appl Phys Lett 101(22):221102

Yang F, Song W, Zhang C, Min C, Fang H, Du L, Yuan X et al (2018) Broadband graphene-based photoacoustic microscopy with high sensitivity. Nanoscale 10(18):8606–8614

Yang F, Guo G, Zheng S, Fang H, Min C, Song W, Yuan X (2021) Broadband surface plasmon resonance sensor for fast spectroscopic photoacoustic microscopy. Photoacoustics 24:100305

Zeng J, Liang DK, Du Y (2007) Prism surface plasmon resonance sensor based on reflection light intensity interrogation. J Optoelectron Laser 18(2):159

Zhang J, Lit JW (1994) Compound fiber ring resonator: theory. JOSA A 11(6):1867–1873

Zhou Q, Lau S, Wu D, Shung KK (2011) Piezoelectric films for high frequency ultrasonic transducers in biomedical applications. Prog Mater Sci 56(2):139–174

Zynio SA, Samoylov AV, Surovtseva ER, Mirsky VM, Shirshov YM (2002) Bimetallic layers increase sensitivity of affinity sensors based on surface plasmon resonance. Sensors 2(2):62–70

Chapter 2
Taper Fiber-Based SPR Sensor

2.1 Introduction

Tapered fiber optical sensors have been used to increase the extension and intensity of the evanescent wave on the outer layer medium (Yuan and Ding 2011). Taper fiber sensors can be fabricated by gradually tapering the waist from its maximum diameter to a minimum diameter and then repeating the reverse process to create the same waist in other directions. The effective refractive index of composite structures can be greatly modulated by continuously and deeply tapering the fiber core so that the field penetrates enough into the sensing material. The concept of producing tapering in simple optical fiber is itself a difficult task, but we can apply our design through software to check the variation and consequences. The allowed wave propagation constant of the fundamental mode can be significantly tailored by the tapering of the optical fiber core. Chemical sensing, gas sensing, biosensing, and refractive index (RI) detection are the basic requirements for any broad outcome in research. The sensors, which can be used for clinical purposes, industrial factories, and security purposes, should also be safe, cost-effective, and compact at the same time. Fiber optic devices have been used for sensors by scholars for more than twenty years, but now some research is focusing on nanoscale particle detection for different kinds of operations. Dispersion-shifted optical fiber (DSF), dispersion compensation fiber (DCF), and dispersion-flattened fiber (DFF) are designed for the transmission of information at high speeds in optical fiber communication systems. These specialty optical fibers are fabricated by doping the core with Ge and are surrounded by a pure silica cladding. The fiber design supports only a single propagating mode per polarization direction, the so-called HE_{11} model. The difference in refractive indices between the core and cladding enables the guiding of light through total internal reflection (TIR). In optical fibers designed as waveguides with TIR, the cladding is thick enough to prevent any evanescent light from penetrating the medium outside the fiber, which means that all the light coupled into the core will reach the end of the fiber. If the taper radius is decreasing due to which field is getting more intense

S. K. Raghuwanshi et al., *Geometric Feature-Based Fiber Optic Surface Plasmon Resonance Sensors*, Springer Tracts in Electrical and Electronics Engineering,
https://doi.org/10.1007/978-981-99-7297-5_2

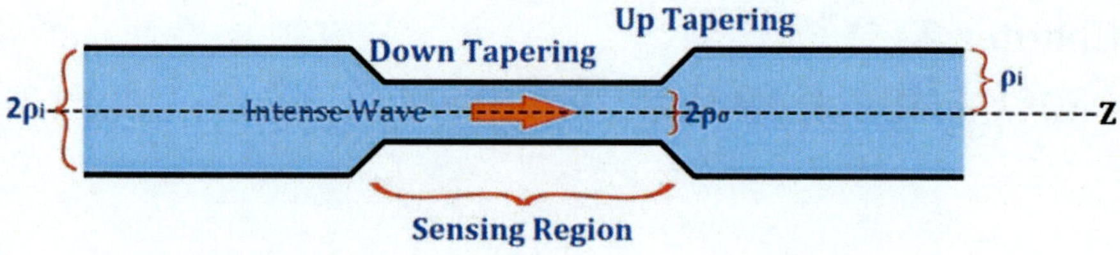

Fig. 2.1 Working methodology of tapered fiber sensors

in the core, this is called a "down taper," and if the taper radius is increasing and the field is not getting intense in the core, this is known as an "up taper." This chapter discusses the details of taper fiber optic sensors in optical fiber communication and photonics systems, device technology, and its effect on optical communication. It also contains complete mathematical analysis with a geometrical interpretation of almost every term. Variation of different parameters in the tapered region is observed in this segment, like being able to understand the taper profile, how the light enters, its number of reflections, and other important measures, as shown in Fig. 2.1.

2.2 Basics Concepts of Taper Fiber

2.2.1 Evanescent Field

Evanescent fields are generated by the total internal reflection phenomenon when light strikes the interface between two mediums with an incident angle greater than the critical angle (Fornel 2001). The evanescent wave forming in an optical fiber can be used in numerous configurations as a sensor.

For optical waveguides, the interaction is proportional to the depth of penetration in the cladding or sensing region. This depth is related to the opto-geometrical parameters of the fibers, which are summed up in the normalized frequency V. The deeper the penetration in the evanescent field is, the easier it is to sense the region, as shown in Fig. 2.2.

2.2.2 Penetration Depth

Light propagation inside the core is tightly confined and guided on axis; however, the evanescent field must be leaked away exponentially toward the cladding region. The evanescent field must decay to $1/e$ of its value at the core–cladding interface after a distance of extension. The corresponding distance is called the penetration depth, as shown in Fig. 2.3.

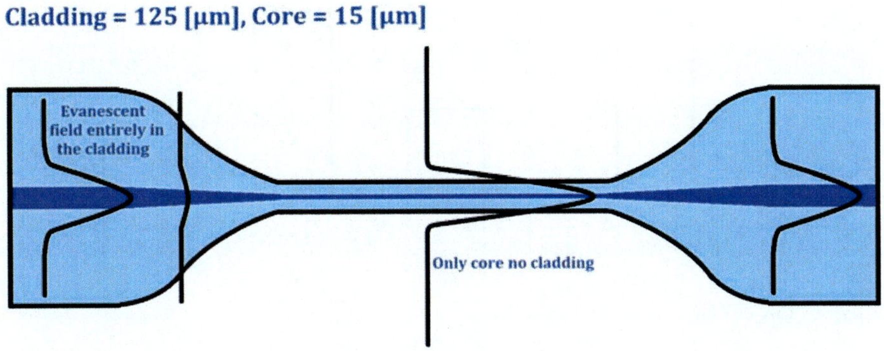

Fig. 2.2 Region showing evanescent fields for different regions

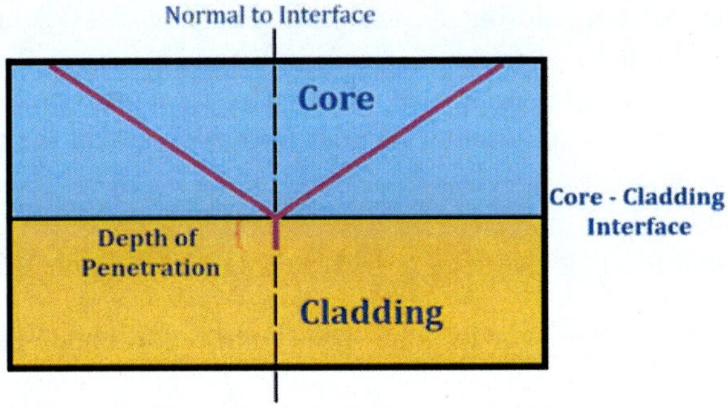

Fig. 2.3 Showing penetration depth at the core–cladding interface

2.2.3 Single-Mode Fibers (SMF-28)

Optical waveguides are much superior to microwave waveguides due to their small dimensions and operating frequencies. This turns out to be a very high bandwidth provided by an optical waveguide with very good stability and reliability of measurements. Presently, we use silica glass (SiO_2) as the backbone of modern communication systems. Earlier, there were losses of 20 dB/km, but presently they are substantially reduced to 0.2 dB/km. Single-mode fibers carry light pulses along a single path only, i.e., they propagate in the lowest order mode (fundamental mode) only, and all higher order modes are not propagating. They have advantages like large information capacity, negligible dispersion, and low degradation. They have a very small core radius and a small difference between the refractive index of the core and the cladding. However, they have some drawbacks, like the difficulty of splicing or joining two single-mode fibers with great precision. Similarly, launching a light with a core of single-mode fiber is difficult with great precision and often requires good couplers, lens assembly, etc. Single-mode fiber can be characterized in several ways, like dispersion-shifted fiber (DSF), viz., double core, double clad fiber, matched clad, depressed core, raided inner cladding, etc., as shown in Fig. 2.4.

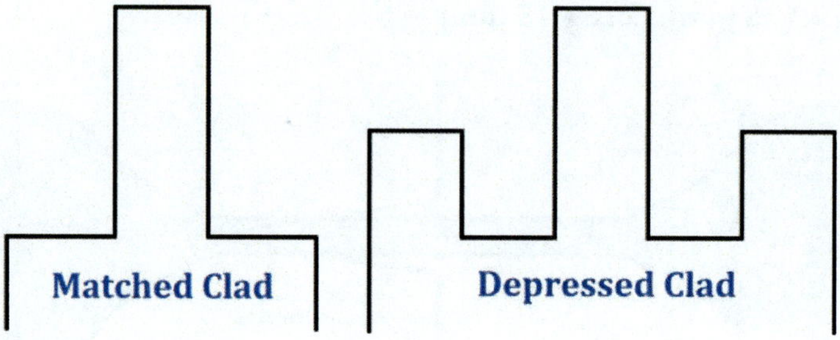

Fig. 2.4 Types of single-mode step-index fiber

Double-core and double-clad fibers may be more useful to tailor the waveguide dispersion. They have specific properties in field conferment and waveguide dispersion tailoring, as discussed in great detail (Fornel 2001).

Model Definition: In the *XY*-plane of the fiber, the mode analysis is made on a cross-section (Gong et al. 2019). The wave propagates in the z direction and has the same form in terms of angular frequency ω and the propagation constant β (Kien et al. 2004),

$$E(x, y, z, t) = E(x, y)e^{-i(\omega t-\beta z)} \tag{2.1}$$

Helmholtz equation has been followed to obtain the eigenvalue equation for the electric field E

$$\nabla \times (\nabla \times E) - {k_0}^2 \eta^2 E = 0 \tag{2.2}$$

The above equation can be solved for eigenvalue $\lambda = -j\beta$.

It is assumed that the field vanishes outside of the cladding region, which is followed by the proper boundary conditions at the interfaces of the core and claddings. The effective mode index of a confined mode is defined by $\eta_{eff} = \frac{\beta}{k_0}$; this is an important criteria to design an efficient taper fiber sensor. Also, we know about normalized frequency of a fiber, i.e.,

$$V = \frac{2\pi\rho}{\lambda_0}\sqrt{n_1^2 - n_2^2} = k_0\rho\sqrt{n_1^2 - n_2^2} \tag{2.3}$$

where ρ is the radius of the core of the fiber.

2.3 Adiabaticity Criteria: Linear Taper Model

If the fiber dimension varies in such a way that the fundamental mode power should remain contained in the waveguide structure taper section without leaking away to cladding modes, this is called adiabatic criteria. Adiabaticity criteria are very important because they provide the opportunity to understand the light propagation properties inside the taper section of multimode fiber. The effective refractive index of a composite structure having a tapered section is a function of the dimensional change in the longitudinal direction of the structure. Hence, it is expected that the fundamental mode can only be coupled to the next higher order mode upon the dimensional change of an axially symmetric composite structure at the critical level. The effective refractive index of composite structures can be as low as the cladding refractive index. In this criteria, coupling due to other higher order modes is generally neglected to minimize the loss in the fundamental mode. Generally, this is a slow tapering process, and since it is a gentle phenomenon, a good number of reflections can happen, and, consequently, a good sensing region can be established. By noticing the change in refractive index, we can measure the electric field or power output variation.

2.4 Key Components Required for Optical Fiber Communication

Key components required for an optical fiber communication system are the optical source that can be modulated at high frequency, single-mode fiber (SMF-28), and optical spectrometer analyzer, as shown in Fig. 2.5 and Table 2.1.

Table 2.1 List of parameters and related optical components required to setup the experiment as in Fig. 2.5

Devices	Parameter	Value
C.W. Laser	Wavelength Optical intensity Laser line width	1550 nm 0 dBm 10 MHz
Fiber (length of 1 and 5 km)	Wavelength Dispersion Second-order dispersion coefficient (β_2) Attenuation	(SMF-28) 1550 nm 17 ps/nm/km $-20\ (\text{ps}^2/\text{km})$ 0.12 dB/km
Optical spectrum analyzer (ANRITSU MS9710B)	Wavelength range	600–1750 nm
Splicing machine	Signal analysis bandwidth	30 MHz

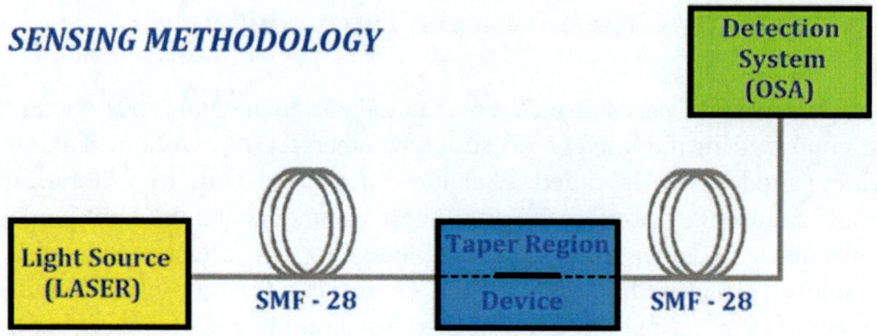

Fig. 2.5 Working methodology of tapered fiber sensor

2.4.1 Optical Source (Laser)

A laser is used as a light source that can transmit the signal at a frequency of around 200 THz. The electric field expression of laser output is $E(t) = E_0 e^{j(2\pi f_0 t + \varphi_0)}$, where E_0 is the amplitude of electric field, and f_0 and φ_0 are frequency and phase of the laser. These three parameters can be modulated.

2.4.2 Optical Spectrum Analyzer (OSA)

To measure the distribution of power from an optical source and also the display of the same over a specified wavelength span, a precision instrument named an optical spectrum analyzer (OSA) is widely used. In an OSA, the wavelength can be monitored on the horizontal scale, and the power can be monitored on the vertical scale.

2.4.3 Fiber Couplers

Coupled mode understanding is required to analyze the behavior of the field inside the tapered fiber. We can find the propagating mode at any point along the taper; these are called local modes. The coupling problem between consecutive segments can be treated by interconnecting many segments with each other's decreased or increasing diameters along the taper sections.

Tapering might be done on optical fibers by heating them under a flame in a controlled manner. Fiber couplers can be easily fabricated in this way without much complexity. The power can be coupled or tuned from one end of a fiber to another, provided that the phase matching condition of the fundamental would be satisfied at the point of coupling region of the individual mode of the composite structure. However, the power will not be coupled between both fibers until the phase condition is satisfied at the point of the tapering section. The tapering section is quite an between both fibers until the phase condition is satisfied at the point of the tapering section. The tapering section is quite sensitive to external perturbations like stress, sound waves, vibration, humidity, etc.

2.5 Taper Fiber Sensor

To understand the various properties of the solid state, a plasma model is required. By applying the boundary condition at interfaces like metal–dielectric, the dispersion relation for surface waves can be used directly. The variation of the electric field associated with surface and plasmon waves across the interface causes an exponential delay at the interface. The main motivation of the chapter was to investigate the parameters and evanescent field in the first place, and then using the in-depth analysis, design and develop a tapered fiber optic sensor. The main contributions of the chapter are to propose and demonstrate some important variations based on its parameters. The derivations are well plotted using the MATLAB tool. Later on, further scope of research work is demonstrated in fiber optic communication utilizing this work.

2.6 Geometrical View of Tapered Fiber and Sensing Principle

2.6.1 Full-Section Geometrical View of Tapered Sensing Region

A schematic diagram representing the geometrical view of full-section tapered sensing is shown in Fig. 2.6, where

$\rho(z)$ Taper radius variation with coordinate z-axis)
ρi Taper radius at input end (at the start of tapered optic fiber)
ρo Taper radius at output end (at the end of tapered optic fiber)
z Propagation direction of light ray-angle from input to output ends
$2L$ Full-sensing length of taper fiber (between both the ends)
L Half-sensing length of taper fiber (between input end and center point)
$\tan\Omega$ Taper angle (angle substended by tapering of core with z-axis).

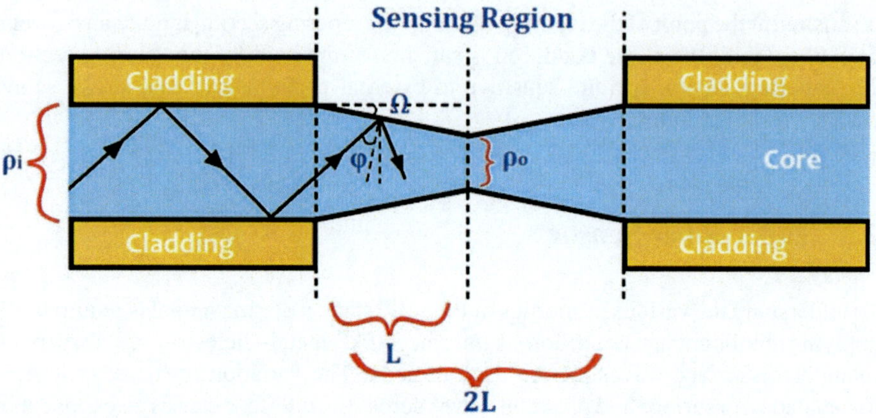

Fig. 2.6 Full-section geometrical view of tapered fiber optic sensing

2.6.2 Half-Section Geometrical View of Tapered Sensing Region

A schematic diagram of the geometrical view of half-section tapered sensing is shown in Fig. 2.7. The sensing length becomes half, i.e., L, for better realization, where

$\rho(z)$	Taper radius variation with coordinate z-axis)
ρi	Taper radius at input end (at the start of tapered optic fiber)
ρo	Taper radius at output end (at the end of tapered optic fiber)
z	Propagation direction of light ray-angle from input to output ends
$2L$	Full-sensing length of taper fiber (between both the ends)
L	Half-sensing length of taper fiber (between input end and center point)
$\tan\Omega$	Taper angle (angle substended by tapering of core with z-axis).

Fig. 2.7 Geometrical view of a half section of taper sensing

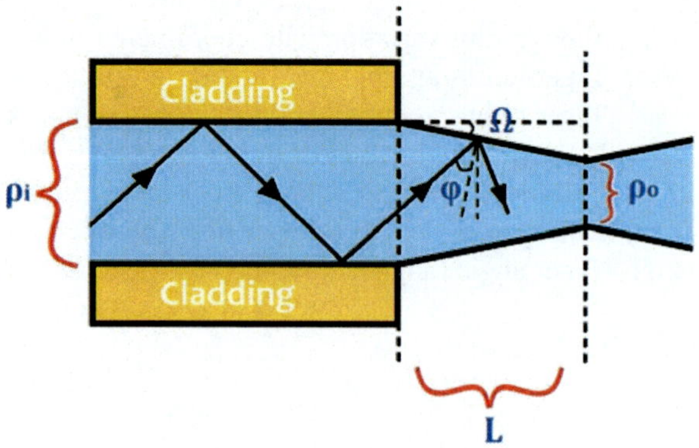

2.6.3 Taper Angle

For simplified analysis, take half the sensing region along the sensing region, as shown in Fig. 2.8. By simplifying the view more, we get the image as follows.

For taper angle calculation, an angle-side relation is required, so we are taking a triangle from its half-sectional region, as shown in Fig. 2.9.

In this triangle, $\tan\Omega = \frac{\rho_i - \rho_o}{L}$. If angles are very small then $\tan\Omega \rightarrow \Omega$ and Ω is considered as taper angle (Jha et al. 2008):

$$\text{Taper angle, } \Omega = \tan^{-1}\left(\frac{\rho_i - \rho_o}{L}\right) \tag{2.4}$$

Taper angle is very dependent on sensing length. As sensing length goes on increasing, taper length increases but decreases as taper ration decreases. For proper sensing operation and transmission, it is very important to select the proper taper angle. The taper angle is dependent on the core radii and sensing length. The taper angle is very small but not zero. If the radius of the taper goes on decreasing, the taper ratio goes on increasing, and the taper angle goes on decaying.

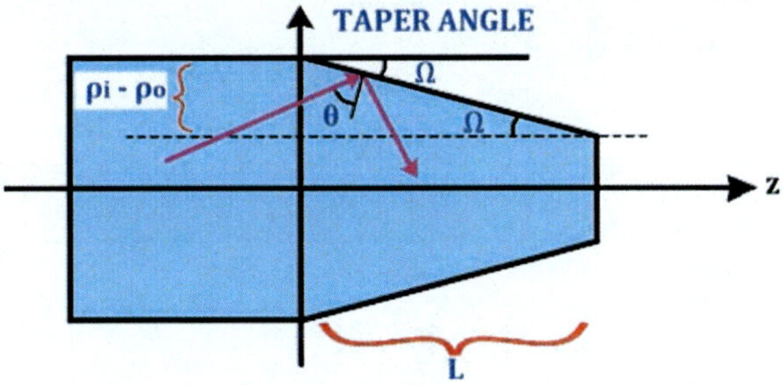

Fig. 2.8 Simplified view of a half-section of taper sensing

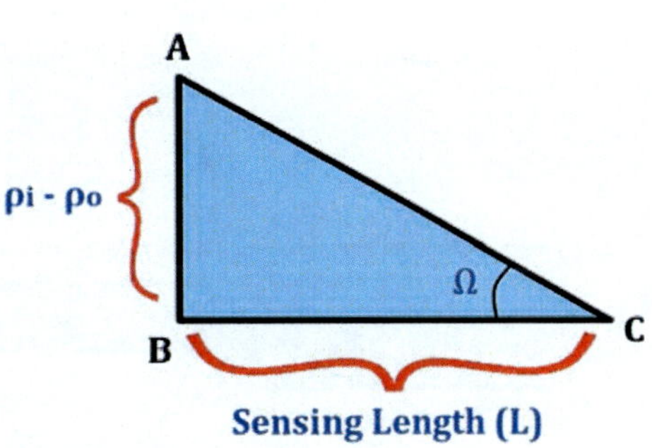

Fig. 2.9 Triangle ABC for the angle-side relationship

2.7 Taper Profile Technique

There are some different approaches to tapering an optical fiber, as shown in Table 2.2. The simplest approach uses a "linear taper profile," which is easier to implement, but sensitivity is between the sensitivities of both an exponential linear taper profile and a parabolic taper fiber profile, as shown in Fig. 2.10. The taper radius variation with z-axis by using all three schemes has been mentioned in the table, but since we are interested in the linear taper fiber profile, we will focus more on that particular scheme. These approaches are listed below in Table 2.2 and have different taper radius variations with axes and characteristics.

In general, if r is radius of core and z is the direction of propagation of beam, i.e., the axis direction then

$$\mathrm{n}\Omega = \frac{d\rho}{dz} \tag{2.5}$$

also,

$$\frac{d\rho}{dz} = \frac{\rho(z) - \rho_i}{z} \tag{2.6}$$

and from geometry,

Table 2.2 Different profile techniques for taper fiber tip as shown in Fig. 2.11

	Taper radius variation with z-axis	Characteristics
Exponential linear taper profile	$\rho(z) = (\rho_i - \rho_o)\left[e^{\left(-\frac{z}{L}\right)} - \frac{z}{L}e^{(-1)}\right] + \rho_o$	Maximum sensitivity but tough to control its radii
Linear taper profile	$\rho(z) = \rho_i - \frac{z}{L}(\rho_i - \rho_o)$	Sensitivity between the two and easier to implement
Parabolic taper profile	$\rho(z) = \left[{\rho_i}^2 - \frac{z}{L}\left({\rho_i}^2 - {\rho_o}^2\right)\right]^{\frac{1}{2}}$	Minimum sensitivity and also tough to control its radii

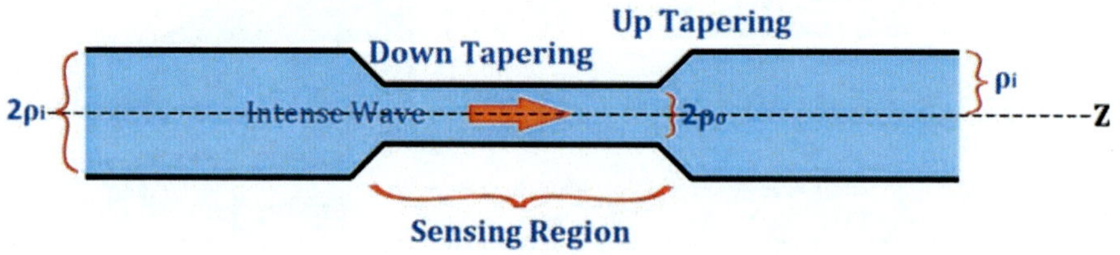

Fig. 2.10 Taper fiber profile

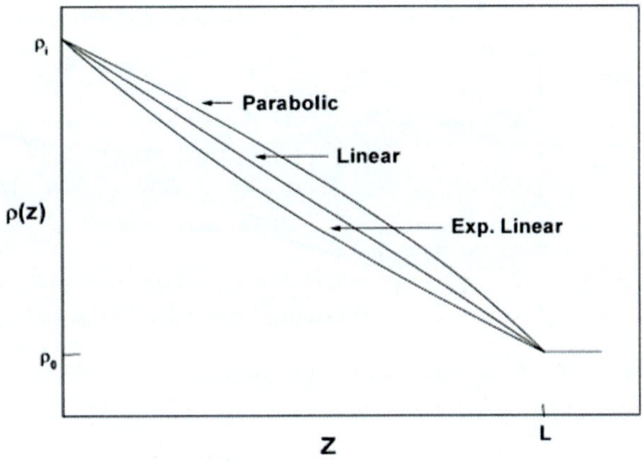

Fig. 2.11 Three different taper profiles show variations in core radius. Reprinted with permission from Optics Communications. Copyright, 2008, Elsevier (Verma et al. 2008)

$$\frac{\rho_i - \rho_o}{L} = \frac{\rho(z) - \rho_i}{z} \tag{2.7}$$

$$\frac{z}{L}(\rho_i - \rho_o) = \rho(z) - \rho_i \tag{2.8}$$

$$\rho(z) = \rho_i - \frac{z}{L}(\rho_i - \rho_o) \tag{2.9}$$

Hence, the taper radius is varying with the coordinate z-axis, as shown in Eq. (2.9).

2.7.1 Variation of Angle Inside the Tapered Region

θ_i the incident angle of the ray with the axis of fiber as shown in Fig. 2.12.
θ_a the angle inside the taper (if considered for any calculation).
θ the angle of the ray after first reflection with normal of fiber.
Ω the taper angle (earlier derived).
$\phi(z)$ the angle subtended with the normal to the interface at starting end of the taper.
$\rho(z)$ taper radii at a distance z from the input end of the taper.
ρ_i taper radii at input end.
ρ_o taper radii at output end.

Snell's law at the outer boundary–core interface.

$$n\sin\theta_i = n_1\sin\left(\frac{\pi}{2} - \alpha\right) \tag{2.10}$$

$$n\sin\left(\frac{\pi}{2} - \theta\right) = n_1\sin\left(\frac{\pi}{2} - \alpha\right) \tag{2.11}$$

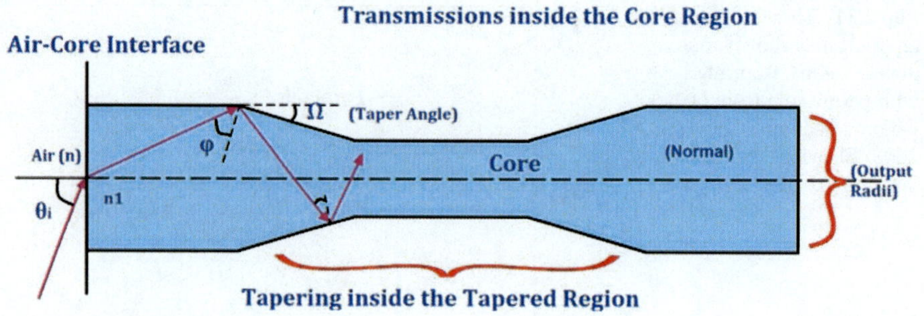

Fig. 2.12 Schematic diagram of transmissions inside the core region

$$n\sin\left(\frac{\pi}{2} - \theta\right) = n_1\cos\alpha \tag{2.12}$$

$$n\cos\theta = n_1\cos\alpha \tag{2.13}$$

$$\left(\frac{n}{n_1}\right)\cos\theta = \cos\alpha \tag{2.14}$$

$$\alpha = \cos^{-1}\left(\frac{n}{n_1}\cos\theta\right) \tag{2.15}$$

Also, $\alpha = \phi + \Omega$

$$\phi = \alpha - \Omega \tag{2.16}$$

Putting the value of taper angle,

$$\Omega = \tan^{-1}\left(\frac{\rho_i - \rho_o}{L}\right) \tag{2.17}$$

$$\phi = \cos^{-1}\left(\frac{n}{n_1}\cos\theta\right) - \tan^{-1}\left(\frac{\rho_i - \rho_o}{L}\right) \tag{2.18}$$

and $\rho(z) = \rho_i - \frac{z}{L}(\rho_i - \rho_o)$.

For linear relation at $z = \text{o}$, $\rho(z) = \rho_i$.

According to Snell's law, we have seen that

$$n\sin\theta_i = n_1\sin\left(\frac{\pi}{2} - \alpha\right) \tag{2.19}$$

so, according to the linearity principle it also satisfies for

$$\rho(z)\sin\theta_a = \rho_i\sin\theta_i$$

Hence,

$$\phi = \cos^{-1}\left(\frac{\rho_i}{\rho(z)} cos\theta\right) - \tan^{-1}\left(\frac{\rho_i - \rho_o}{L}\right) \tag{2.20}$$

In the tapered region, the angle range of rays vary from θ_{critical} to $\pi/2$, and the coordinate z alters to $\phi = \phi_1(z)$ and $\phi_1(z)$

$$\rho(z) = z \text{ then } \theta_1 = \theta_{\text{critical}} \tag{2.21}$$

$$\phi = \cos^{-1}\left(\frac{n}{n_1}\cos\theta\right) - \tan^{-1}\left(\frac{\rho_i - \rho_o}{L}\right) \tag{2.22}$$

$$\rho(z) = \infty \text{ then } \theta_2 = \frac{\pi}{2}$$

$$\phi = \left(\frac{\pi}{2}\right) - \tan^{-1}\left(\frac{\rho_i - \rho_o}{L}\right) \tag{2.23}$$

2.7.2 Number of Reflections Inside Tapered Region

The light is incident from the air into the core of the optical fiber, where it progresses forward in the z-direction. It undergoes many reflections in its path, as shown in Fig. 2.13.

Take its geometrical view for the derivation of the number of reflections. Take the small section where the first collision is taking place as the distance b, as shown in Fig. 2.14.

Here, θ is the angle of incidence at the interface or striking angle. For a single reflection, distance covered by single ray is b. For N number of reflections, total length covered will be $L = NB$ from geometry, $\tan\theta = \frac{b}{2\rho}$

$$2\rho\tan\theta = b \tag{2.24}$$

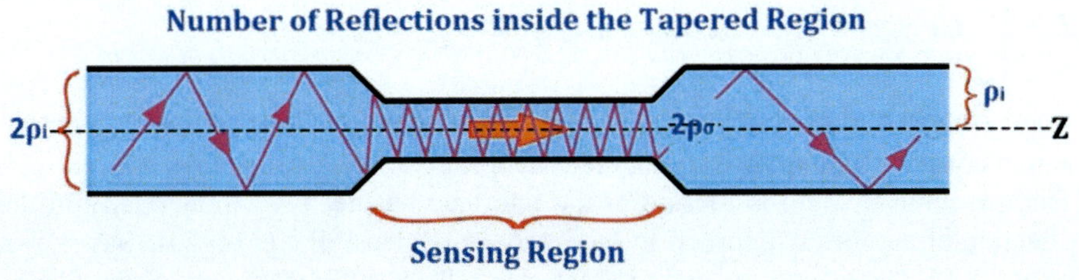

Fig. 2.13 Number of reflections inside the tapered region

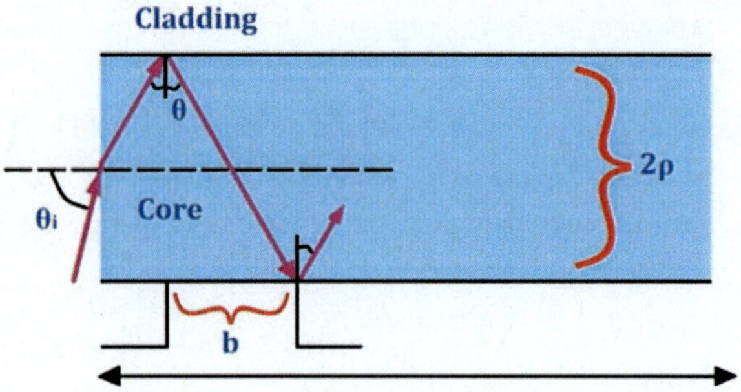

Fig. 2.14 Geometrical view for the derivation of number of the reflections

$$L = 2\rho N \tan\theta \tag{2.25}$$

$$N = \frac{L}{2\rho\tan\theta} = \frac{L\cot\theta}{2\rho} \tag{2.26}$$

Putting, $\theta = \phi + \Omega$

$$N = \frac{L}{2\rho\tan(\theta + \Omega)} = \frac{L\cot(\theta + \Omega)}{2\rho} \tag{2.27}$$

at air–core interface.

From Snell's law (taking $\eta = 1$ for air at air–core interface)

$$n_1 \sin(90 - \theta) = 1.\sin\theta_i$$

The total number of reflections inside the tapered region is dependent upon sensing length (Jha et al. 2008). For a larger number of reflections, we get an angle with the normal to the axis of each ray that goes on decreasing inside the taper region, resulting in a smaller number of reflections. Another important measure can be analyzed by varying the taper ratio for the number of reflections. More reflections for continuously decreasing core radii or a greater taper ratio For the satisfaction of TIR, the angle inside the tapered section is kept between the critical angle and ninety degrees.

2.7.3 *Concept of Penetration Depth*

Depth of penetration, $\delta = \frac{1}{\alpha} = \frac{1}{\beta}$, where α = attenuation constant and β = propagation constant. For space varying electric field, $E = E_0 \exp(-z/\delta) = E_0 \exp^{-\alpha z}$. The maximum E-field is located at the interface but decays exponentially in the direction of an outward normal to the interface (Balan Pillai et al. 2012) as shown in Fig. 2.15. Here, $n_1 > n_2$ and total internal reflection (TIR) occurs. From Snell's law, $n_1\sin\theta_1 = n_2\sin\theta_2$. Assuming light is incident at critical angle, $\theta_1 = \theta_c$:

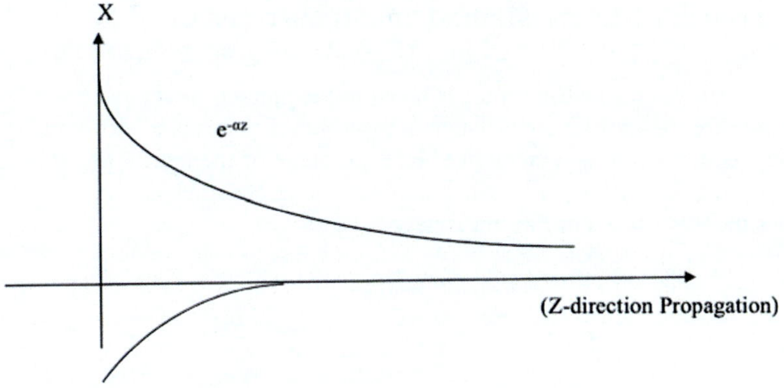

Fig. 2.15 Depth of penetration shown at the interface

$$n_1 \sin\theta_c = n_2 \sin\frac{\pi}{2} \tag{2.28}$$

$$n_1 \sin\theta_c = n_2 = M(let)$$

$$n_1^2 \sin^2\theta_c - n_2^2 = M^2 \tag{2.29}$$

$$M = n_1\left(\sqrt{n_1^2 \sin^2\theta_c - \frac{n_2^2}{n_1^2}}\right) \tag{2.30}$$

$$M = n_1\left(\sqrt{n_1^2 \sin^2\theta_c - n_{21}^2}\right) \tag{2.31}$$

where $n_{21} = \frac{n_2}{n_1}$

Now, depth of penetration, $\delta = \frac{1}{\alpha} = \frac{1}{\beta}$.
Where $\alpha =$ attenuation constant and $\beta =$ propagation constant.
Also, $\beta = \frac{2\pi}{\lambda} M$

$$\delta = \frac{1}{\beta} = \frac{\lambda}{2\pi M} \tag{2.32}$$

Putting the value of M,

$$\delta = \frac{\lambda}{2\pi M} = \frac{\lambda}{2\pi n_1\left(\sqrt{n_1^2 \sin^2\theta_c - n_{21}^2}\right)} \tag{2.33}$$

Depth of penetration is typically less than λ.

2.8 Transfer Matrix Method and Transmission

At the interface, the electric vector is taken in the plane of incidence (p-polarized). Keep our case for homogeneous and non-absorbing dielectrics as shown in Fig. 2.16.

The electric vector is assumed to be in the plane of incidence, i.e., p-polarized. Here, plane $x = 0$, which represents the interface of two dielectrics; since the field satisfies the wave equation, we must have

$$\beta_1^2 = \omega^2 \varepsilon_1 \mu_1 \tag{2.34}$$

$$\beta_2^2 = \omega^2 \varepsilon_2 \mu_2 \tag{2.35}$$

$$\beta_3^2 = \omega^2 \varepsilon_3 \mu_3 \tag{2.36}$$

Here, we have two cases: (1) the E-vector lies in the plane of incidence, and (2) the E-vector lies in the perpendicular plane of incidence (Zvyagin and Ohtsu 1997). By using boundary conditions, if the E-field vector associated with an incident plane wave lies in the plane of incidence, then the electric vector associated with reflected and refracted waves also lies in the plane of incidence (Avendaño-Alejo et al. 2007). The same will also hold for an electric vector lying in the perpendicular plane of incidence (Zvyagin and Ohtsu 1997). Electric vector fields can be represented as

$$\vec{E} = E_1 e^{j(\omega_1 t - \beta_1 z)} \tag{2.37}$$

$$\vec{E} = E_2 e^{j(\omega_2 t - \beta_2 z)} \tag{2.38}$$

$$\vec{E} = E_3 e^{j(\omega_3 t - \beta_3 z)} \tag{2.39}$$

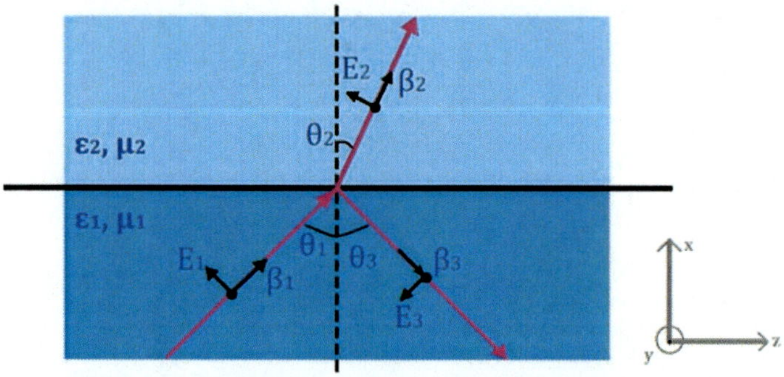

Fig. 2.16 Reflection and refraction of plane waves incident at the interface of two dielectrics

All the fields are assumed to lie in the plane of incidence. Resolving the electric field vector E along the x- and z-axes, since the z-component is tangential to the surface, we must have E_z continuous across the interface:

$$E_{1z} + E_{3z} = E_{2z} \tag{2.40}$$

$$E_1 e^{j(\omega_1 t - \beta_1 z)} \cos\theta_1 + E_3 e^{j(\omega_3 t - \beta_3 z)} \cos\theta_3 = E_2 e^{j(\omega_2 t - \beta_2 z)} \cos\theta_2 \tag{2.41}$$

where $\beta_Z = \beta_x x + \beta_y y + \beta_z z$.

For plane, $x = 0$

$$\beta_Z = \beta_y y + \beta_z z \tag{2.42}$$

Snell's law of refraction

$$n_1 \sin\theta_1 = n_2 \sin\theta_2 \tag{2.43}$$

For a two-layer model, the transfer matrix method has been widely used in optics to analyze the propagation of electromagnetic waves through a stratified (layered) medium (Dehdashti et al. 2013). Dielectric mirrors and anti-reflective coatings have been analyzed by the same method. The Fresnel equation is used to study the reflection and transmission of light from a multilayer structure (Mohammed 2019). To derive the reflection and transmission coefficients, the field continuity condition should be satisfied at each interface, as shown in Fig. 2.17.

Tangential field of electric field

$$E_i = a_{il} + b_{il} \tag{2.44}$$

Tangential field of magnetic field

$$H_i = \frac{a_{il} - b_{il}}{z_i} \tag{2.45}$$

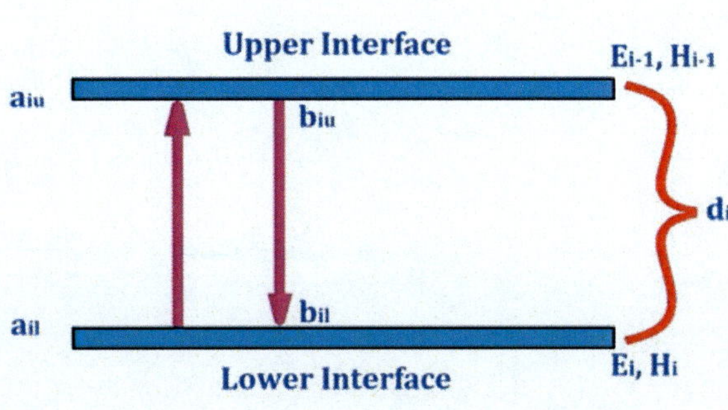

Fig. 2.17 Two-interface, three-layer model representing the transfer matrix (TMM)

Now,

$$E_{i-1} = a_{iu} + b_{iu} \tag{2.46}$$

$$E_{i-1} = a_{il}e^{-j\beta_i d_i} + b_{il}e^{+j\beta_i d_i} \tag{2.47}$$

$$H_{i-1} = \frac{a_{il}e^{-j\beta_i d_i} - b_{il}e^{+j\beta_i d_i}}{z_i} \tag{2.48}$$

From Eq. (2.48),

$$H_i z_i = a_{il} - b_{il} \tag{2.49}$$

Equations (2.46) + (2.49):

$$2a_{il} = E_i + H_i z_i \tag{2.50}$$

$$a_{il} = \frac{E_i + H_i z_i}{2} \tag{2.51}$$

Equations (2.46)–(2.51);

$$2b_{il} = E_i - H_i z_i \tag{2.52}$$

$$b_{il} = \frac{E_i - H_i z_i}{2} \tag{2.53}$$

Putting the value of a_{il} and b_{il} in equations, we get

$$E_{i-1} = \frac{E_i + H_i z_i}{2} e^{-j\beta_i d_i} + \frac{E_i - H_i z_i}{2} e^{+j\beta_i d_i} \tag{2.54}$$

$$e^{-j\beta_i d_i} = \cos(\beta_i d_i) - j\sin(\beta_i d_i) \tag{2.55}$$

$$e^{+j\beta_i d_i} = \cos(\beta_i d_i) + j\sin(\beta_i d_i) \tag{2.56}$$

$$E_{i-1} = E_i\cos(\beta_i d_i) - j(H_i z_i)\sin(\beta_i d_i) \tag{2.57}$$

$$H_{i-1} = \frac{\frac{E_i + H_i z_i}{2} e^{-j\beta_i d_i} - \frac{E_i - H_i z_i}{2} e^{+j\beta_i d_i}}{z_i} \tag{2.58}$$

$$H_{i-1} = \frac{-j}{2i}\sin(\beta_i d_i) + H_i\cos(\beta_i d_i) \tag{2.59}$$

Representing equations E_{i-1} and H_{i-1} in matrix form, we get

$$\begin{bmatrix} E_{i-1} \\ H_{i-1} \end{bmatrix} = \begin{bmatrix} \cos(\beta_i d_i) & -jz_i \sin(\beta_i d_i) \\ \frac{-j\sin(\beta_i d_i)}{z_i} & \cos(\beta_i d_i) \end{bmatrix} \begin{bmatrix} E_i \\ H_i \end{bmatrix} \tag{2.60}$$

$$\begin{bmatrix} E_{i-1} \\ H_{i-1} \end{bmatrix} = M_i \begin{bmatrix} E_i \\ H_i \end{bmatrix} \tag{2.61}$$

where

$$M = \begin{bmatrix} M_{11} & M_{12} \\ M_{21} & M_{22} \end{bmatrix} \tag{2.62}$$

$$M_i = \begin{bmatrix} \cos(\beta_i d_i) & -jz_i \sin(\beta_i d_i) \\ \frac{-j\sin(\beta_i d_i)}{z_i} & \cos(\beta_i d_i) \end{bmatrix} \tag{2.63}$$

For p-polarized wave at kth boundary,

$$z_k = \left(\frac{\mu_k}{\varepsilon_k}\right)^{1/2} \cos\theta_k \tag{2.64}$$

$$\mu_k = 1$$

$$z_k = \left(\frac{1}{\varepsilon_k}\right)^{1/2} \cos\theta_k \tag{2.65}$$

$$z_k = \frac{\left(\varepsilon_k - n_1^2 \sin^2\theta_{i1}\right)^{1/2}}{\varepsilon_k} \tag{2.66}$$

For our case at 2nd interface, i.e., $k = 2$,

$$z_2 = \frac{\left(\varepsilon_2 - n_1^2 \sin^2\theta_{i1}\right)^{1/2}}{\varepsilon_2} \tag{2.67}$$

Propagation constant,

$$\beta_k = \frac{2\pi}{\lambda} \eta_k \cos\theta_k (z_k - z_{k-1}) \tag{2.68}$$

$$\cos\theta_k = \left(1 - n_{12}^2 \sin^2\theta_k\right)^{1/2} \tag{2.69}$$

$$\cos\theta_k = \left(\frac{n_2^2 - n_1^2 \sin^2\theta_k}{n_2^2}\right)^{1/2} \tag{2.70}$$

and $\varepsilon_2 = n_2^2$

$$\cos\theta_k = \frac{\left(\varepsilon_2 - n_1^2 \sin^2\theta_k\right)^{1/2}}{\sqrt{\varepsilon_2}} \tag{2.71}$$

$$\beta_k = \frac{2\pi d_k}{\lambda}\left(\varepsilon_k - \eta_1^2 \sin^2\theta_1\right)^{1/2} \tag{2.72}$$

The reflection coefficient for p-wave TM is (Gupta and Sharma 2005)

$$r_p = \frac{(M_{11} + M_{12}q_3)q_1 - (M_{21} + M_{22}q_3)}{(M_{11} + M_{12}q_3)q_1 + (M_{21} + M_{22}q_3)} \tag{2.73}$$

Reflectivity is (Gupta and Sharma 2005)

$$R_p = \left|r_p\right|^2 \tag{2.74}$$

On further analysis from Fig. 2.10,

$$E_1 e^{j(\omega_1 t - \beta_1 y - \beta_1 z)}\cos\theta_1 + E_3 e^{j(\omega_3 t - \beta_3 y - \beta_3 z)}\cos\theta_3 = E_2 e^{j(\omega_2 t - \beta_2 y - \beta_2 z)}\cos\theta_2 \tag{2.75}$$

$$E_1\cos\theta_1 + E_3\cos\theta_3 = E_2\cos\theta_2 \tag{2.76}$$

As $\beta_1, \beta_2, \beta_3$ will all be parallel to XZ-plane, also

$$\beta_1\sin\theta_1 = \beta_2\sin\theta_2 = \beta_3\sin\theta_3 \tag{2.77}$$

Since $\beta_{1y} = \beta_{2y} = \beta_{3y}$ and $\beta_{1z} = \beta_{2z} = \beta_{3z}$, and $\beta_1\sin\theta_1 = \beta_3\sin\theta_3$ and $\beta_1 = \beta_3$, therefore, $\sin\theta_1 = \sin\theta_3$, we get $\theta_1 = \theta_3$. Normal components of electric field is continuous:

$$D_{1x} + D_{3x} = D_{2x} \tag{2.78}$$

$$\varepsilon_1 E_1\sin\theta_1 - \varepsilon_1 E_3\sin\theta_3 = \varepsilon_2 E_2\sin\theta_2 \tag{2.79}$$

$$E_1\cos\theta_1 + E_3\cos\theta_3 = \frac{\cos\theta_2(\varepsilon_1 E_1\sin\theta_1 - \varepsilon_1 E_3\sin\theta_3)}{\varepsilon_2\sin\theta_2} \tag{2.80}$$

$$\varepsilon_2 E_1\cos\theta_1\sin\theta_2 + \varepsilon_2 E_3\cos\theta_3\sin\theta_2 = \cos\theta_2(\varepsilon_1 E_1\sin\theta_1 - \varepsilon_1 E_3\sin\theta_3) \tag{2.81}$$

Already, we have proved that $\theta_1 = \theta_3$, thus putting this in the Eq. (2.81), we get

$$\varepsilon_2 E_1\cos\theta_1\sin\theta_2 + \varepsilon_2 E_3\cos\theta_1\sin\theta_2 = \cos\theta_2(\varepsilon_1 E_1\sin\theta_1 - \varepsilon_1 E_3\sin\theta_1) \tag{2.82}$$

$$\varepsilon_1 E_1 \sin\theta_1 \cos\theta_2 - \varepsilon_2 E_1 \cos\theta_1 \sin\theta_2 = (\varepsilon_1 E_3 \sin\theta_1 \cos\theta_2 + \varepsilon_2 E_3 \cos\theta_1 \sin\theta_2) \quad (2.83)$$

$$E_1(\varepsilon_1 \sin\theta_1 \cos\theta_2 - \varepsilon_2 \cos\theta_1 \sin\theta_2) = E_3(\varepsilon_1 \sin\theta_1 \cos\theta_2 + \varepsilon_2 \cos\theta_1 \sin\theta_2) \quad (2.84)$$

$$\frac{E_3}{E_1} = \frac{\varepsilon_1 \sin\theta_1 \cos\theta_2 - \varepsilon_2 \cos\theta_1 \sin\theta_2}{\varepsilon_1 \sin\theta_1 \cos\theta_2 + \varepsilon_2 \cos\theta_1 \sin\theta_2} \quad (2.85)$$

$$n_1 = \frac{c}{v_1} = \sqrt{\frac{\varepsilon_1 \mu_1}{\varepsilon_0 \mu_0}} \quad (2.86)$$

$$n_2 = \frac{c}{v_2} = \sqrt{\frac{\varepsilon_2 \mu_2}{\varepsilon_0 \mu_0}} \quad (2.87)$$

Putting above parameters, we get

$$\Gamma_p = \frac{E_3}{E_1} = \frac{n_1^2 \sin\theta_1 \cos\theta_2 - n_2^2 \cos\theta_1 \sin\theta_2}{n_1^2 \sin\theta_1 \cos\theta_2 + n_2^2 \cos\theta_1 \sin\theta_2} \quad (2.88)$$

Reflection coefficient, $\Gamma_p = \frac{E_3}{E_1}$ and reflectivity, $R_p = |\Gamma_p|^2$.

Considering the situation for only single-mode propagation in fiber and keeping our rays completely inside the core, taking $n_1 = n_2$ and Snell's law at the interface

$$n_1 \sin\theta_1 = n_2 \sin\theta_2 \quad (2.89)$$

i.e., $\sin\theta_1 = \sin\theta_2$. But this case is possible only when $\theta_1 = \theta_2$. Since for no reflection and total transmission $\Gamma_p = 0$,

$$\frac{n_1^2 \sin\theta_1 \cos\theta_2 - n_2^2 \cos\theta_1 \sin\theta_2}{n_1^2 \sin\theta_1 \cos\theta_2 + n_2^2 \cos\theta_1 \sin\theta_2} = 0 \quad (2.90)$$

$$n_1^2 \sin\theta_1 \cos\theta_2 = n_2^2 \cos\theta_1 \sin\theta_2 \quad (2.91)$$

If we consider propagation inside the core, only then incident light can be represented in terms of θ and n_1 as $n_1^2 \sin\theta_1 \cos\theta_2$ and since $\theta_1 = \theta_2 = \theta$ then it can be represented as $n_1^2 \sin\theta \cos\theta$. As we are giving rays inside by some means of source like led or laser, it can be also verified by "Lambert's cosine law" which says that the total radiant power observed from a surface is directly proportional to the cosine of an angle θ between the normal to surface and observer's line of sight. Total luminous intensity: $\phi \propto \sin\theta \cos\theta$. Again, as both the angle maximum sum up to 90 degrees, i.e., $\theta_1 + \theta_2 = \frac{\pi}{2}$,

$$\theta_1 = \theta_2 = \frac{\pi}{4} \quad (2.92)$$

$$T = \frac{4n_1 n_2 \cos\theta \cos\theta_T}{|n_1 \cos\theta + n_2 \cos\theta_T|^2} \quad (2.93)$$

where θ_T is the refracted angle

$$\cos\theta_T = \left(1 - \left(\frac{n_1^2}{n_2^2}\right)\sin^2\theta\right)^{1/2} \quad (2.94)$$

$$\cos\theta_T = \left(1 - n_{12}^2 \sin^2\theta\right)^{1/2} \quad (2.95)$$

2.9 Taper Fiber-Based Sensors

Goswami et al.'s (2016) proposal of a surface plasmon resonance (SPR) configuration using a tapered fiber structure and a radially polarized light beam represents a novel and novelly analyzed approach to improving the sensitivity of SPR-based fiber optic sensors. Here, a 40 nm thick Ag layer, a 10 nm thick Au layer, and a sensing layer with a refractive index of 1.3331.353 are deposited on the taper waist area of an optical fiber with a diameter of roughly 330 m to create a bimetal-coated taper fiber optic sensor in SPR mode. The higher excitation of surface plasmon waves is a result of the radially polarized light's distinctive radial field distribution and cylindrical symmetry. When compared to a p-polarized light beam, the output response of a fiber optic sensor is 10 times more sensitive when using a wavelength interrogation approach and 2.307 times more sensitive when using an intensity interrogation technique. Our sensitivity study, which takes into account the impact of temperature on the suggested taper bimetallic structure, offers support for pursuing a novel approach to increasing the excitation n of SPR, which opens up novel possibilities for using radially polarized light in sensor applications. The proposed sensor setup is shown in Fig. 2.18.

The surface plasmon resonance (SPR)-based tapered fiber optic sensor was developed by Vikas and Verma (2018), who began by covering a tapered fiber core with a thin film of platinum, then added monolayers of graphene and molybdenum disulfide (MoS_2). Performance metrics, including sensitivity and detection accuracy, are measured for four taper profiles, including exponential—linear, linear, parabolic, and quadratic, and the taper ratio, to determine the impact of graphene–MoS_2 over layers. With a higher taper ratio, the detection accuracy drops monotonically, but the sensitivity rises. The proposed sensor setup is shown in Fig. 2.19.

The ability to determine saltwater salinity with great precision has significant practical implications. Tapered fiber optic SPR sensors offer great sensitivity to refractive index and find applications in a variety of scientific disciplines. By modifying the taper length, waist diameter, and coating thickness, Wei et al. (2023) were able to build an optimal tapered fiber optic SPR sensor. Using simulations, the impact of these variables on sensor performance was investigated. Experiments led to the

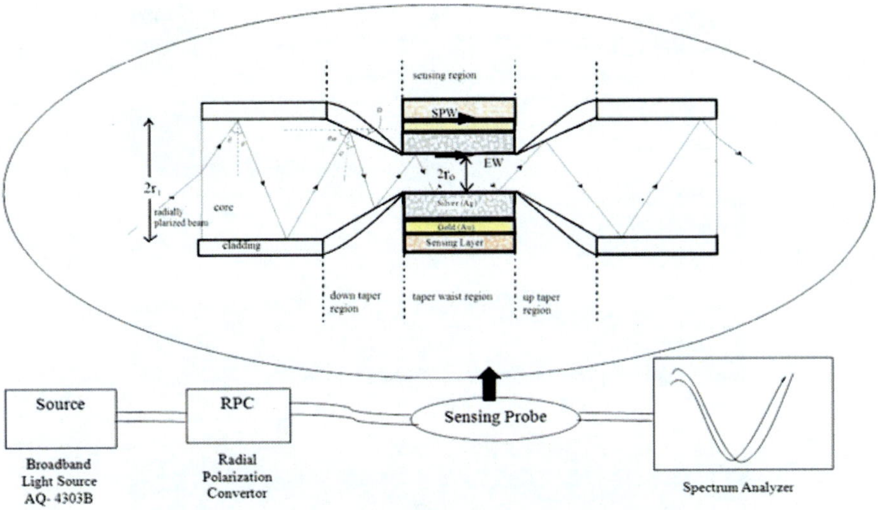

Fig. 2.18 Sensor setup schematic and ray tracing for a radially polarized, SPR-equipped, tapered, multimode fiber (MMF) with an Ag–Au bimetal coating. Reprinted with permission from Optics Communications. Copyright, 2016, Elsevier (Goswami et al. 2016)

Fig. 2.19 Schematic diagram of a tapered fiber optic sensor. Reprinted with permission from Optik. Copyright, 2019, Elsevier (Vikas and Verma 2018)

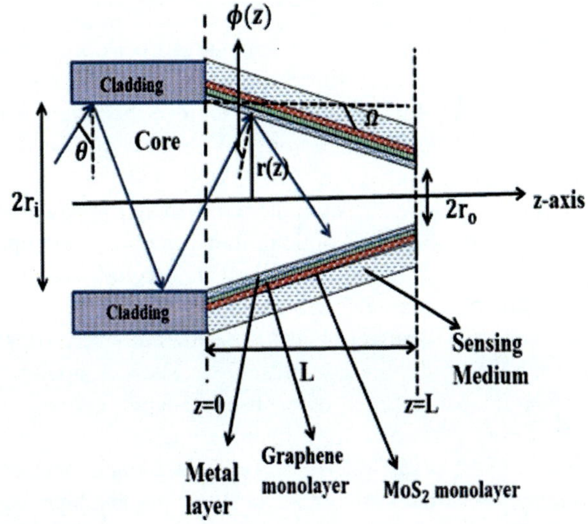

selection of a tapered fiber optic SPR sensor with a 27-mm taper length, a 25 μm waist diameter, and a 40-nm (Au) coating thickness, all of which were judged to be optimal on the basis of the quality factor. The salinity of seawater was detected with this sensor and high linearity. The experiment also demonstrated the sensor's durability. The proposed sensor setup is shown in Fig. 2.20.

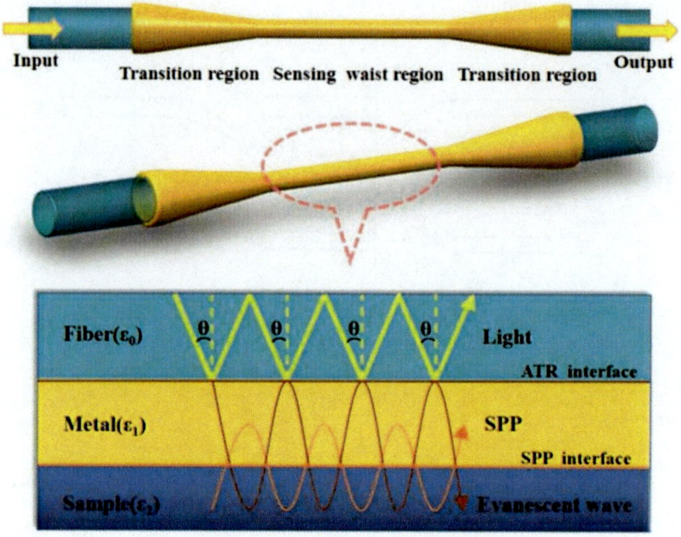

Fig. 2.20 SPR salinity sensor structure and sensing technique utilizing a tapered optical fiber. Reprinted with permission from Optical Fiber Technology. Copyright, 2023, Elsevier (Wei et al. 2023)

The trend in future development of fiber optic sensors is toward miniaturization, high sensitivity, and multifunctionality to meet the demands of various emerging fields (Agrawal et al. 2020). To achieve this, there is a need for the integration and miniaturization of tapered optical fiber (TOF) sensors, making them more portable and versatile (Kumar et al. 2019; Yang et al. 2020a, b). Additionally, there is potential for further improvement in the performance of TOF sensors by optimizing the structural design of tapered optical fibers, refining the preparation process, and exploring new surface modification methods (Zhu et al. 2021). Exploring new fiber materials and structures can also lead to better sensitivity and resolution, catering to diverse applications (Wang et al. 2021a; Zhu et al. 2022; Singh et al. 2019, 2020).

Innovative fiber optic structures, such as dual-periodic tapered, quad-periodic tapered, taper-in-taper, taper-in-multi-taper, humanoid-shaped, and dual-periodic S-taper configurations, have been extensively investigated to enhance sensor performance (Wang et al. 2021b; Zhu et al. 2020; Kumar et al. 2022; Li et al. 2022). Furthermore, novel fiber structures like W-shapes and taper-in-taper with core mismatch are also being explored, offering promising possibilities for various applications (Liu et al. 2023; Singh et al. 2023).

Despite the broad application prospects of TOF sensors, they do face certain challenges. Large-scale preparation remains a challenge, and there is a need to develop efficient and controllable processes to meet industrial production demands. The stability of tapered fiber optic sensors during long-term use is another important issue. The fragile nature of the sensor structure makes it susceptible to environmental disturbances, potentially leading to structural deformation or damage over time, which must be addressed to ensure prolonged and reliable sensor performance.

2.10 Summary

Geometrical optics theory is presented to describe the tapered fiber optic sensor. As for a proper in-depth study of the modeling of the taper fiber sensor, it is very much required to correlate it with the simulating result. Firstly, taper radii variation has been highlighted, and then taper angle and taper length have been formulated. Some derivations have been done in order to understand the propagation and tapering of the model. For proper sensing operation and transmission, it is very important to select the proper taper angle. The taper angle is dependent on the core radii and sensing length. The taper angle is very small but not zero. If the radii of the taper go on decreasing, the taper ratio goes on increasing, and the taper angle goes on decaying. When the incident is closer to the critical angle, the penetration depth increases and thus the sensitivity increases. Depth of penetration is a very important tool for measuring sensing capability; it increases with an increase in operating wavelength. It decreases with the incident angle. The propagation constant is maximum at the critical angle and zero at ninety degrees. It decreases to a greater extent above the critical angle. The propagation constant is also observed with respect to wavelength between the critical angle and ninety degrees and is found to decrease with wavelength.

References

Agrawal N, Zhang B, Saha C, Kumar C, Kaushik BK, Kumar S (2020) Development of dopamine sensor using silver nanoparticles and PEG-functionalized tapered optical fiber structure. IEEE Trans Biomed Eng 67:1542–1547

Avendaño-Alejo M, Moreno I, Stavroudis O (2007) Minimum deviation angle in uniaxial prisms. J Opt Soc Am A 24(8):2431. https://doi.org/10.1364/JOSAA.24.002431

Balan Pillai A, Varghese B, Madhusoodanan KN (2012) Design and development of novel sensors for the determination of fluoride in water. Environ Sci Technol 46(1):404–409. https://doi.org/10.1021/es2028718

Dehdashti S, Roknizadeh R, Mahdifar A (2013) Analogue special and general relativity by optical multilayer thin films: the Rindler space case. J Mod Opt 60(3):233–239. https://doi.org/10.1080/09500340.2013.769638

De Fornel, F. (2001). Evanescent waves: from Newtonian optics to atomic optics (Vol. 73). Springer Science & Business Media.

Gong W et al (2019) Experimental and theoretical investigation for surface plasmon resonance biosensor based on graphene/Au film/D-POF. Opt Express 27(3):3483. https://doi.org/10.1364/oe.27.003483

Goswami N, Chauhan KK, Saha A (2016) Analysis of surface plasmon resonance based bimetal coated tapered fi ber optic sensor with enhanced sensitivity through radially polarized light. Opt Commun 379:6–12. https://doi.org/10.1016/j.optcom.2016.05.047

Gupta BD, Sharma AK (2005) Sensitivity evaluation of a multi-layered surface plasmon resonance-based fiber optic sensor: a theoretical study. Sens Actuators B Chem 107(1 SPEC. ISS.):40–46. https://doi.org/10.1016/j.snb.2004.08.030

Jha R, Verma RK, Gupta BD (2008) Surface plasmon resonance-based tapered fiber optic sensor: sensitivity enhancement by introducing a teflon layer between core and metal layer. Plasmonics 3(4):151–156. https://doi.org/10.1007/s11468-008-9068-9

Kumar S, Kaushik BK, Singh R, Chen N-K, Yang QS, Zhang X, Wang W, Zhang B (2019) LSPR-based cholesterol biosensor using a tapered optical fiber structure. Biomed Opt Express 10:2150–2160

Kumar S, Wang Y, Li M, Wang Q, Malathi S, Marques C, Singh R, Zhang B (2022) Plasmon-based tapered-in-tapered fiber structure for p-cresol detection: from human healthcare to aquaculture application. IEEE Sens J 22:18493–18500

Le Kien F, Liang JQ, Hakuta K, Balykin VI (2004) Field intensity distributions and polarization orientations in a vacuum-clad subwavelength-diameter optical fiber. Opt Commun 242(4–6):445–455. https://doi.org/10.1016/j.optcom.2004.08.044

Li G, Xu Q, Singh R, Zhang W, Marques C, Xie Y, Zhang B, Kumar S (2022) Graphene oxide/multiwalled carbon nanotubes assisted serial quadruple tapered structure-based LSPR sensor for glucose detection. IEEE Sens J 22:16904–16911

Liu F, Zhang W, Lang X, Liu X, Singh R, Li G, Xie Y, Zhang B, Kumar S (2023) Development of taper-in-taper-based optical fiber sensors for chemical and biological sensing. Photonics

Mohammed ZH (2019) The Fresnel coefficient of thin film multilayer using transfer matrix method TMM. IOP Conf Ser Mater Sci Eng 518(3):9. https://doi.org/10.1088/1757-899X/518/3/032026

Singh L, Zhu G, Singh R, Zhang B, Wang W, Kaushik BK, Kumar S (2020) Gold nanoparticles and uricase functionalized tapered fiber sensor for uric acid detection. IEEE Sens J 20:219–226

Singh R, Wang Z, Marques C, Min R, Zhang B, Kumar S (2023) Alanine aminotransferase detection using TIT assisted four tapered fiber structure-based LSPR sensor: from healthcare to marine life. Biosens Bioelectron 236:115424

Singh L, Singh R, Zhang B, Cheng S, Kumar Kaushik B, Kumar S (2019) LSPR based uric acid sensor using graphene oxide and gold nanoparticles functionalized tapered fiber. Opt Fiber Technol 53:102043

Verma RK, Sharma AK, Gupta BD (2008) Surface plasmon resonance based tapered fiber optic sensor with different taper profiles. Opt Commun 281(6):1486–1491. https://doi.org/10.1016/j.optcom.2007.11.007

Vikas, Verma RK (2019) Design considerations of a surface plasmon resonance (SPR) based tapered fiber optic bio-sensing probe with graphene-MoS2 over layers. Optik (Stuttg) 180:330–343. https://doi.org/10.1016/j.ijleo.2018.11.081

Wang Y, Zhu G, Li M, Singh R, Marques C, Min R, Kaushik BK, Zhang B, Jha R, Kumar S (2021a) Water pollutants p-Cresol detection based on Au-ZnO nanoparticles modified tapered optical fiber. IEEE Trans Nanobiosci 20:377–384

Wang Z, Singh R, Marques C, Jha R, Zhang B, Kumar S (2021b) Taper-in-taper fiber structure-based LSPR sensor for alanine aminotransferase detection. Opt Express 29:43793–43810

Wei X, Peng Y, Chen X, Zhang S, Zhao Y (2023) Tapered optical fiber spr sensor for salinity measurement with high sensitivity. SSRN Electron J 78:103309. https://doi.org/10.2139/ssrn.4333815

Yang Q, Zhu G, Singh L, Wang Y, Singh R, Zhang B, Zhang X, Kumar S (2020a) Highly sensitive and selective sensor probe using glucose oxidase/gold nanoparticles/graphene oxide functionalized tapered optical fiber structure for detection of glucose. Optik 208:164536

Yang Q, Zhang X, Kumar S, Singh R, Zhang B, Bai C, Pu X (2020b) Development of glucose sensor using gold nanoparticles and glucose-oxidase functionalized tapered fiber structure. Plasmonics 15:841–848

Yuan Y, Ding L (2011) Theoretical investigation for excitation light and fluorescence signal of fiber optical sensor using tapered fiber tip. Opt Express 19(22):21515. https://doi.org/10.1364/oe.19.021515

Zhu G, Agrawal N, Singh R, Kumar S, Zhang B, Saha C, Kumar C (2020) A novel periodically tapered structure-based gold nanoparticles and graphene oxide—immobilized optical fiber sensor to detect ascorbic acid. Opt Laser Technol 127:106156

Zhu G, Singh L, Wang Y, Singh R, Zhang B, Liu F, Kaushik BK, Kumar S (2021) Tapered optical fiber-based LSPR biosensor for ascorbic acid detection. Photonic Sens 11:418–434

Zhu G, Wang Y, Wang Z, Singh R, Marques C, Wu Q, Kaushik BK, Jha R, Zhang B, Kumar S (2022) Localized plasmon-based multicore fiber biosensor for acetylcholine detection. IEEE Trans Instrum Meas 71:1–9

Zvyagin A, Ohtsu M (1997) Near-field optical microscope for true surface topography: theoretical study. Opt Commun 133(1–6):328–338. https://doi.org/10.1016/S0030-4018(96)00452-X

Chapter 3
U-shape Fiber Optic-Based SPR Sensor

3.1 Introduction

Surface plasmon resonance (SPR) occurs when an incident light hits the electron sea on the metal-thin surface at a certain angle; some part of the incident light energy gets absorbed by the electrons, and a wave parallel to the surface is established due to the collective oscillation of the excited electrons. This collective oscillation of electrons is referred to as a "surface plasmon," and the propagating wave is called a "surface plasmon wave" (SPW). These surface plasmons are transverse magnetic (TM) or p-polarized and exponentially decay along the direction of propagation. The absorption of the particular energy depends on the refractive index (RI) of the metal–dielectric interface, and therefore a wavelength shift is observed when the dielectric or its RI changes and is used to measure molecular adsorption, thickness change, density, etc.

The propagation constant for SPW can be derived using Maxwell's equation and is given by (Brier and Jayanti 2020)

$$\beta = \frac{\omega}{c}\sqrt{\frac{\epsilon_d \epsilon_m}{\epsilon_d + \epsilon_m}} = \frac{2\pi}{\lambda}\sqrt{\frac{\epsilon_d \epsilon_m}{\epsilon_d + \epsilon_m}} \tag{3.1}$$

And for dielectric media, the propagation constant is given by (Brier and Jayanti, 2020)

$$\beta = \frac{\omega}{c}\sqrt{\epsilon_s} \tag{3.2}$$

The propagation constant for SPWs is higher than that of the light in dielectric media, and it also requires the polarization state of SPWs to be matched with that of the incident light; thus, the surface plasmons cannot be excited using normal light or by directly exposing the metal–dielectric surface. Instead, coupling devices,

S. K. Raghuwanshi et al., *Geometric Feature-Based Fiber Optic Surface Plasmon Resonance Sensors*, Springer Tracts in Electrical and Electronics Engineering,
https://doi.org/10.1007/978-981-99-7297-5_3

or couplers, are used to achieve the correct match to optically excite the electrons and generate surface plasmon waves. The most common coupling devices used are a prism coupler, a waveguide coupler, and a grating coupler. Traditional prism couplers and Kretschmann and Otto's configurations involving the principle of attenuated total reflection (ATR) have a wide application in brewery, food processing, drug discovery, clinical diagnosis, and automobile. In the present scenario, many of the researchers are working on fiber optic-based SPR sensors for different applications.

3.1.1 Fiber Optic-Based SPR Sensors

Traditional prism coupler-based SPR sensors have several limitations due to their bulkiness and moving optical and mechanical parts and are thus not suitable for remote applications. The optical fiber-based SPR sensors involve the propagation of light by TIR inside the core. Some portions of the silica cladding of the core are replaced with a thin layer of plasmonic materials coated with the dielectric sensor. Light launched from one side, guided by TIR, generates EWs at the core–metal interface, which couples with the surface plasmon in the metal to generate a surface plasmon wave at the metal–dielectric interface. The evanescent waves (EWs) generated decay exponentially along the depth of the interface. The exponential relationship between the electric field and depth is given by (Tan et al. 2021)

$$E = E_0 e^{-\frac{z}{d_p}} \tag{3.3}$$

where z is the radial distance from the core (z = 0 at the core–metal interface), d_p is the penetration depth of the evanescent field, and E_0 is the maximum amplitude. A typical optical fiber-based SPR sensor is shown in Fig. 3.1.

Several factors like fiber parameters (single-mode or multimode), a geometrical configuration like straight, tapered, D-shaped, or U-shaped, tilted fiber Bragg gratings (TFBG), long-period fiber gratings (LPFG), specialty fibers (microstructure, photonic crystal fibers, photonic crystal, etc.), and numerical aperture determine the extent of coupling of the evanescent wave with surface plasmon. In contrast to prism configurations, a greater number of reflections per unit length occurs in fiber

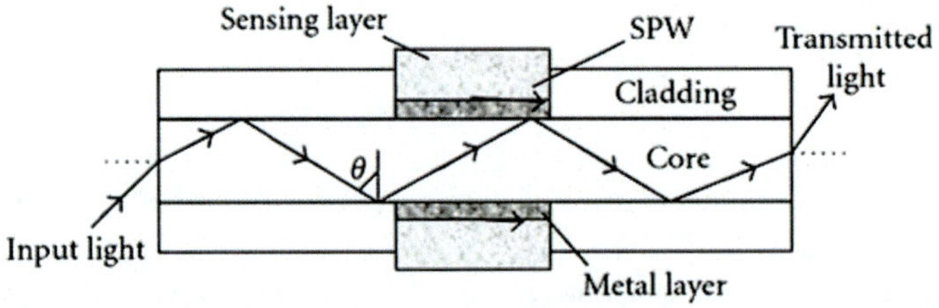

Fig. 3.1 A typical optical fiber-based SPR sensor. Adapted by (Gupta and Verma 2009)

optic-based ones, and the SPR curve width (which depicts the accuracy of the sensor) depends on the number of reflections, which further depends on the core diameter and sensing length of the metal–core interface (Gupta and Verma 2009). The net transmitted power at the detector end is proportional to the magnitude of the reflection coefficient. The coefficient of reflection can be determined using the transfer matrix method. The reflection coefficient for p-polarized light can be calculated using the N-layer model depicted in the picture for the SPR probe. The P polarization of light is the only one capable of excitation by the surface plasmon. Assume the layers are stacked perpendicular to the fiber core axis along the z-axis. The relationship of the tangential field at the first boundary $Z = Z_1 = 0$ with the final boundary $Z = Z_{N-1}$ is given by Jiang et al. (2017).

$$\begin{bmatrix} U_1 \\ V_1 \end{bmatrix} = M \begin{bmatrix} U_{N-1} \\ V_{N-1} \end{bmatrix} \tag{3.4}$$

The characteristic matrix of the combined structure, M, is given by Eq. 3.4 where U_1 and V_1 are the transverse and magnetic components of the electric field in the first and second layers, U_{N-1} and V_{N-1} are in the (N-1)th and Nth layers, and so on:

$$M = \prod_{k=2}^{N-1} M_k = \begin{bmatrix} M_{11} & M_{12} \\ M_{21} & M_{22} \end{bmatrix} \tag{3.5}$$

where

$$M_k = \begin{bmatrix} \cos\beta_k & (-i\sin\beta_k)/q_k \\ -iq_k\sin\beta_k & \cos\beta_k \end{bmatrix} \tag{3.6}$$

and,

$$q_k = \left(\frac{\mu_k}{\varepsilon_k}\right)^{\frac{1}{2}} \cos\theta_k = \frac{\left(\varepsilon_k - n_1^2 \sin^2\theta_1\right)^{\frac{1}{2}}}{\varepsilon_k} \tag{3.7}$$

$$\beta_k = \frac{2\pi}{\lambda} n_k \cos\theta_k (z_k - z_{k-1}) = \frac{2\pi\, d_k}{\lambda}\left(\varepsilon_k - n_1^2 \sin^2\theta_1\right)^{\frac{1}{2}} \tag{3.8}$$

When incident light is p-polarized, the reflection coefficient (rp) has the amplitude (Jiang et al. 2017)]:

$$r_p = \frac{(M_{11} + M_{12}q_N)q_1 - (M_{21} + M_{22}q_N)}{(M_{11} + M_{12}q_N)q_1 + (M_{21} + M_{22}q_N)} \tag{3.9}$$

and reflection coefficient intensity, or reflectance, for a single reflection is given by Jiang et al. (2017)

$$R_p = |r_p|^2 \tag{3.10}$$

Effective transmitted power can be computed by raising R_p by the number of reflections a ray takes to reach the detector end. Also, the prism-based couplers are more focused on angular interrogation in which the angle is varied while keeping the wavelength constant, and the sharp dip corresponds to a resonant angle, while the fiber-based couplers use spectral interrogation where the wavelength is varied (i.e., a polychromatic light is used) while keeping the angle constant and the sharp dip appears at a resonant wavelength. The resonant wavelength varies with the refractive index of the dielectric or analyte under test. A large shift in resonant wavelength with a small variation in RI, i.e., high sensitivity and also high SNR, are the desirable characteristics of an SPR sensor. Since the resonant wavelength $\delta\lambda$ is shifted in spectral interrogation when the RI is increased by δn, the sensitivity is defined as (Gupta and Verma 2009)

$$S_n = \frac{\delta n}{\delta\lambda} \tag{3.11}$$

And the detection accuracy, or SNR, with spectral interrogation is given by (Gupta and Verma 2009)

$$SNR = \frac{\delta\lambda}{\delta\lambda_{0.5}} \tag{3.12}$$

where $\delta\lambda_{0.5}$ is the spectral width at 50% reflectance. In addition to signal-to-noise ratio (SNR), another typical statistic used to assess an SPR sensor's efficiency is the figure of merit (FOM), which is calculated as

$$\text{F.O.M} = \frac{S}{FWHM} \tag{3.13}$$

where S is the sensitivity and FWHM is the resonance dip's full width of half maxima, which depends upon the reflectance minima R_{min} and local reflectance maxima R_{max}. A higher FOM or smaller FWHM is the desired performance of an SPR sensor.

3.1.2 Probe Design (Material and Geometry)

Glass and plastic fibers are the two most common types. Plastic fiber offers more advantages over glass fiber, like higher mechanical stability and strength, a lower possible bend radius and cost, and more robustness and flexibility. It also offers a higher numerical aperture (NA) than glass fibers, which leads to high sensitivity. Many researchers have used plastic optical fiber to develop SPR sensors. For example, a side-polished POF-based SPR sensor has been proposed by Cennamo et al. in

2011, a D-shaped microstructured plastic optical fibers (POF) has been proposed by Gasior et al. and a highly sensitive SPR sensor based on perfluorinated POF has been proposed by S. Cao et al. Having several advantages over glass fiber, plastic fiber limits its use when the operating temperature is high. The highest operating temperature for plastic optical fiber is between 80 °C and 100 °C. Operating at high temperatures results in attenuation in the final signal due to quality deterioration of the plastic fiber. Also, high humidity causes abrupt optical transmission loss as it absorbs humidity from the surrounding air. Thus, based on the application of the sensor, a fiber of interest can be used. Various plasmonic materials such as silver (Ag), aluminum (Al), copper (Cu), graphene on top of silver, gold, or copper, niobium (Nb), and indium tin oxide (ITO) are fabricated to develop the probe and improve its sensitivity. Among them, the most commonly used plasmonic material is gold due to its chemical inertness, biocompatibility, and long-term stability. Also, it has no oxidation issues and is easy to structure. But due to optical damping and a large resonance wavelength peak, it gives the wrong positive analyte detection. Silver (Ag), with its sharp resonance peak and highly sensitive behavior, is also a good candidate as a plasmonic sensor. It has no inter-band transition loss and has low optical damping. But it suffers from oxidation issues, which can be avoided using a layer of gold, graphene, or Ti_2O_5 (titanium oxide). Using an additional layer increases the fabrication complexity and hampers the performance of the sensor. Aluminum (Al) has a high electron density and moderate damping loss, but it also suffers from oxidation issues. Moreover, these plasmonic materials require an additional adhesion layer to be fabricated with the optical fibers, which brings in more damping loss and causes the performance to degrade significantly. Niobium, having strong adhesion with silica glass, high mechanical strength, and a continuous layer even below the ten nanometer thickness, in contrast to gold, is a novel plasmonic material. Indium tin oxide (ITO), which has the same optical damping as gold and silver, can also be a potential candidate because of its low bulk plasma frequency. Apart from the choice of metals, the SPR condition also depends on the RI of the core of the fiber. In General, core of the fiber is silica, however, its RI can be changed by adding some impurities, such as oxides of Boron, Germanium, and Phosphorous. The concentration of these dopants changes the sensitivity of SPR as well as signal to noise ratio, as depicted in Fig. 3.2 (Gupta and Verma 2009).

Sensitivity enhancement can also be achieved using a 2D (e.g., coating graphene around gold) model. When graphene is used to coat gold metal, the electric field around the sensor surface is increased, leading to greater sensitivity than when utilizing gold metal alone. About 2.5 and 25 percent of the enhancement can be obtained using a monolayer and 10 layers of graphene on gold, respectively, using a conventional Kretschmann arrangement. Enhanced sensitivity of 15 μM/RIU, for the U, bent fiber SPR sensor, which is 424.32% concerning straight and 381.33% for the U bent fiber sensor without graphene coating, is obtained.

Several geometries of SPR probes—straight, tapered, D-shaped, and U-shaped—have been reported to date, and some of them have been used for different applications. The sensitivity of a sensor depends upon the extent of coupling of the EWs with SPWs or how deep the EWs penetrate the metal thickness. And the extent of

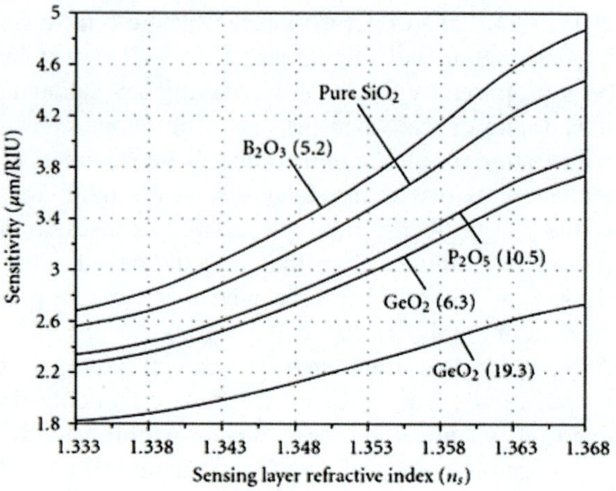

Fig. 3.2 Sensitivity variation for different dopants with different concentrations or RI. Adapted from (Gupta and Verma 2009)

coupling highly depends on the geometry of the probe. For example, in a tapered probe, different penetration of EWs along the length of the sensing region is observed, while in a straight probe, EWs penetrate uniformly (Sharma et al. 2007). Evanescent field absorption also depends on the radius of the core of the fiber.

3.2 Basic Concepts of U-type Fiber Optic Sensor

The occurrence of this phenomenon in U-shaped macro-bend fibers is a function of the light incident on and transmitted through the straight waveguide. At the macroscopic level, light from the emitter penetrates the cladding at the U-shaped bend. The light within the cladding is reflected at the cladding/air interface due to the whispering gallery effect. Variations in bending radius, temperature, and the RI of the fiber's surface all affect the effective refractive index of the cross-section of the fiber. Radiation coupling to the cladding causes some of the energy in the fiber core to be dissipated while some of it is transported along the fiber's outer wall (Gupta and Verma 2009). As a result, the curved fiber causes new interference patterns. Round off the diameters at the tight spots where the pipe bends. Light is reflected and transmitted along the cladding–air interface when the fiber is bent in the following ways.

A U-shaped probe is made by heating and bending the unclad, or sensing length, fabricated with a metallic coating of fiber in the middle, as shown in Fig. 3.3. It shows high absorption sensitivity in comparison to straight probes because of the conversion of the evanescent wave from lower modes to higher order modes. It offers high sensitivity—about 10 times more than that of a straight probe—less fragility, easy fabrication, and utilization as a point sensor.

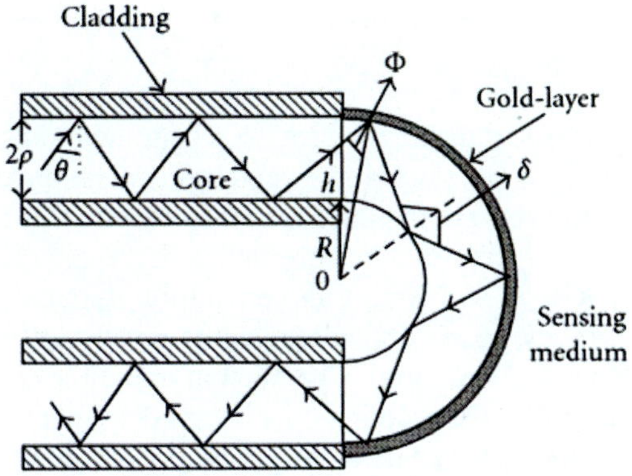

Fig. 3.3 A typical U-shaped probe. Adapted from (Gupta and Verma 2009)

A unique advantage that the U-shaped probe offers over other geometries is that it can be used as a dip-type probe, which facilitates efficient interaction with the analyte with a small volume. The probe's sensitivity is determined by how well the evanescent field is absorbed and how well the EW couples to the surface plasmon. And the evanescent absorption coefficient (γ) depends on two factors: the number of reflections per unit length (N) and the evanescent field's penetration depth (d_p):

$$N = \frac{\cot\theta}{2\rho} \tag{3.14}$$

$$d_p = \frac{\lambda}{2\pi n_1\sqrt{sin^2\theta - \sin^2\theta_c}} \tag{3.15}$$

In a straight probe, for a given angle of reflection, θ which remains the same throughout the guided path, the number of reflections per unit length decreases with the increase in core radius. But penetration depth is not affected; thus, in this case, the absorption coefficient of the evanescent field only depends on the core radius, and it decreases with an increase in the core radius. And for a given core radius, a decrease in the angle increases the value of N and d_p as mentioned in Eqs. 3.17 and 3.18.

A U-shaped probe, on the other hand, alters the interface angle from θ to ϕ and δ while entering the bent region, as shown in Fig. 3.3. The process by which lower order modes are transformed into higher order ones is known as "mode conversion." For a meridional ray, the angle of incidence at the boundary of the curved sensing zone is

$$\phi = \sin^{-1}\left[\left(\frac{R+h}{R+2\rho}\right)\sin\theta\right] \tag{3.16}$$

And the incident ray's angle on the inner surface is calculated as

$$\delta = \sin^{-1}\left[\left(\frac{R+h}{R}\right)\sin\theta\right] \tag{3.17}$$

where R is the sensing region's bending radius, h is the height from the inner core–cladding interface at the bent region's entrance ($0 \leq h \leq 2\rho$), and θ is the ray angle in the straight region. From this equation, it can be observed that for given R, θ, and h, increasing the core radius decreases the value of ϕ, and the angle (ϕ) value results in an increase in penetration depth, as can be seen from Eq. 3.19. Therefore, the absorption coefficient is mostly affected by the depth of penetration. The evanescent absorption coefficient of the U-shaped probe increases as the core radius increases. Considering only the guided light transmits through the fiber, the value of the incidence angle in the straight part will range from $\theta_{cr} to \pi/2$.

At the outer surface of the bending region, the corresponding transformed angle will range from $\phi_1 to \phi_2$, and it is given by Xie et al. (2021)

$$\phi_1(h) = \sin^{-1}\left[\left(\frac{R+h}{R+2\rho}\right)sin\theta_{cr}\right] \tag{3.18}$$

$$\phi_2(h) = \sin^{-1}\left[\left(\frac{R+h}{R+2\rho}\right)sin\frac{\pi}{2}\right] \tag{3.19}$$

The incident power profile $P(\theta)$ is given by

$$P(\theta) = \frac{n_1^2 \sin\theta\cos\theta}{\left(1 - n_1^2 cos^2\theta\right)^2} \tag{3.20}$$

Transmission power normalized at the detector end of a U-shaped cable is given by Eq. 3.24 Xie et al. (2021).

$$P_{trans} = \frac{\int_0^{2\rho} dh \int_{\phi_1(h)}^{\phi_2(h)} R_p^{N_{ref}} \frac{n_1^2 \sin\theta\cos\theta}{\left(1-n_1^2 cos^2\theta\right)^2} d\theta}{\int_0^{2\rho} dh \int_{\phi_1}^{\phi_2} \frac{n_1^2 \sin\theta\cos\theta}{\left(1-n_1^2 cos^2\theta\right)^2} d\theta} \tag{3.21}$$

where R_p is the single reflectance of p-polarized light in the sensing region, and N_{ref} is the total number of reflection-guided light events that occur at the outer surface of the bent region and is given by Eq. 3.25 (Xie et al. 2021):

$$N_{ref} = \frac{L}{8\rho}\left[\cot\theta + cot\left(\left(\frac{R+2\rho}{R}\right)\theta\right)\right] \tag{3.22}$$

where sensing region length is denoted by L. The further effect of the numerical aperture for the U-shaped probe is similar to that for the straight probe. An increase in numerical aperture increases evanescent wave absorption and thus sensitivity as well. The ratio of optical power in the environment to the sum of core and cladding power is the definition of EW power:

$$\frac{P_{sm}}{P} = \frac{4\sqrt{2}}{3}\frac{\lambda}{2\pi r\sqrt{n_{co}^2 - n_{sm}^2}} \tag{3.23}$$

And the absorption coefficient is defined as

$$\gamma = \alpha\frac{P_{sm}}{P} \tag{3.24}$$

where α is absorbing medium concentration. The bending radius is also a major factor that affects the evanescent field depth penetration and, thus, γ. Absorption coefficient γ variation with a bending radius for NA 0.17 and a core radius of 100 μm for 0.05 M cobalt chloride solution is reported by Khijwania and Gupta (1999).

The bend region contributes highly to RI sensitivity, as shown by Jiang et al. (2017) where it has been observed that RI sensitivity increases as the bend diameter is reduced from 1.7 to 0.75 m. And further reduction of the bend diameter results in decreasing sensitivity. This reduction in sensitivity can be attributed to two reasons:

(1) Light is coupled directly between the two straight arms before the bend region, resulting in cross-coupling (or overlapping) of the evanescent fields of the two unclad fiber arms.
(2) Incident angles ϕ and δ become less than the required critical angle θ_c for total internal reflection (TIR) due to a small bending radius, and thus propagating light refracts and gets lost.

Therefore, the highest sensitivity can be achieved at some optimum bend radius. The core diameter and numerical aperture of the fiber both play a role in determining this optimal bend radius. For NA with a 0.17 and 600 μm core diameter, the optimum bend radius is observed to be 0.12 cm, whereas the same NA but a 200 μm core diameter shows an optimum bend radius below 0.9 cm (Khijwania and Gupta 2000). Thus, the sensitivity of a U-shaped optical fiber can be evaluated by the dimensionless number R/ρ (ratio of bend radius and core radius), also called bend ratio. The diameter of fiber cores are 200, 400, and 600 μm which show optimum RI sensitivity at a bend diameter of 1.4 mm. Sai and co-workers have shown bend ratio as a potential factor to achieve the optimum geometry of a U-shaped fiber optic sensor. They have also demonstrated that a bend ratio of fewer than 5 shows high sensitivity to refractive index changes (Danny et al. 2020).

A graphene layer coating also enhances the sensitivity of a probe due to an improvement in the electric field around the sensing material. A monolayer coating of graphene on an Au-based SPR sensor gives 4.13 μm/RIU sensitivity, and a 5-layer coating gives 4.31 μm/RIU sensitivity, which are 37.2% and 43.2% enhanced sensitivity, as compared to a straight fiber sensor. Although the sensitivity enhances with the increased layer of graphene, the FWHM also increases, which results in a reduced FOM (Eq. 3.9). A single-layer coating shows 34.10 RIU^{-1}, whereas a 5-layer coating shows 33.49 RIU^{-1} of FOM (Xie et al. 2021).

RI sensitivity and FOM variation with a thickness of Au illustrate that both parameters increase with the increasing thickness of the Au film. However, this increase also results in a broadened and shallow dip in the SPR curve of transmitted light, which hampers detection accuracy. Therefore, sensitivity to depth (StD), another parameter that can be used to achieve optimized thickness, is given by (Xie et al. 2021)

$$StD = \Delta R_p \times S \tag{3.25}$$

where $\Delta R_p = R_{max} - R_{min}$.

Reflectance minima (R_{min}) against thickness gives a minimum value at 43 nm, which corresponds to a sensitivity of 4.09 μm/RIU, whereas StD plotted against thickness gives it maxima at 48 nm, which corresponds to a sensitivity of 4.12 μm/RIU, greater than that obtained by considering R_{min} only at 43 nm thickness. Further enhancement in sensitivity can be achieved by using more layers of graphene. A 20-layer graphene coating gives 5.1 μm/RIU sensitivity (Xie et al. 2021).

A longer sensing region results in a large number of reflections and thus more coupling of an evanescent field with the surface plasmon. However, it is found that an optimum sensing length exists, and increasing sensing length decreases sensitivity due to a higher loss of power in propagating light to the cladding material.

3.3 U-type of Fiber Optic SPR Sensor Structure and Operation Principle

To demonstrate how the whispering gallery effect (WGE) can be used for sensing displacement and temperature, the author employs an asymmetric U-shaped tapered fiber. The suggested sensor can measure displacements on the micron scale with high precision, reaching a maximum sensitivity of 38 pm/m and a linear sensitivity of 0.048 dB/m throughout a displacement range of 10–60 μm at ambient temperature. The asymmetrical U-type fiber structure is depicted in microphotograph and 3D model form as in Fig. 3.4a. The smallest diameter is 2.3 μm, the greatest is 8 μm, and both the symmetric and asymmetric regions are larger than the central position. The U-shaped whispering gallery and the stimulated higher order modes in the bending area are transmitted jointly, thanks to the structure's bilateral asymmetry. The fiber pull-cone method is used to construct this unit. The machinery used to produce tapered fibers is depicted in Fig. 3.4b. To heat the regular SMF once the coating has been stripped off, use a flame generated by the burning of hydrogen fuel. When the fiber is in its melting stage, the motor clamps down on both ends. To make a tapered fiber (TF), two motors rotate at the same speed in opposite directions. Figure 3.4c is a diagrammatic representation of the tapered fiber. Two transition zones and a waist zone make up the tapering SMF. An adiabatic pull-taper fiber was used to study the aftereffects of the U shape and to eliminate any interference caused by the cone area

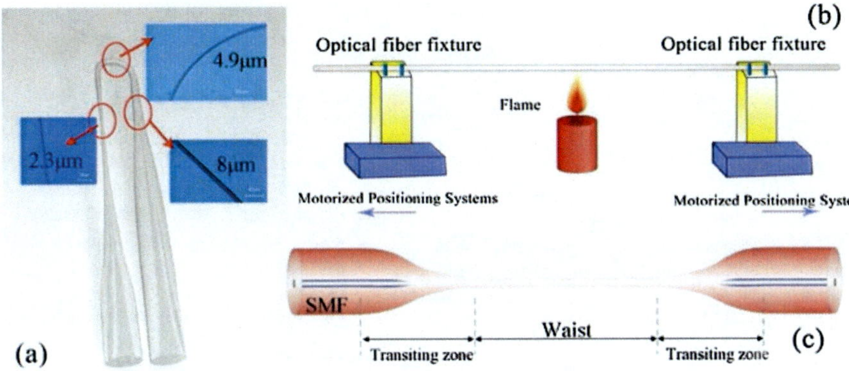

Fig. 3.4 **a** A microphotograph and 3D model of the asymmetrical U-shaped fiber sensing head. Figure **b** The process of making a tapered fiber. **c** Illustration of a tapered fiber. Reprinted with permission from Optik. Copyright, 2021, Elsevier (Guo et al. 2021)

of the fiber. All parameters are now free to accommodate the interference mode's wavelength change.

In Teng et al. (2022), U-shaped tapered plastic optical fibers (TPOFs) are proposed to be used as the basis for SPR sensors used in the measurement of RI. Experimentally explore the effect of Au-film coating on the performance of the U-shaped probes by comparing the RI sensing performances of straight and U-shaped TPOFs. In testing, a U-shaped probe by a taper ratio of 6.7 and a vertical Au-film coating has a sensitivity of 1534.53 nm/RIU in the RI detecting range of 1.335–1.41.

The SPR effect is activated, and an SPR peak is produced at a specific wavelength in the transmission spectrum. When the evanescent wave's wave vector is equal to the surface plasmon wave's wave vector, an evanescent wave is produced. Figure 3.5 is a schematic depiction of both a straight TPOF and a U-shaped TPOF.

To improve the sensor, this research coats a bare-bend loss-sensitive single-mode fiber with India ink on the inside and nickel on the outside (Peng et al. 2017). After the buffer coatings are stripped away, the cladding is smeared with India ink to cancel out the WG modes within the fiber. A chemical-electroplating procedure is used to generate the protective nickel coating, which then hardens into a U-shaped SMF. A ratio power measuring method and an FBG temperature calibration device are used to get even more accurate values of bend loss and temperature. The sensor is protected against corrosion in high-temperature and high-pressure settings by a nickel coating. The fiber's bend loss roughly linearly correlates to the sensor temperature range of 0 to 80 °C. The suggested sensor has a 0.5 °C temperature resolution and a 0.023 dB/°C bend loss response. The outer nickel coating on the U-shaped fiber has a bend diameter of roughly 22 mm. Figure 3.6 depicts the U-shaped fiber that has a nickel coating on the exterior and an India coating on the inside (Peng et al. 2017).

U-shaped zero-knotted and double-knotted spiral fiber optical probes with a 0.6 mm bending radius were studied for glucose solution monitoring (Hou et al. 2020). The fiber optical probe is used to measure the effective refractive index changes

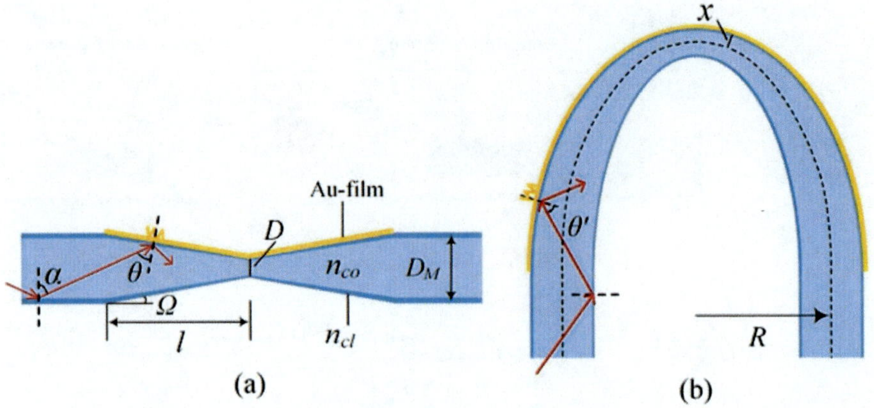

Fig. 3.5 **a** Schematics of a straight TPOF-based SPR sensor probe and **b** a U-shaped probe. Reprinted with permission from Optik. Copyright, 2022, Elsevier (Teng et al. 2022)

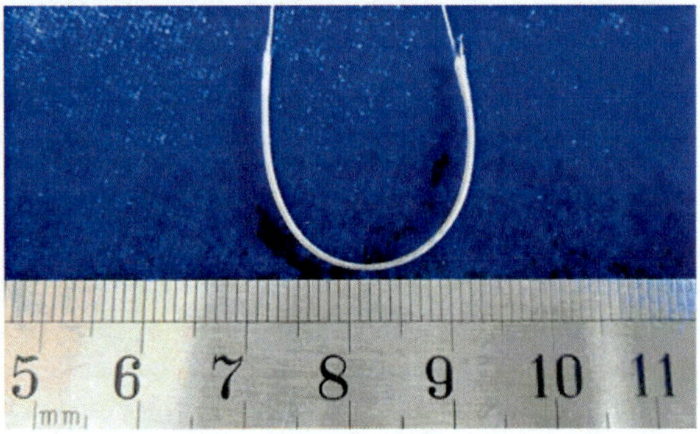

Fig. 3.6 The U-shaped fiber with an exterior nickel coating and an interior layer of Indian ink. Reprinted with permission from Optik. Copyright, 2017, Elsevier (Peng et al. 2017)

in a glucose solution at varying concentrations. Whispering-gallery mode (WGM) theory is the foundation of the probe's design. The maximum linearity achieved in experiments was 0.99, while the sensitivity of the double-knotted transmission loss was found to be as high as 0.061 dB/%. The results demonstrated that a double-knotted U-shaped spiral optical fiber probe with a limited radius provided accurate readings of glucose concentration. Figure 3.7 shows the formation of the WGM and a simplified depiction of the relevant optical transmission channel (Hou et al. 2020).

It was found that by changing the radius of curvature of the U-shaped fiber structure, the working band of the multi-channel fiber SPR sensor could be fine-tuned (Wang et al. 2022a, b). The U-shaped structure with curvature radii of 1 mm and 5 mm was used to create and test a dual-channel fiber SPR, with a sensitivity to changes in the RI of 1414 nm/RIU and 3687 nm/RIU, respectively. Both glucose and sucrose were detected simultaneously utilizing this dual-channel fiber SPR, with a detection sensitivity of 0.172 and 0.738 nm/μg/ml, respectively. The proposed sensor

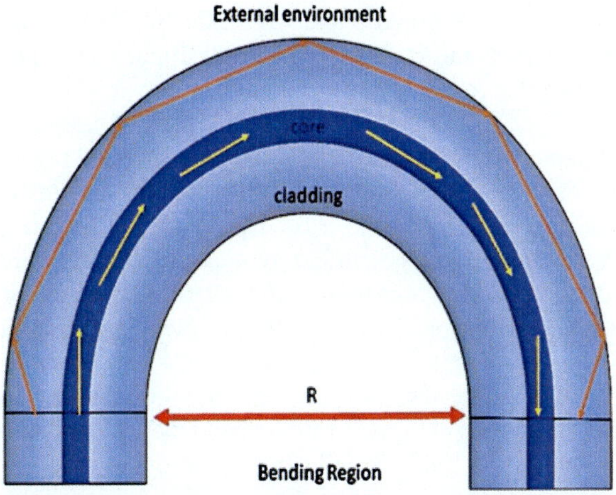

Fig. 3.7 The optical signal path for a U-shaped fiber optic probe is depicted schematically. Reprinted with permission from Optik. Copyright, 2020, Elsevier (Hou et al. 2020)

was used for the cascade design of dual-channel sensors to achieve maximum separation between the operating bands and therefore implement wavelength division multiplexing. Figure 3.8 depicts the sensing probe's structure.

To get over the issue of resonance superposition, the U-shaped and tapered core mismatch SPR sensor was demonstrated in (Ren et al. 2023). Because some higher order modes can be cut off, thanks to the tapered structure, the full width at half maximum (FWHM) is reduced. Additionally, tests demonstrated that after incorporating SMF and a tapered structure, the thickness of a gold film, in conjunction with the degree of macro-bending, has an ideal value. After fine-tuning the sensor's parameters, it outperforms the U-shaped multimode fiber sensor by a factor of 2.5 on the complete evaluation indicator (CEI).

Sensors of varying curvature diameters (5.5 mm, 5 mm, 4 mm, and 2.5 mm) are depicted in Fig. 3.9a. All of these sensor probes have an SMF length of 1 cm, a gold film thickness of roughly 69.5 nm, and a total of 1 cone arm. The sensor's curvature

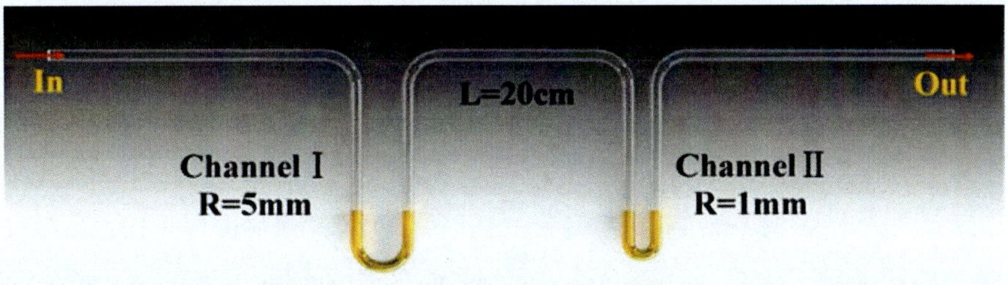

Fig. 3.8 A dual-channel sensing probe with a U-shaped structure. Reprinted with permission from Optik. Copyright, 2022, Elsevier (Wang et al. 2022a, b)

was modified by varying the duration and size of the SMF's heated zone (Ren et al. 2023). Figure 3.9b depicts the resonance spectrum. Figure 3.9c shows transmission spectrum enhancement at resonance wavelength, and Fig. 3.9d depicts the sensitivity fitting curves.

Transmission patterns in the sensing region were obtained using NaCl solutions with RI values between 1.3324 and 1.3376. For RI sensing testing, only a gold layer is plated on the sensor because it focuses on structural changes. The solute of interest can be identified and quantified by measuring its RI. Since the RI of the medium around the sensor changes when the solution contains different amounts of biomass, the suggested sensor may detect multiple analytes by just having their respective antibodies fixed to its surface. Table 3.1 shows the experimental outcomes, which detail how many times the sensor was tested with various solutions of known refractive indices after being equipped with several tapered arms in various positions (as indicated in Table 3.1). Table 3.1 illustrates that the number of cone arms that affects the FWHM more than the sensitivity and RVD values. The presence of the tapered arms narrows the FWHM because they decrease the light's transmission

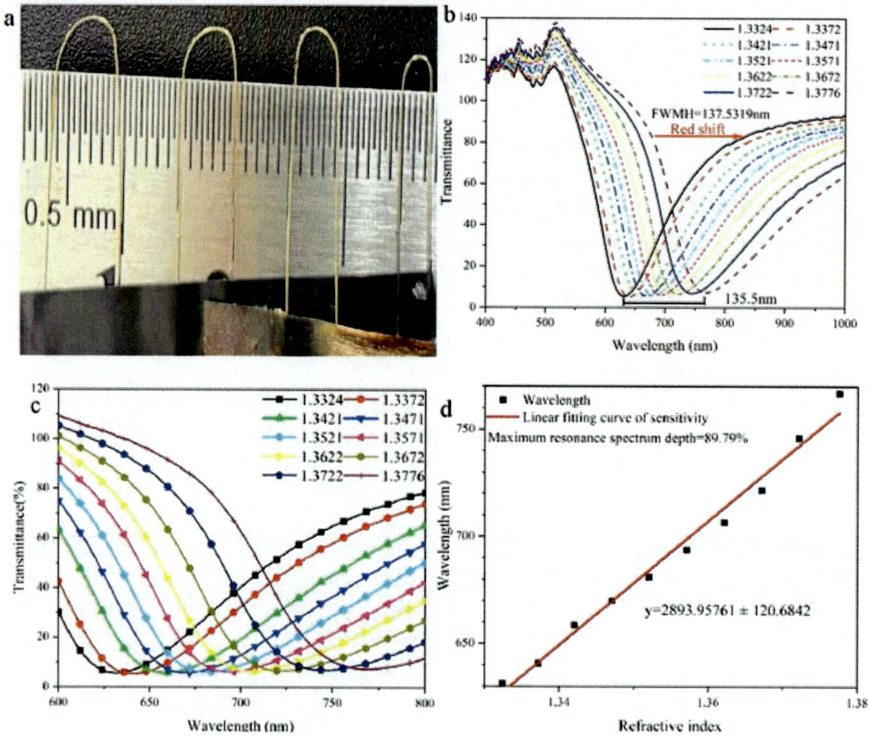

Fig. 3.9 **a** Sensors for different radii of curvature, **b** 5 mm spectrum, **c** transmission spectrum with magnification at resonance wavelength, and **d** optimal sensor sensitivity fitting curves. Reprinted with permission from Measurement. Copyright, 2022, Elsevier (Ren et al. 2023)

pattern through the cladding and decrease the gap between the resonant wavelengths produced by the various cladding modes (Ren et al. 2023). Table 3.1 show the sensing performance of different types of U-shape fiber optic structure.

After making U-bent probes in sizes ranging from 250 to 1000 m, some studies have been carried out to determine what size bend was best for each fiber diameter (Gowri and Sai 2016). Probe bend diameters between 2 and 3 times the fiber diameter were most sensitive. Probes with a 500 m core and a 1.25 mm bent diameter had the maximum visible spectrum sensitivity to RI variations from 1.33 to 1.47, with accuracy greater than 1 milli RI units (Gowri and Sai 2016).

Figure 3.10 shows how bending and pressing fibers into a capillary, followed by a simple heat treatment, yielded U-bent POF probes with varying bent diameters (D). Bending fibers of a specific diameter required the use of special glass capillaries with the right inside diameter.

Table 3.1 Diameters of varying curvature yield varying values for each parameter. Reprinted with permission from measurement. Copyright, 2022, Elsevier (Ren et al. 2023)

Number of cone arms	Illustrations	Sensitivity	FWHM	RVD	CEI
0		3121.31793	186.622168869	82.324	13.76897056
1		2969.17623	177.2549720644	81.25	13.61036203
2		2893.95761	137.5309700136	89.79	18.89367545
3		2324.69621	122.4272692629	82.247	15.61734473

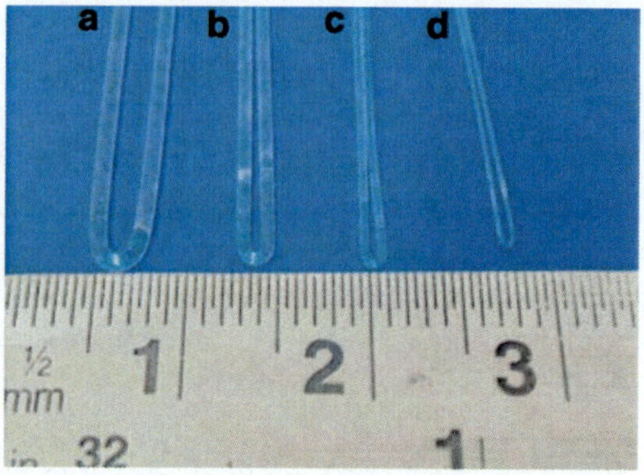

Fig. 3.10 The picture of U-shaped POF probes **a** D = 3 mm at 1000 m **b** D = 1.25 mm **c**, 500 m, D = 1.25 mm **d** D = 0.75 mm. Reprinted with permission from Sensors and Actuators B: Chemical, Copyright, 2016, Elsevier (Gowri and Sai 2016)

3.4 Recent Advances in U-type Fiber-Based Plasmon Sensors

The core ideas and crucial factors that must be taken into account when developing U-shaped FOSs are already well known. Recent developments in U-shaped optical fiber sensors encompass a wide range of mechanisms and fields. For various applications, numerous researchers have recently proposed various U-shaped sensor techniques, including evanescent wave absorption (EWA), LSPR, SPR, fluorescence, and interferometry-based U-shaped FOSs (Tan et al. 2021).

U-shaped FOSs that rely on evanescent waveguides operate on the principle that light refracts outward as evanescent waves from the fiber core at the bending region (Sharma et al. 2019). An absorbing material then absorbs the evanescent waves, and the amount of absorption is correlated with the concentration of the absorbing medium. In U-shaped EWA-based FOS, the sensing element is typically a bare fiber coating. The former is a form of direct sensing because the analyte itself is interacting with the evanescent waves. In indirect sensing, the analyte binds to the transducer coating and changes its dielectric characteristics. The presence of the analyte, either free or bound to the detecting coating, alters the wave characteristic, and this alteration can be measured and associated with the analyte concentration at the sensor. This bending of the fiber resulted in a change in penetration of 362.14 nm. As shown in Fig. 3.11, the absorbance of the signal at the output was caused by proteins present in E. coli cells that had been adsorbed on the surface with the aid of an antibody (Sharma et al. 2019).

When evanescent waves of light from an external source strike a substance, it excites another transverse wave called a surface plasmon. This process is known as SPR, or propagating surface plasmon resonance (PSP). Both in the direction of the metal film and the direction of the surrounding medium, the free electron oscillations decay exponentially (Luo et al. 2016). To excite SPR, typically thin coatings of noble metals are deposited onto the exposed fiber core. Another layer may be put on top of the metal film to increase its sensitivity and selectivity, although this depends on the specific use case. Gold (Au) coating of varied thickness was deposited onto the fiber probe using the sputtering technique. This study involved the detection

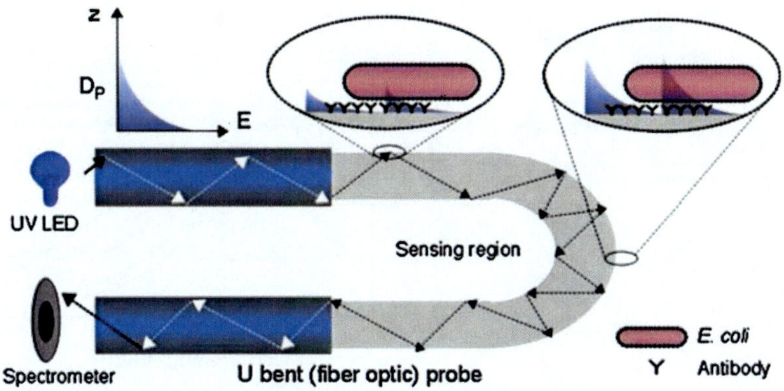

Fig. 3.11 Fiber U-bending increases the depth at which E. coli can be detected. Reprinted with permission from Optik. Copyright, 2019, Elsevier (Sharma et al. 2019)

of tiny compounds using aptamers and unmodified gold nanoparticles (AuNPs). When exposed to high concentrations of salt, the AuNPs/ssDNA complexes remain stable. As can be seen in Fig. 3.12, a U-shaped fiber optic probe (Thorlabs, with a 0.37 numerical aperture) was used to focus the light from a tungsten lamp onto its input ends. Light with a plasmon resonance wavelength was absorbed by the SPR probe on its way there due to intense vibration on the silver film. The signal-channel spectrometer picked up the corresponding light signal.

This technique depends on the sensing principle of salt-induced aggregation of AuNPs. Figure 3.13a is a diagrammatic representation of salt-induced aggregation. As can be seen in Fig. 3.13b, a UV–vis spectrometer (Lambda 25 PerkinElmer,

Fig. 3.12 The measurement setup for the U-shaped FOSPR probe. Reprinted with permission from Sensors and Actuators B: Chemical, Copyright, 2016, Elsevier (Luo et al. 2016)

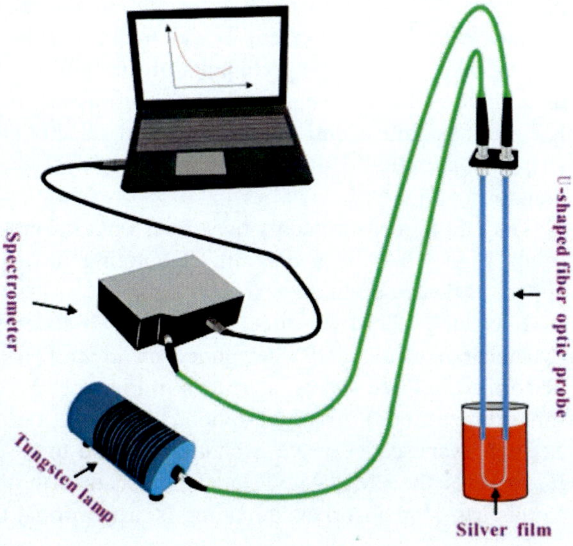

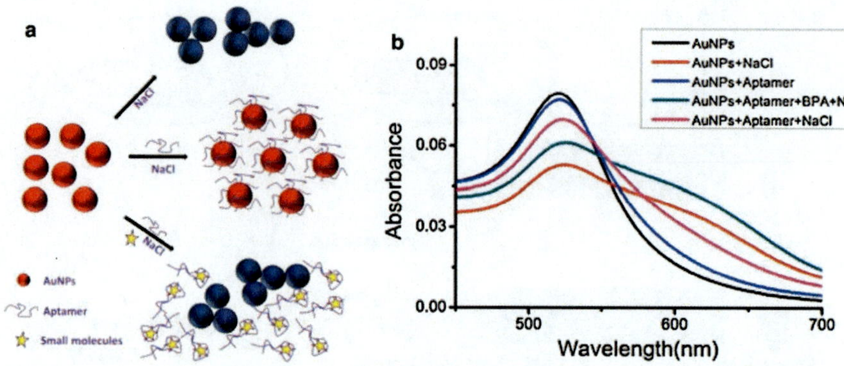

Fig. 3.13 **a** An aptamer-associated salt-induced aggregation of unmodified AuNPs is used to illustrate the detection of tiny molecules. **b** UV–vis absorption spectra of AuNPs at room temperature under various experimental circumstances. Reprinted with permission from Sensors and Actuators B: Chemical, Copyright, 2016, Elsevier (Luo et al. 2016)

USA) was used to track the AuNPs' signature absorption peak. In short, the typical absorption peak of unaltered AuNPs will shift to red due to the aggregation induced by high salt conditions via the electrostatic screening effect. The size of the AuNPs in the solution affects the shape of the characteristic absorption peak (Luo et al. 2016).

U-shaped fiber optic LSPR cytosensing detects cancer cells and analyzes cell surface N-glycans (Luo et al. 2019). U-shaped-LSPR fiber optics are sensitive to refraction variations. U-shaped fiber optic LSPR cytosensing was ultrasensitive for cancer cell detection with a LOD of 30 cells/mL and strong linearity in a broad range of 1102–1106 cells/mL under optimal conditions of modified AuNPs size and Con A concentration. Straight fiber optic LSPR cytosensing has a detection limit 29 times higher than U-shaped fiber optics. In addition, the built-in U-shaped cytosensing was utilized to evaluate the extent to which cancer cells' surface N-glycan expression was influenced by varying amounts of the inhibitor tunicamycin (TM). U-shaped fiber optic LSPR cytosensing was used to assess N-glycan expression in six cell lines. Repeatability, anti-interference, and selectivity were excellent. Thus, the U-shaped fiber optic LSPR is useful for biophysical investigations and clinical cancer diagnostics.

The LSPR measurements over fiber optics were performed using a makeshift setup. Figure 3.14 is a diagram elaborating on the U-shaped fiber optic LSPR apparatus described in the text.

To create a label-free (direct) and sandwich-style plasmonic biosensor, the author (Manoharan et al. 2019) developed an idea. The plasmonic sandwich assay is performed in three stages, as shown in Fig. 3.15a. To observe LPS binding in real time (direct mode), first hydrophobically trap it on an OTS-functionalized UFOP. Second, evanescence wave excitation is used to assess the plasmonic absorbance response of the PMB-AuNP labels attached to the bound LPS (sandwich assay). In this case, the absorbance reading is proportional to the amount of LPS present.

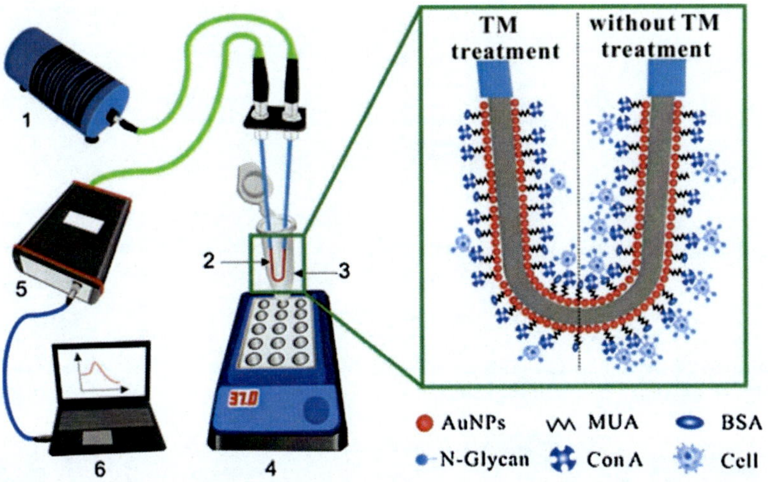

Fig. 3.14 A light source, optical fiber probe, sample cell, thermostat, spectrometer, and computer for a U-shaped fiber optic LSPR setup. Reprinted with permission from Sensors and Actuators B: Chemical, Copyright, 2019, Elsevier (Luo et al. 2019)

Lastly, the assay's sensitivity is increased by a factor of ten with the addition of a straightforward silver reduction step at the conclusion. Here, this assay was built from the ground up and shows how it may be used as a simple and portable sensor for LPS detection in aqueous samples.

Based on the SMS structure's advantages and the U-shaped taper NCF's high evanescent field, a sensor for measuring the salinity and temperature of seawater is investigated (Li et al. 2023). The SMS sensing structure is made up of a segment of UTNCF that has been joined to two ends of regular SMF. Since seawater's RI varies linearly with salinity and temperature, an optical SMS structure based on UTNCF can be used to determine the RI of a water sample. The experimental setup attempts to replicate the variation in RI seen in the natural marine environment by altering the salinity and temperature of the salt solution. Other marine microorganisms' potential impact is disregarded. The transfer matrix approach is used to address the salinity–temperature cross-sensitivity. The temperature sensitivity is—0.11 nm/°C and the sensitivity to salinity is 0.69 nm/%. The sensor's great sensitivity and straightforward production are two of its best features. In addition, the sensor will help make seawater temperature and salinity sensors more commercially viable in the long run.

Figure 3.16 is a diagrammatic representation of the SMS structure, which consists of a segment of UTNCF joined at both ends to regular SMF. The TIR of higher order modes is broken in the UTNCF. As a result, the NCF loses more energy, which boosts the evanescent field and improves sensitivity even more. However, UTNCF's sensor performance will degrade with increasing bending angles. Since the UTNCF's evanescent field grows stronger as the bending angle gets smaller, the sensing system's sensitivity improves while the interference spectrum's output power and extinction ratio go down. The sensor's capabilities are diminished as a

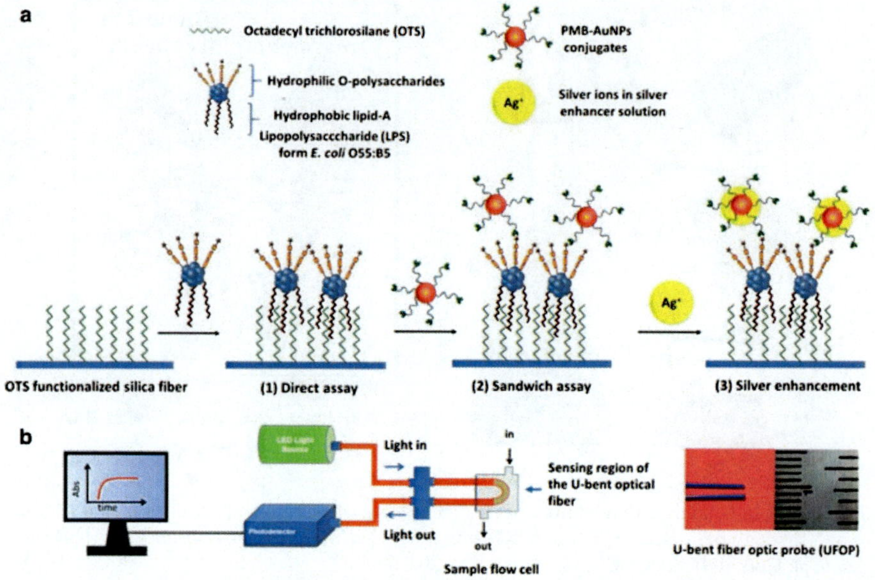

Fig. 3.15 **a** LPS bioassay on an OTS-functionalized silica UFOP surface, **b** an optical setup for LPS detection, **c** illustrated with an actual image of a 1 mm-bent UFOP. Reprinted with permission from Biosensors and Bioelectronics. Copyright, 2019, Elsevier (Manoharan et al. 2019).

result. The bending angle of 60° was chosen after considering these considerations and the practicability of the experimental operation. Figure 3.17 displays schematic and scanning electron micrographs of the three sensor probes. NCF has a length of 4.4 cm and a width of 125 μm. UTNCF has a bending angle of 60 degrees, while the minimum diameter of tapered NCF is 20 μm.

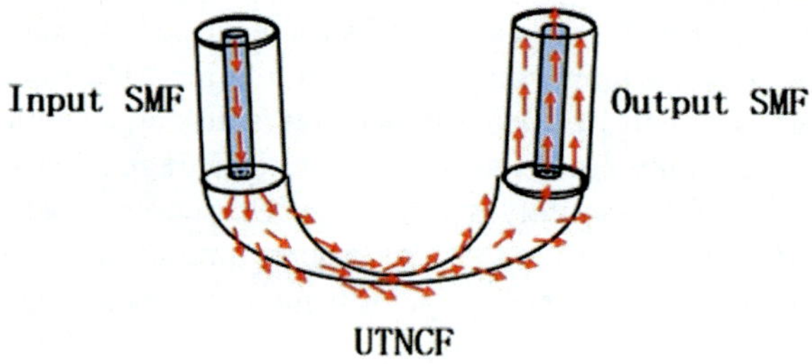

Fig. 3.16 SMS structure schematic based on UTNCF. Reprinted with permission from Infrared Physics & Technology. Copyright, 2023, Elsevier (Li et al. 2023)

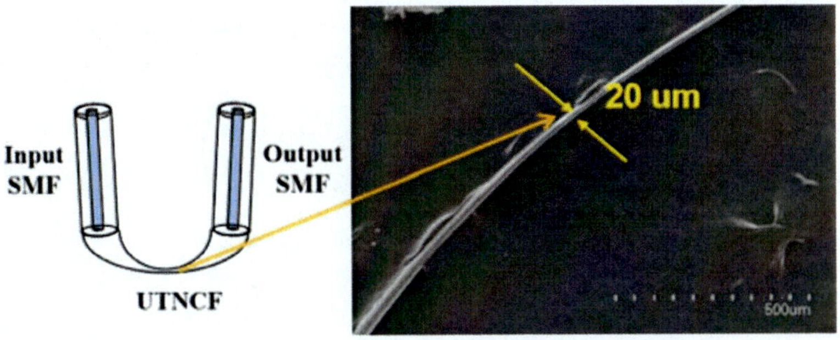

Fig. 3.17 Sensing probe UTNCF structures and SEM images. Reprinted with permission from Infrared Physics & Technology. Copyright, 2023, Elsevier (Li et al. 2023)

3.5 Probe Design of U-type Fiber Optic Sensor

The preparation of a U-shaped probe involves two steps: de-cladding the straight fiber and bending. Further, on the bent shape, plasmonic material can be fabricated to form a U-shaped SPR sensor. A straight fiber is declared by removing the cladding material from the central portion of a particular length either by mechanical striping or by chemical etching. Mechanical stripping is done with a sharp razor or surgical blade to remove polymer coatings and buffer coatings. Chemical etching is common for all polymer-based fibers, where chemical etching is done by using acetone, ethyl acetate, and isopropyl alcohol, and HF is used by many to declare all silica single-mode fibers (SMF). To preserve the remaining cladding, it is covered with Tygon tubing, and Teflon-based spray is used to seal the gaps. Then the fiber is dipped into a 48% HF solution to remove the cladding of the uncovered part and rinsed with deionized water.

After de-cladding, the fiber is then heated and bent at some temperature to form a U, with the declared part in a semi-circular shape. For glass fiber, direct heating like a candle flame is used to achieve a high temperature, while for plastic fiber, indirect heating (ambient heating) is performed to facilitate smooth bending. Different flames are used to achieve different bending diameters. This method prevents black carbon deposition on the flame. Bending of the straight fiber with a uniform refractive index causes inhomogeneity of the RI of the bent part of the fiber due to the photoelastic effect. The RI of the inner bent region becomes high due to compression, whereas the RI of the outer bent region becomes low due to tension. RI profile variation due to bending is observed for 200-m core diameter fused silica fiber (Xie et al. 2021). This increases the field intensity at the outer part of single-mode and multimode fibers. A variation of RI from 1.49 to 1.44 at the outer part and from 1.49 to 1.54 is observed for a core diameter of 0.5 mm and a bend radius of 1 mm for a plastic optical fiber.

Bending changes mechanical properties like Poisson's ratio and elasticity coefficient (Danny et al. 2020). But these changes are negligible in comparison to RI inhomogeneity caused by bending, also called bend-induced RI (BIRI) inhomogeneity. The tensile strain at the outer curvature and the compressive strain developed at its inner curvature produce light propagation changes that can be evaluated using the two equations below (Danny et al. 2020):

$$\Delta\varphi = \frac{d\varphi}{d\epsilon}\varepsilon + \frac{1}{2}\left(\frac{d^2\varphi}{d\epsilon^2}\right)\epsilon^2 + \frac{1}{6}\left(\frac{d^3\varphi}{d\epsilon^3}\right)\epsilon^3 \quad (3.26)$$

$$\Delta\varphi = \frac{2\pi}{\lambda}L(\Delta n + \epsilon n) \quad (3.27)$$

The optical fiber core's RI is n, phase retardation is φ, strain is ε, and length is L. Substituting the bending stress in the above equation results in Neumann–Maxwell stress equations of refractive index variation in the outer bending region, which are given by (Danny et al. 2020)

$$\Delta n_{ij} = P_{ijkl}\epsilon_{kl} \quad (3.28)$$

$$n_{bent}(x) = n_{co}\left[1 - \frac{n_{co}^2 x}{2R}[P_{12} - \upsilon(P_{11} + P_{12})]\right] \quad (3.29)$$

where Δn is the RI tensor, P is the Pockels tensor, P_{11} and P_{12} are the core material's photoelastic coefficients or Pockels, x is the distance from the outer bent region's center of curvature, and υ is Poisson's ratio. A decrease in the bend radius causes asymmetric propagation of light along the axis of the fiber in the bending region for graded index fiber. Greater radiation loss occurs in comparison to step-index fiber as the Fresnel transmission coefficient approximates 1 at the core–cladding interface. NA also varies as RI varies between the core and cladding, or the metal. NA in the bent region is given by

$$NA(R, a) = \sqrt{n_{co}^2 - n_{sm}^2\left(\frac{R + r}{R + a}\right)^2} \quad (3.30)$$

where a is the distance from the center of the core, where $-r \leq a \leq r$. It was demonstrated that the NA of the curved region grows further from the center of curvature. From Eq. 3.33, it is given that the angle at the inner surface is greater than the angle at the outer bent surface. Thus, the chances of TIR occurring at the inner surface are higher, and thus more light is lost at the surface of the outer portion. The minimum bend radius of the fiber below which no transmission of light occurs is known as the critical radius of curvature, which is given by Eq. 3.31 for SMF and Eq. 3.32 for MMF:

$$R_{c(SMF)} = \frac{20\lambda}{(n_{co} - n_{cl})^{\frac{3}{2}}}\left[2.748 - 0.996\frac{\lambda}{\lambda_c}\right]^{-3} \quad (3.31)$$

$$R_{c(MMF)} = \frac{3n_{co}^2\lambda}{4\pi(NA)^3} \tag{3.32}$$

As NA and the difference between core and cladding RI become larger, the critical radius of curvature lowers. And thus, a plastic fiber with a large NA allows transmission without loss in a fiber with a small bend radius. The refractive loss induced due to RI inhomogeneity and fiber geometry can be analyzed using either the wave optics model or the ray optics model. The wave optics model involves computational limitations, and the analytical model for computing bending loss is limited to single and few-mode fibers. Thus, for multimode fiber and ease of computation, the ray optics model is preferable. Considering input parameters (such as material, geometric, and simulation parameters), sensitivity, guided and lost power, width, and linearity of the propagating light can be obtained (Danny et al. 2020). Refractive loss and sensitivity are given by

$$\text{Normalized Refractive Loss } (dB) = 10\log\left(\frac{P_{ref}}{P_{med}}\right) \tag{3.33}$$

$$\text{Sensitivity} = \frac{\Delta Loss(dB)}{\Delta RI} \tag{3.34}$$

where Pref is the reference medium power and Pmed is the output power at the detector end for the respective RI of the surrounding medium. There are several other numerical analyses, like the beam propagation method (BPM), the finite element method (FEM), and the finite difference time domain (FDTD), used to analyze the effect of the bending region on the transmission of light and the loss that occurred due to bending. BPM and FEM involve high computation costs as they require meshing and mode-solving to achieve high accuracy. Similar reasons force FDTD to incur a higher cost margin. The isotropic, non-magnetic, and uniform behavior of stacked layers in the Transfer Matrix Method is assumed for stacked layers. This assumption might incur a discrepancy between experimental and simulation results.

The effective refractive index of the inner and outer curvatures of the core of the plastic optical fiber varies with the bend ratio. It deviates prominently from that of straight fiber below the bend ratio of 7. The BIRI inhomogeneity effect on light propagation is negligible in the geometric region (bend ratio greater than 17). Its effect on RI sensitivity varies drastically in a region with a bend ratio less than 7 (the plastic region), and further reduction of the bend ratio results in a significant loss of sensitivity.

The ideal bend ratio, represented by the highest point of the sensitivity curve, is determined by the RI of the surrounding medium and the optical fiber material. For example, the surrounding medium with RI 1.36 shows maximum sensitivity at a bend ratio of 1.69 and zero sensitivity at a 1.40 bend ratio. And for silica fiber, peak and zero values are obtained at bend ratios of 3.4 and 2.5, respectively (Danny et al. 2020).

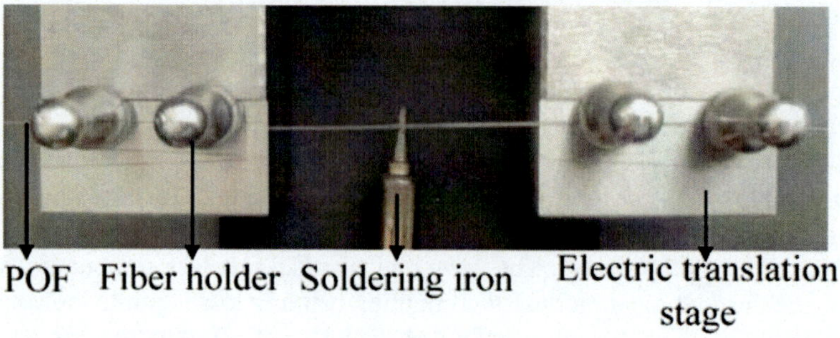

Fig. 3.18 Image of a tapering instrument. Reprinted with permission from Optik. Copyright, 2022, Elsevier (Teng et al. 2022)

Researchers (Teng et al. 2022) advised for the production of the TPOF probe, a step-index POF was used. The cladding is made of a fluorinated polymer with a reflectance index of 1.41. The RI of the PMMA used to construct the building is 1.49. The POF core was 490 μm in diameter, and the cladding was 5 μm thick. Figure 3.18 is a photo of the tapering apparatus used to create the TPOF via a heating and drawing operation. Once the POF has reached its softening point, you can take it off the soldering iron and simultaneously draw both ends of the POF using the electric translation steps. The soldering iron's temperature and the electric translation stages' positioners allowed for precise assembly, and the TPOF with a variety of taper ratios can be manufactured.

The manufactured TPOF was rolled onto a cylindrical metal mold with a macro-bending radius of 4 mm and then immersed in hot water at around 80° C to create a stable U-shaped TPOF probe. The final step involved depositing an Au coating on one side of the probe using a plasmon sputtering device (SD-900 M, Vision Precision Instruments). Because the Au-film distribution and evanescent field power will be different at the curved top and lateral directions, the U-shaped TPOFs were coated using either the vertical or horizontal Au-film coating procedures, as shown in Fig. 3.19a, b; U-shaped TPOF probes with and without Au-film coating are shown in Fig. 3.19c, d, respectively (Teng et al. 2022).

Similarly, another group (Hou et al. 2020) suggested to making the probe by cutting a 40-cm-long portion of fiber, removing about 4 cm of its intermediate protective layer using a wire stripper, and then cleaning the exposed fiber end with alcohol. After attaching the tube and optical fiber to a micro-motion platform, the latter pulled the fiber back inside the quartz glass tube. When the glass temperature hit 800 °C–1000 °C, the optical fiber structure instantly changed to a small U-shaped curved structure (Fig. 3.20).

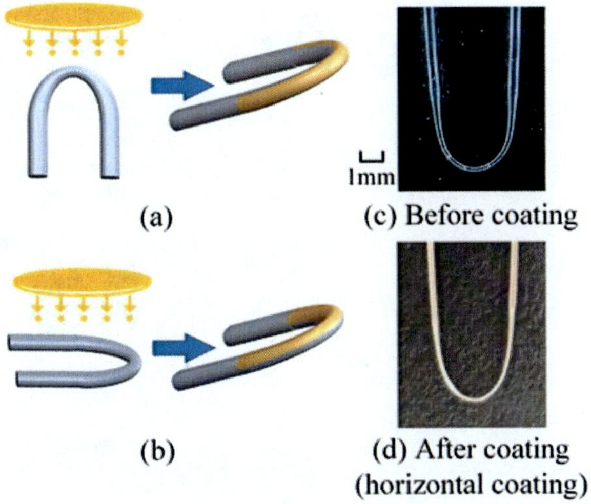

Fig. 3.19 **a** and **b** schematic diagrams of vertical and horizontal Au-film coating, and **c** and **d** photos of the U-shape TPOF probe before and after Au-film coating. Reprinted with permission from Optik. Copyright, 2022, Elsevier (Teng et al. 2022)

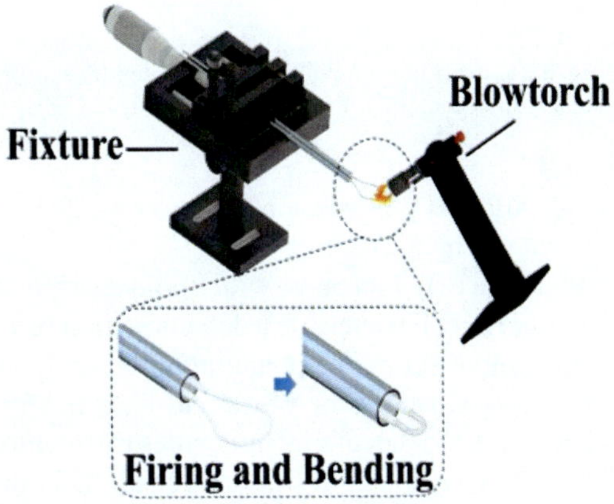

Fig. 3.20 Flame-made U-shaped fiber probe. Reprinted with permission from Optik. Copyright, 2020, Elsevier, Elsevier (Hou et al. 2020)

3.6 Application of a U-type Fiber Optic Surface Plasmon Resonance Sensor

The first U-shaped fiber optic sensor was made almost four decades ago to measure the RI of the surrounding medium (Hou et al. 2020). Lamp oil is used as a testing medium to obtain experimental and analytical observations. In 1992, a CO_2 laser source was used for heating to bend and fabricate a U-shaped fiber optic sensor. In 1996, Gupta et al. (Mason et al. n.d.) used 2D geometrical analysis to study evanescent wave absorption on U-shaped plastic-clad silica (PCS) fiber. Khijwania and Gupta (2000) later compared the sensitivity of the U-shaped sensor based on numerical aperture, core radius, and bend radius. A U-shaped SPR sensor finds its

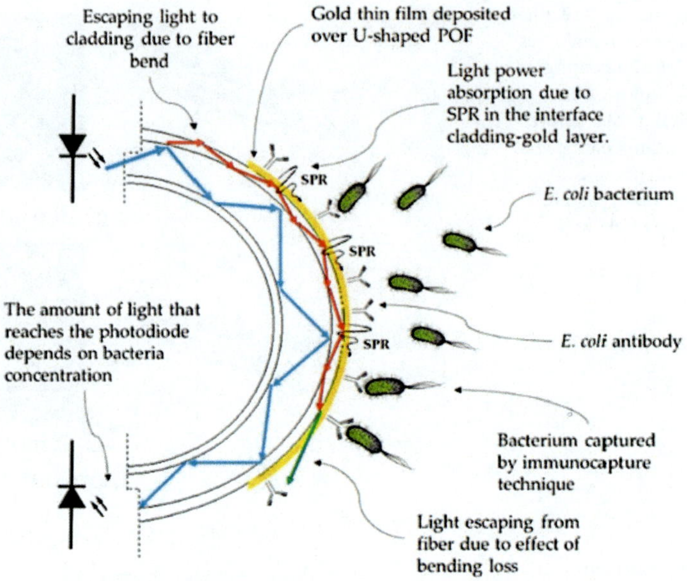

Fig. 3.21 Working principle of gold-coated biosensor. Reprinted with permission from Measurement Copyright, 2022, Elsevier (Zhou et al. 2023)

application as a biosensor (Arcas et al. 2018) to analyze the biological analyte, as shown in Fig. 3.21.

Using a POF biosensor with a gold coating, Khijwania and Gupta (2000) were able to detect E. coli bacteria at a detection limit of 1.5×10^3 colony-forming units (CFU)/mL, demonstrating the potential of this technology as an affordable replacement for current approaches for water and food quality analysis. The U-shaped sensor can also be used to detect tiny molecules by monitoring changes in their refractive index. The refractive index of the solution shifts in the presence of bisphenol A (Luo et al. 2016) as a result of the aggregation of unaltered AuNp (gold nanoparticles) under high salt conditions.

In Duarte et al. (2015), a U-shaped sensor is shown being used differently, this time as a pressure sensor implanted in a block of silicon rubber. The benefit of optical fiber has not been fully utilized because most SPR-based prism configurations have been employed for gas pressure sensors with a resolution of 16 kPa, while Fabry–Perot interferometers are used in pressure sensors based on optical fiber. Even if 0.02 kPa was the finest resolution device available, making such a structure can be difficult. In Mi et al. (2013), researchers proposed a piezo sensor based on SPR phenomena that communicates via optical fiber cables. External pressure was employed to create a refractive index gradient, and a fiber tip coated with a polymer and metal layer was used to measure it. This resulted in a sensitivity of 1.75×10^3nm/MPa. However, this setup has drawbacks, such as the need for a highly precise and controlled etching process on the cladding.

Kumar et al. (2016) have proposed a U-shaped multimode optical fiber SPR probe embedded in a silicon rubber block, with high sensitivity coated with 50 nm thick metal. It is possible to quantify the pressure-induced shift in SPR resonance because of the pressure-induced change in the bending radius. To create a reliable pressure monitoring system, it is proposed in this study to embed a cladded fiber probe in a block of silicon rubber. This protects the metal layer, which opens the door to the use of metals with lesser chemical stability, such as copper. Srinivasulu and Venkateswara Rao (2021) presented a cheap, straightforward, small, dependable, robust, miniaturized, and sturdy refractometer. Two multimode PCS fibers, each 50 cm in length, are used in this setup; one is connected to the 660 nm light source, and the other is attached to the light detector. To create the sensing zone depicted in Figs. 3.22 and 3.23, the remaining ends of two multimode fibers. This rod was the same length as the rest of the fibers. Toluene and t-butanol were kept in the detector as binary liquids in the sensing zone at various temperatures to track the light emitted from the source. The sensor's index of refraction–power output relationship is optimized for use between 10 °C and 60 °C, and its maximum resolution is on the order of 10^{-5} in the working range of 1.36312 nD to 1.50915 nD (Srinivasulu and Venkateswara Rao 2021).

A U-shaped-wound fiber bending based loss crack sensor has been discussed by (Cheng et al. 2020). The results show that the optical splitter improves the reliability of the U-shaped-wound fiber bending loss crack sensor by minimizing the impact of temperature on the light source and the fiber link between them.

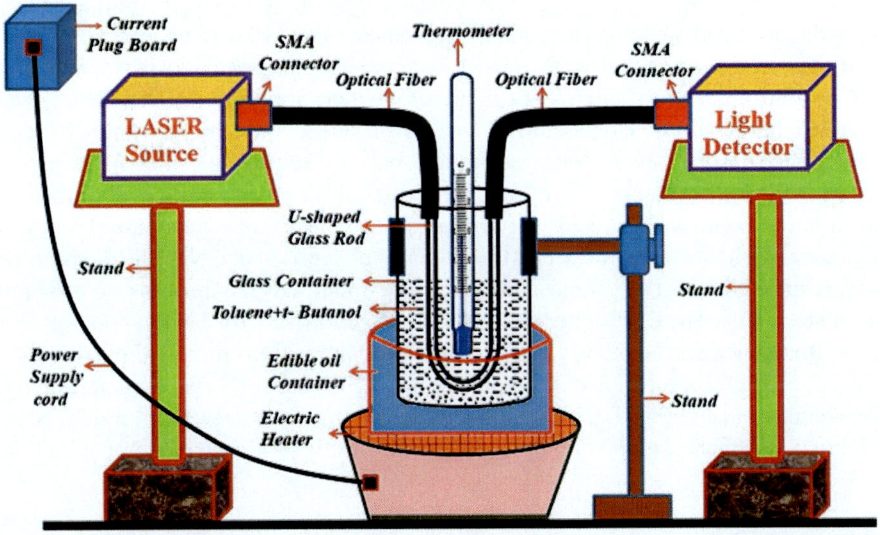

Fig. 3.22 Fiber optic U-shaped sensor used in an experimental setting to heat a chemical reaction. Reprinted with permission from Materials Today. Copyright, 2022, Elsevier (Srinivasulu and Venkateswara Rao 2021)

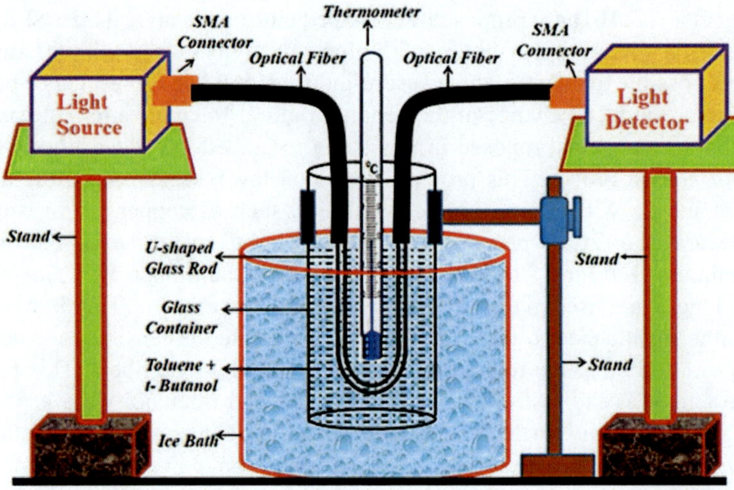

Fig. 3.23 The temperature of the chemical combination was lowered in an experiment using a fiber optic U-shaped glass sensor. Reprinted with permission from Materials Today. Copyright, 2022, Elsevier (Srinivasulu and Venkateswara Rao 2021)

As illustrated in Fig. 3.24a, U-wound fiber optic macro-bending crack sensors have a light source, fiber, optical power meter, fiber winding shaft, top cover, sensor base, moveable rod, and fixed rack. Figure 3.24c shows that the fiber winding shaft has a gear at both ends. See Fig. 3.24b, where the two gears mesh simultaneously with the drive rod's sliding rack and the top cover's fixed rack (Cheng et al. 2020).

LSPR of gold nanoparticles (GNP) detects vapor-phase explosives, including 2,4,6-trinitrotoluene (TNT) and cyclotrimethylenetrinitramine (RDX) (Bharadwaj and Mukherji 2014). GNP localized surface plasmons were trapped on a U-bend fiber optic sensor probe for evanescent field-based stimulation. 4-mercaptobenzoic acid (4-MBA), l-cysteine, and cysteamine were added to immobilize GNP to bind nitro-based explosive chemicals. After explosive analytes bound nanoparticle surface moieties, GNP RI changes were found. This changed the GNP-LSPR spectra's absorbance features. L-Cysteine and cysteamine-modified GNP probes also had high TNT selectivity. The LSPR fiber optic probe's detection limit for TNT vapors was low ppb (Bharadwaj and Mukherji 2014). Gold–thiol interactions modified nanoparticle surfaces with receptor molecules, enabling the LSPR sensor's chemical selectivity. As demonstrated in Fig. 3.25, the explosive detection receptors were immobilized on GNP-coated fibers after 60 min of incubation in freshly produced receptor solutions.

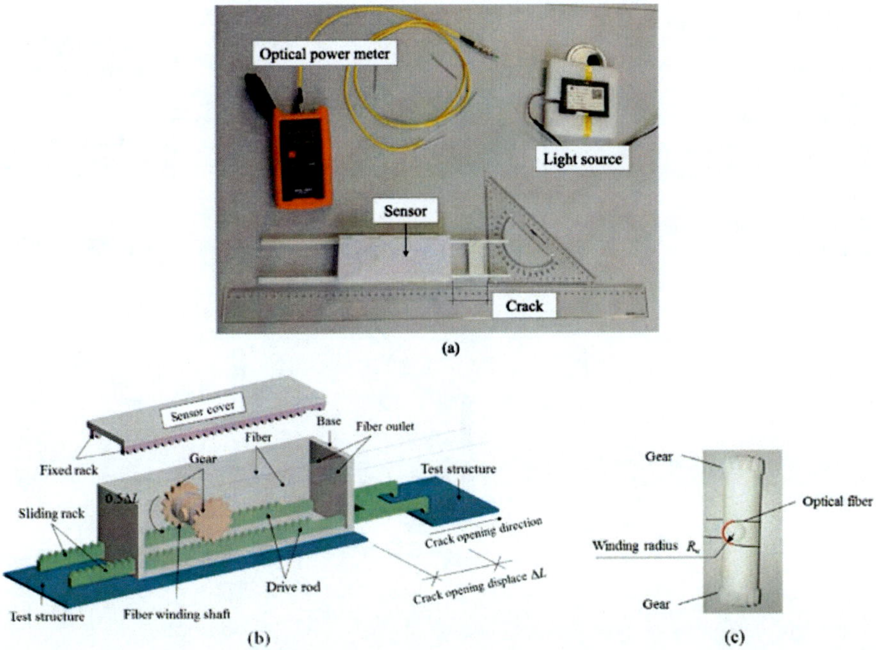

Fig. 3.24 Sensor architecture is shown in schematic form. **a** U-wound fiber optic macro-bending crack sensor experiment as viewed through a simplified model of the experimental setup, **b** the principle of displacement transfer, and **c** The axis of a fiber winding. Reprinted with permission from Optical Fiber Technology. Copyright, 2020, Elsevier (Cheng et al. 2020)

3.7 Future Prospects of U-shaped SPR Sensor

There are many de facto advantages relating to U-shaped SPR sensors. They support various complex industrial applications such as husbandry, environmental hazard monitoring, and medical applications, among others. Cascading other shaped fiber sensors, such as U-bend and D-shaped, can not only improve the performance of the overall sensor but can also incur a loss of power due to multiple bending. Therefore, it is worth researching how to analyze and fabricate such an SPR or any fiber optic sensor. OTDR techniques for implementing U-bend fiber probes also possess potential for remote sensing applications, and since there are few publications in this regard, these areas will also require some research in the future. Based on the application and resources available, various numerical analyses among BPM, FEM, ray tracing methods, etc. to compute power loss are worth researching. The heating temperature required for the bending of the fiber involves precise bending forces to maintain the uniformity of the core of the fiber. Different core diameter fibers have different optimum bending temperatures and bending forces, and the guidelines for optimum bending temperatures are not available and require more studies. Also, irrelevant parameters causing false readings obtained from U-shaped sensors

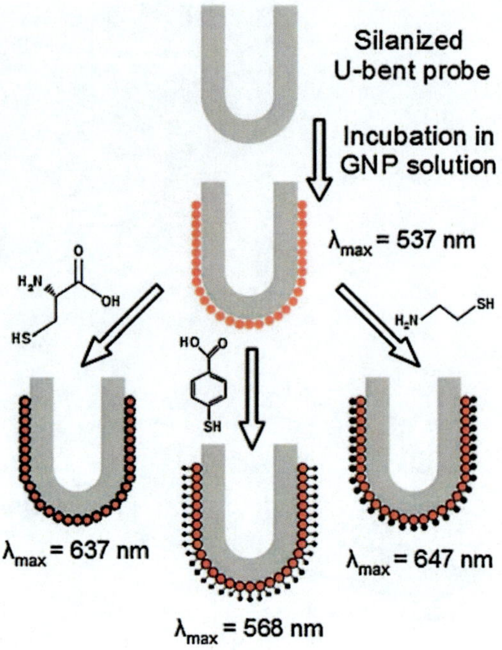

Fig. 3.25 This schematic depicts the functionalization of the U-bend fiber probes' surfaces using gold nanoparticle coatings. Reprinted with permission Sensors and Actuators B: Chemical. Copyright, 2014, Elsevier (Bharadwaj and Mukherji 2014)

are obstructions for field deployment and thus required to be minimized. These false readings can be prevented with the help of machine learning tools and principal component analysis (PCA). Effects of environmental perturbations such as humidity, vibration, temperature, and noise on instruments also cause performance errors and are thus required to be minimized. The needs of the consumer will drive the future application of SPR sensors; thus, these sensors are required to be consumer-friendly. Commercialization of SPR sensors is also an important aspect of their development and use outside of a laboratory (Wang et al. 2022a, b). Low fabrication cost, rapid detection, robustness, high specificity, user friendliness, etc. are the main areas of continuous research (Pandey et al. 2022a, b; Shadab et al. 2022).

Nanomaterials play a crucial role in optical sensing due to their unique properties and the ability to manipulate light–matter interactions at the nanoscale (Li et al. 2022). They offer significant advantages in enhancing the sensitivity, selectivity, and performance of optical sensors (Kumar and Singh 2021, Kaur et al. 2023; Kumar et al. 2022a, b). Here are some key aspects of the role of nanomaterials in optical sensing: (1) Nanomaterials, such as nanoparticles, nanowires, and nanotubes, possess a large surface-to-volume ratio, enabling a higher probability of interaction with target analytes. This increased surface area allows for more efficient capture and detection of analyte molecules, leading to improved sensitivity in optical sensing applications (Wang et al. 2021). (2) Certain nanomaterials, like gold or silver nanoparticles, exhibit unique optical properties related to localized surface plasmon resonance (Agrawal et al. 2022; Kumar et al. 2023a, b; Kumari et al. 2023).

When these nanoparticles are appropriately sized and shaped, they can interact with specific wavelengths of light, resulting in a distinctive optical response (Kumar et al. 2022a, b; Uniyal et al. 2022a, b). This phenomenon has been extensively utilized in various optical sensing techniques, such as surface plasmon resonance spectroscopy, for label-free detection of biomolecules and chemical analytes (Kumar et al. 2022a, b; Pandey et al. 2022a, b). (3) Nanomaterials can serve as signal amplifiers in optical sensing systems. For instance, nanomaterials with high fluorescence quantum yields can be used as labels or tags in fluorescence-based sensors, enabling the detection of low-concentration analytes (Kumari et al. 2022). Additionally, nanomaterials like quantum dots or upconversion nanoparticles, can emit light at specific wavelengths, offering improved signal-to-noise ratios and enabling multiplexed detection in optical sensing platforms (Subbanna et al. 2022). (4) Nanomaterials can be engineered and tailored to exhibit specific optical properties by controlling their size, shape, composition, and surface chemistry. This tunability allows for the optimization of nanomaterials to match the desired sensing requirements, such as specific analyte detection, wavelength range, and sensitivity level (Singh and Kumar 2022; Kumar et al. 2023a, b). (5) Nanomaterials can be integrated with various optical devices, such as optical fibers, waveguides, and microfluidic channels, to enhance their sensing capabilities (Uniyal et al. 2022a, b). By incorporating nanomaterials into these platforms, it becomes possible to enhance light–matter interactions, increase light confinement, and improve signal transduction efficiency, leading to improved sensing performance. Overall, nanomaterials offer unique opportunities to revolutionize optical sensing by enabling high sensitivity, selectivity, and versatility (Singh and Kumar 2020). Their unique properties and controllable synthesis make them valuable tools for developing next-generation optical sensors for applications ranging from healthcare diagnostics and environmental monitoring to food safety and security.

The U-shape fiber plays a significant role in wearable sensors, providing several advantages and functionalities (Lang et al. 2023). Here are some key roles of U-shape fiber in wearable sensors: (1) Comfortable and flexible design: The U-shape fiber allows for a more comfortable and flexible design of wearable sensors. Its curved shape conforms well to the contours of the body, ensuring a better fit and improved wearer comfort. This is particularly important for sensors that are worn for extended periods, such as fitness trackers or health monitoring devices. (2) Enhanced signal stability: The U-shape fiber helps in maintaining the stability of the sensor signals. The curved shape reduces the chances of sensor displacement or movement during physical activities, preventing signal artifacts or fluctuations. This is crucial for accurate and reliable data collection in wearable sensors. (3) Improved signal sensitivity: The U-shape fiber can enhance the signal sensitivity of wearable sensors. By curving the fiber, the effective sensing area can be increased, allowing for improved interaction between the sensor and the target analyte. This leads to enhanced sensitivity in detecting physiological parameters, such as temperature, pressure, and strain. (4) Versatile sensor placement: The U-shape fiber provides flexibility in sensor placement on different parts of the body. Its curved design enables easy attachment to curved surfaces or body parts, such as wrists, arms, and even clothing. This versatility allows for the integration of sensors into various wearable devices, accommodating different applications and user preferences. (5) Integration with other components: The U-shape fiber can be easily integrated with other components of wearable sensors, such as electronic circuits, microcontrollers, and wireless modules. Its flexible nature facilitates the integration process and ensures seamless connectivity between different sensor elements, enabling real-time data monitoring and transmission. (6) Durability and reliability: The U-shape fiber offers improved durability and reliability for wearable sensors. Its curved structure provides mechanical strength, preventing fiber breakage or damage during everyday movements. This durability ensures the longevity of wearable sensors, allowing for long-term usage without compromising performance. Overall, the U-shape fiber plays a crucial role in wearable sensors by providing comfort, stability, sensitivity, versatility, and integration capabilities (Kaur et al. 2023; Kumar et al. 2023a, b). Its unique design features contribute to the development of wearable devices that are user-friendly, accurate, and capable of monitoring various physiological parameters for health care, sports, and other applications.

3.8 Summary

This chapter is devoted to a detailed discussion of SPR-based FOS, with a focus on the U-shaped SPR sensor. This chapter describes the basic principles of the SPR technique and traditional configurations like prism and otto, and then urge the need for and advantages of optical fiber-based sensors in comparison to traditional ones. Performance-deciding parameters, especially the sensitivity of the sensor, factors affecting the sensitivity, enhancement of sensitivity, etc., have been discussed thoroughly. Various numerical analyses, such as the ray tracing method and transfer matrix method, are expanded to analyze the loss of power due to different factors. Parameters like bend radius, core diameter, numerical aperture, sensing region length, metal thickness, etc. are discussed for optimizing the sensor sensitivity. Performance limiting factors during fabrication like RI inhomogeneity, its cause and effect, optimum bending temperature, bending force, etc. are vital in developing the sensor of interest. The U-shaped SPR sensor is known for its abundant benefits like high sensitivity, easy fabrication, utilization as a point sensor, and simple experimental setup over another geometrical configuration. Based on the application of interest, different materials for fiber, such as glass or plastic, can be used. For operating at high temperatures, glass fibers are more suitable, while plastic fibers provide more mechanical strength. The U-shaped sensor has wide applications such as a biosensor, chemical analyte sensor, and pressure sensor.

References

Agrawal N, Saxena R, Singh L, Saha C, Kumar S (2022) Recent advancements in plasmonic optical biosensors: a review. ISSS J Micro Smart Syst 11(1):31–42

Arcas AdaS, Dutra FdaS, Allil RCSB, Werneck MM (2018) Surface plasmon resonance and bending loss-based U-shaped plastic optical fiber biosensors. Sensors (Switzerland) 18(2):1–16. https://doi.org/10.3390/s18020648

Bharadwaj R, Mukherji S (2014) Gold nanoparticle coated U-bend fibre optic probe for localized surface plasmon resonance based detection of explosive vapours. Sens Actuators, B Chem 192:804–811. https://doi.org/10.1016/j.snb.2013.11.026

Brier J, Jayanti LD (2020) No Analysis of the co-dispersion structure of health-related indicators in the center of the subjective sense of health. Title 21(1). http://journal.um-surabaya.ac.id/index.php/JKM/article/view/2203

Cheng L, Song F, Zhang K, Li Y, Yang J (2020) A U-shaped-wound fiber macro-bending loss crack sensor improved by an optical splitter. Opt Fiber Technol 58:102259. https://doi.org/10.1016/j.yofte.2020.102259

Danny CG, Danny Raj M, Sai VVR (2020) Investigating the refractive index sensitivity of U-bent fiber optic sensors using ray optics. J Lightwave Technol 38(6):1580–1588. https://doi.org/10.1109/JLT.2019.2958044

Duarte DP, Alberto N, Bilro L, Nogueira R (2015) Theoretical design of a high sensitivity SPR-based optical fiber pressure sensor. J Lightwave Technol 33(22):4606–4611. https://doi.org/10.1109/JLT.2015.2477353

Gowri A, Sai VVR (2016) Development of LSPR based U-bent plastic optical fiber sensors. Sens Actuators, B Chem 230:536–543. https://doi.org/10.1016/j.snb.2016.02.074

Guo Y, Zhang Y, Yi G (2021) Displacement and temperature sensor based on whispering gallery effect within unsymmetrical U-type fiber structure. Optik 238:166459. https://doi.org/10.1016/j.ijleo.2021.166459

Gupta BD, Verma RK (2009) Surface plasmon resonance-based fiber optic sensors: principle, probe designs, and some applications. J Sensors 2009. https://doi.org/10.1155/2009/979761

Hou CH, Chiang CC, Hsieh TS, Li YL, Tsai L, Hsu CY (2020) Double-knotted U-shaped spiral optical fiber probe for measuring glucose solutions. Optik 207. https://doi.org/10.1016/j.ijleo.2019.163765

Jiang S, Li Z, Zhang C, Gao S, Li Z, Qiu H, Li C, Yang C, Liu M, Liu Y (2017) A novel U-bent plastic optical fibre local surface plasmon resonance sensor based on a graphene and silver nanoparticle hybrid structure. J Phys D: Appl Phys 50(16). https://doi.org/10.1088/1361-6463/aa628c

Kaur B, Kumar S, Kaushik BK (2022) Recent advancements in optical biosensors for cancer detection. Biosens Bioelectron 197:113805

Kaur B, Kumar S, Kaushik BK (2023) Novel wearable optical sensors for vital health monitoring systems—a review. Biosensors 13(2):181

Khijwania SK, Gupta BD (1999) Fiber optic evanescent field absorption sensor: effect of fiber parameters and geometry of the probe. Opt Quant Electron 31(8):625–636. https://doi.org/10.1023/A:1006956824722

Khijwania SK, Gupta BD (2000) Maximum achievable sensitivity of the fiber optic evanescent field absorption sensor based on the U-shaped probe. Opt Commun 175(1):135–137. https://doi.org/10.1016/S0030-4018(00)00465-X

Kumar S, Singh R (2021) Recent optical sensing technologies for the detection of various biomolecules: review. Opt Laser Technol 134:106620

Kumar V, Raghuwanshi SK, Kumar S (2022a) Advances in nanocomposite thin-film-based optical fiber sensors for environmental health monitoring—a review. IEEE Sens J 22(15):14696–14707

Kumar V, Raghuwanshi SK, Kumar S (2022b) Recent advances in carbon nanomaterials based SPR sensor for biomolecules and gas detection—a review. IEEE Sens J 22(16):15661–15672

Kumar S, Singh R, Wang Z, Li M, Liu X, Zhang W, Zhang B, Li G (2023a) (Invited) Advances in 2D nanomaterials-assisted plasmonics optical fiber sensors for biomolecules detection. Results Opt 10:100342

Kumar S, Wang Z, Zhang W, Liu X, Li M, Li G, Zhang B, Singh R (2023b) Optically active nanomaterials and its biosensing applications—a review. Biosensors 13(1):85

Kumar M, Raghuwanshi SK, Palodiya V (2016) The design and analysis of a noble surface plasmon resonance-based pressure sensor. Plasmon: Des, Mater, Fabr, Charact, Appl XIV 9921:99212V. https://doi.org/10.1117/12.2237728

Kumari A, Vyas V, Kumar S (2022) Advances in electrochemical and optical sensing techniques for vitamins detection: a review. ISSS J Micro Smart Syst 11(1):329–341

Kumari A, Vyas V, Kumar S (2023) Synthesis, characterization, and applications of gold nanoparticles in development of plasmonic optical fiber-based sensors. Nanotechnology 34(4):042001

Lang X, Liu X, Zhang W, Singh R, Li G, Xie Y, Zhang B, Kumar S (2023) Homemade low-cost fabrication technique and stability analysis of a U-shaped fiber sensor structure. Appl Opt 62(18):4753–4758

Li M, Singh R, Wang Y, Marques C, Zhang B, Kumar S (2022) Advances in novel nanomaterial-based optical fiber biosensors—a review. Biosensors 12(10):843

Li P, Chen Y, Hu J, Zhang G, Cui B, Meng L, Lv M (2023) Simultaneous measurement of salinity and temperature of seawater based on U-shaped tapered no-core fiber. Infrared Phys Technol 130:104617. https://doi.org/10.1016/j.infrared.2023.104617

Luo Z, Zhang J, Wang Y, Chen J, Li Y, Duan Y (2016) An aptamer based method for small molecules detection through monitoring salt-induced AuNPs aggregation and surface plasmon resonance (SPR) detection. Sens Actuators, B Chem 236:474–479. https://doi.org/10.1016/j.snb.2016.06.035

Luo Z, Wang Y, Xu Y, Wang X, Huang Z, Chen J, Li Y, Duan Y (2019) Ultrasensitive U-shaped fiber optic LSPR cytosensing for label-free and in situ evaluation of cell surface N-glycan expression. Sens Actuators, B Chem 284:582–588. https://doi.org/10.1016/j.snb.2019.01.015

Manoharan H, Kalita P, Gupta S, Sai VVR (2019) Plasmonic biosensors for bacterial endotoxin detection on biomimetic C-18 supported fiber optic probes. Biosens Bioelectron 129:79–86. https://doi.org/10.1016/j.bios.2018.12.045

Mason A, Chandra S, Krishanthi M, Jayasundera P (n.d.) Smart sensors, measurement and instrumentation 11. Sens Technol: Curr Status Futur Trends III. http://www.springer.com/series/10617

Mi H, Wang Y, Jin P, Lei L (2013) Design of an ultrahigh-sensitivity SPR-based optical fiber pressure sensor. Optik 124(21):5248–5250. https://doi.org/10.1016/j.ijleo.2013.03.097

Pandey PS, Raghuwanshi SK, Kumar S (2022a) Recent advances in two-dimensional materials-based Kretschmann configuration for SPR sensors: a review. IEEE Sens J 22(2):1069–1080

Pandey PS, Raghuwanshi SK, Shadab A, Ansari MTI, Tiwari UK, Kumar S (2022b) SPR based biosensing chip for COVID-19 diagnosis—a review. IEEE Sens J 22(14):13800–13810

Peng X, Cha Y, Zhang H, Li Y, Ye J (2017) Light intensity modulation temperature sensor based on U-shaped bent single-mode fiber. Optik 130:813–817. https://doi.org/10.1016/j.ijleo.2016.11.003

Ren ZH, Wang Q, Cong XW, Zhao WM, Tang JR, Wang L, Yan X, Zhu AS, Qiu FM, Chen BH, Zhang KK (2023) A fiber SPR sensor with high comprehensive evaluation indicator based on core mismatched U-Shaped and tapered arm. Meas: J Int Meas Confed 206:112248. https://doi.org/10.1016/j.measurement.2022.112248

Shadab A, Raghuwanshi SK, Kumar S (2022) Advances in micro-fabricated fiber Bragg grating for detection of physical, chemical, and biological parameters—a review. IEEE Sens J 22(16):15650–15660

Sharma AK, Jha R, Gupta BD (2007) Fiber-optic sensors based on surface plasmon resonance: a comprehensive review. IEEE Sens J 7(8):1118–1129. https://doi.org/10.1109/JSEN.2007.897946

Sharma AK, Gupta J, Sharma I. (2019). Fiber optic evanescent wave absorption-based sensors: a detailed review of advancements in the last decade (2007–18). Optik 183:1008–1025. https://doi.org/10.1016/j.ijleo.2019.02.104

Singh R, Kumar S (2020) Chapter nine—applications of graphene-based sensors for biomedical industries. Compr Anal Chem. C. M. Hussain, Elsevier. 91:201–233

Singh R, Kumar S (2022) Cancer Targeting and diagnosis: recent trends with carbon nanotubes. Nanomaterials 12(13):2283

Srinivasulu S, Venkateswara Rao S (2021) Application of U-shaped hybrid fiber optic sensor to determine the temperature dependent variation of refractive index of binary liquids. Mater Today: Proc 51:31–37. https://doi.org/10.1016/j.matpr.2021.04.358

Subbanna BB, Choudhary K, Singh S, Kumar S (2022) 2D material-based optical sensors: a review. ISSS J Micro Smart Syst 11(1):169–177

Tan AJY, Ng SM, Stoddart PR, Chua HS (2021) Trends and applications of U-shaped fiber optic sensors: a review. IEEE Sens J 21(1):120–131. https://doi.org/10.1109/JSEN.2020.3014190

Teng C, Li M, Cheng Y, Peng H, Deng S, Deng H, Yuan L, Chen M (2022) Investigation of U-shape tapered plastic optical fibers based surface plasmon resonance sensor for RI sensing. Optik 251:168461. https://doi.org/10.1016/j.ijleo.2021.168461

Uniyal S, Choudhary K, Sachdev S, Kumar S (2022a) Recent advances in K-SPR sensors for the detection of biomolecules and microorganisms: a review. IEEE Sens J 22(12):11415–11426

Uniyal S, Choudhary K, Sachdev S, Kumar S (2022b) Synthesis and characterization of gold nanoparticles and graphene oxide for the development of optical fiber biosensors. Appl Opt 61(26):7618–7624

Wang Y, Huang Y, Bai H, Wang G, Hu X, Kumar S, Min R (2021) Biocompatible and biodegradable polymer optical fiber for biomedical application: a review. Biosensors 11(12):472

Wang R, Liu C, Wei Y, Jiang T, Liu C, Shi C, Zhao X, Li L (2022a) Research and application of multi-channel SPR sensor cascaded with fiber U-shaped structure. Optik 266:169603. https://doi.org/10.1016/j.ijleo.2022.169603

Wang Z, Zhang W, Liu X, Li M, Lang X, Singh R, Marques C, Zhang B, Kumar S (2022b) Novel optical fiber-based structures for plasmonics sensors. Biosensors 12(11):1016

Xie T, He Y, Yang Y, Zhang H, Xu Y (2021) Highly sensitive surface plasmon resonance sensor based on graphene-coated U-shaped fiber. Plasmonics 16(1):205–213. https://doi.org/10.1007/s11468-020-01264-x

Zhou Y, Zhang YN, Han B, Cheng L, Li D, Zheng W, Zhao Y (2023) Biochemical sensor based on functional material assisted optical fiber surface plasmon resonance: a review. Meas: J Int Meas Confed 207:112353. https://doi.org/10.1016/j.measurement.2022.112353

Chapter 4
Cascaded Fiber Optic SPR Sensor

4.1 Introduction

SPR-based fiber optic sensors have a number of benefits, including immunity to electromagnetic interference, quick response, high sensitivity, and good reproducibility. Furthermore, they demonstrate probe miniaturization and remote sensing capabilities (Tabassum and Kant 2020a). Multi-channel fiber optic sensing schemes that cascade two (or more) sensing regions that are clearly separated on a single fiber optic probe have been investigated to allow the simultaneous optical detection of several analytes (Homola et al. 2001; Sun et al. 2018). Cascaded channel fiber optic sensors improve on the benefits of single-channel fiber optic SPR sensors by allowing for the simultaneous detection of multiple analytes. Additionally, a cascaded channel fiber optic sensor's two (or more) independent and well-separated sensing zones can be set up in various manners to find the identical analyte. Therefore, instead of using two (or more) separate single-channel fiber optic probes, a single sensing probe can be used to compare the performance of several sensing channels for detecting a particular analyte. Also, one of the detecting channels of these cascaded sensors can be designated as a "reference" channel, which is an efficient method of reducing noise introduced by the RI background or the ambient temperature. Cascaded channel fiber optic sensors are highly adapted for the quantitative inspection of truly complex sample combinations, such as those present in petroleum products, clinical evaluations, biological systems, and pharmaceutical formulations, due to their aforementioned properties. Studies on the feasibility of a fiber optic SPR sensing platform for the simultaneous detection of many analytes have been extensive. To argue for the relevance and use of this possible field of study, it is crucial to comprehend the historical development of multi-channel fiber optic sensors. Using optical fibers, multi-channel different geometrical configurations of sensors have been developed, such as D-shaped fiber (Xia et al. 2012), hetero-core structured fiber shaped like a circular truncated cone (Liu et al. 2017), and cascading of multimode and single-mode fibers (Wei et al. 2016). The knowledge of fiber optic sensors with numerous

S. K. Raghuwanshi et al., *Geometric Feature-Based Fiber Optic Surface Plasmon Resonance Sensors*, Springer Tracts in Electrical and Electronics Engineering,
https://doi.org/10.1007/978-981-99-7297-5_4

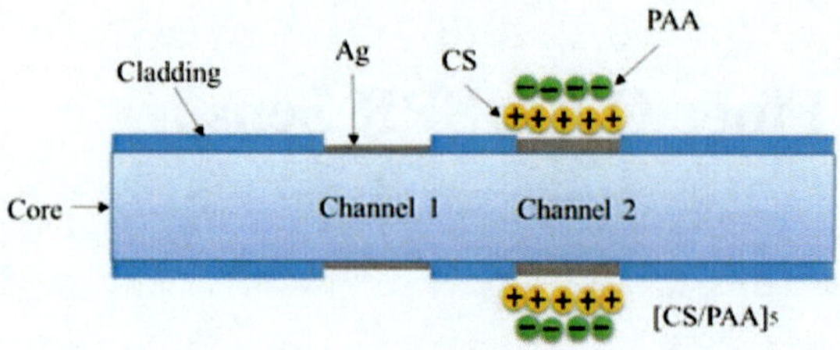

Fig. 4.1 The structure description of the dual-channel optical fiber heavy metal ion sensor. Reprinted with permission from Optics Communications. Copyright, 2020, Elsevier (Zhang et al. 2019)

channels is essential in order to comprehend the idea of multi-channel detection for a variety of analytes.

Using surface plasmon resonance, Zhang et al. (2019) suggested an experimentally novel dual-channel optical fiber sensor. First, two resonance dips with RI sensitivities of 1730.88 and 1334.56 nm/RIU are created by sputtering Ag onto two fiber channels of varying thicknesses. Identifying metal ions in polluted water is made possible by coating the channel with polyacrylic acid (PAA) and chitosan (CS) in a layer-by-layer approach. The channel that contains the thinner Ag can then be used as a reference in order to determine if the change in the preceding channel was brought about by heavy metal ions or the local refractive index. The CS/PAA-functionalized channel causes a resonance dip at low ion concentrations, and this dip is redshifted when Cu^{2+} and other metal ion concentrations increase in the experimental sample. According to experimental findings, Cu^{2+} binds the CS/PAA layer more strongly when the concentration is below 80 nM, reaching 0.249 nm/M, indicating that this sensor is highly sensitive to modest concentrations of Cu^{2+} (Fig. 4.1).

By adjusting the thickness of the gold film coating, Zhang et al. (2017) suggested and demonstrated a technique for distributed multi-channel SPR sensing in a fiber. Resonance frequencies, dynamic ranges, and testing sensitivities of an SPR sensor based on a multimode fiber might vary depending on the thickness of the gold film used. A two-channel multimode fiber SPR sensor was therefore created, with a gold layer thickness of 30 nm in the first sub-channel and 60 nm in the second sub-channel (both cascaded for optimal efficiency). The values of 1712 and 3038 nm/RIU are employed in the testing, where RI is between 1.333 and 1.385. As shown by the experiments, for RIs with gold film thicknesses between 20 and 60 nm, the resonance dip is deeper, the testing range is larger, and the testing sensitivity is lower. Despite its distributed nature, the distributed multi-channel fiber SPR sensor's testing sensitivity and dynamic range are preserved by the cascade connection of several sensing sub-channels. The sensitivity of each of the fiber SPR channels was independently controlled by adjusting the gold layer thickness (Fig. 4.2).

Using Ag, Cu, and Au thin films as plasmonic metals and ZnO and Si high-RI overlayers, Tabassum and Kant (2020b) studied the functionality of SPR-based RI sensors in cascaded dual- and triple-channel fiber optic devices. The use of a multimode optical fiber with two physically independent detecting channels made

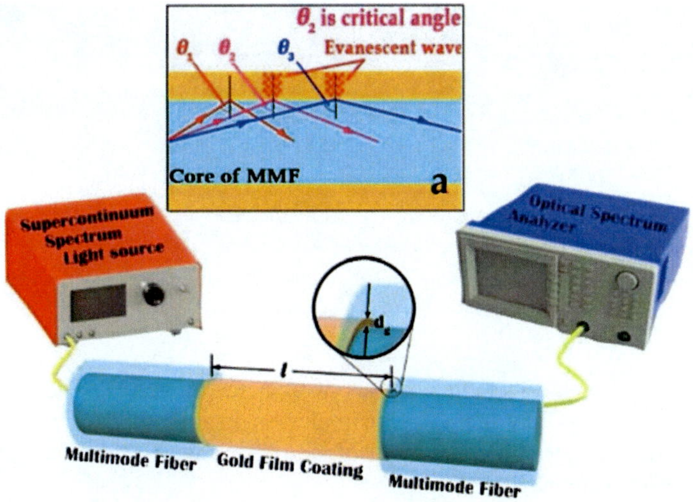

Fig. 4.2 The illustration shows the structure of the sensor probe. Reprinted with permission from Sensors and Actuators A: Physical. Copyright, 2017, Elsevier (Zhang et al. 2017)

of Ag and Cu/ZnO layers simulates a dual-channel sensor. Silver, copper/zinc oxide, and gold/silicon are all tested as potential components of a triple-channel sensor. Different SPR spectra are acquired when the sensor response is studied by varying the RIs of the analytes that enclose the various sensing zones of the cascaded fiber optic probe. This is because Ag, Cu, and Au all have somewhat different SPR resonance wavelengths. This makes it possible to quickly and simply modify the sensor's working range to fit the needs of a wide variety of applications. The results provided provide encouraging evidence for the use of in-line sensing technologies to analyze components of complex mixtures such as those found in clinical samples, petroleum products, and a wide range of other biomolecules.

A new fiber-based dual-channel SPR refractive index sensor with a hetero-core design has been proposed and proven by Liu et al. (2017). The standard hetero-core fiber SPR sensor does not allow for tuning of the resonance frequency range. That issue was fixed by reshaping the standard hetero-core structure fiber into a truncated cone with a circular cross-section. They proved that shifting the angle at which the fiber is polished can change the resonance wavelength range. Resonance wavelengths shift toward the red when fiber polishing angles rise. When the refractive index is between 1.333 and 1.385 and the fiber polishing angle is 14 degrees, the resonance wavelengths are 754 and 965 nm; this is distinct from the typically used hetero-core structure fiber SPR sensor (600–700 nm). By employing wavelength division multiplexing, they were able to achieve dual-channel sensing.

Multi-channel fiber SPR sensors with a U-shaped fiber topology were proposed by Wang et al. (2022). The dual-channel fiber SPR was constructed using a U-shaped construction with a 1 and 5 mm radius of curvature for testing. Glucose

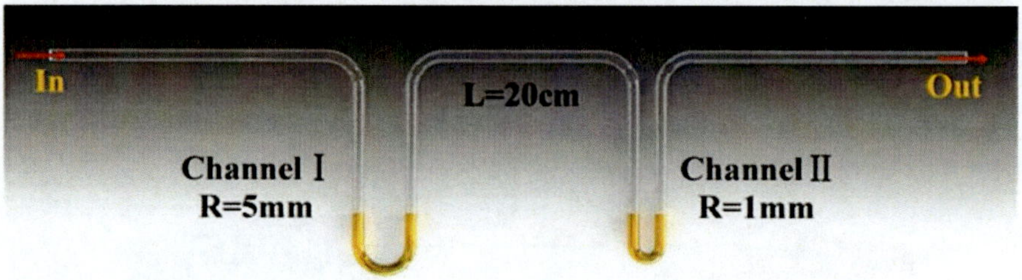

Fig. 4.3 U-shaped dual-channel sensing probe structure illustration. Reprinted with permission from Optik. Copyright, 2022, Elsevier (Wang et al. 2022)

and sucrose were both detectable with this two-channel fiber SPR. The proposed U-shaped construction is easy and simple to implement using a multi-channel fiber SPR sensor. For simultaneous measurements of many parameters, it can be put directly into a confined location (Fig. 4.3).

Using the electric field coupling between SPR and LSPR, Wang et al. (2019) suggested a dual-channel fiber optic biosensor. For identifying human IgG that has been marked with Au nanoparticles (Au NPs), a graphene oxide/gold bilayer with a surface that has been changed by goat anti-human immunoglobulin G (IgG) that is put on one of the sensing channels. The graphene oxide layer close to coupled electric fields speeds up both the immune response and the biological material's adsorption to a sensor surface. When compared to a traditional LSPR sensor, this sandwich design makes optimal use of the electric field created by the combination of the Au NPs and Au film. To lessen the impact of temperature cross-sensitivity and non-specific binding, a thin Ag film is only applied to one of the sensing channels, serving merely as a reference channel (Fig. 4.4).

They carried out corresponding experiments after numerically analyzing the SPR/electric LSPR's field coupling strength. Excellent temperature insensitivity, high sensitivity, and accuracy are all features of the suggested biosensor.

An extremely sensitive temperature-compensated cascaded fiber SPR sensor was proposed by Tian et al. (2022). Three layers of MoS_2 and gold and silver films of varying thicknesses are recommended for use on the surfaces of the two sensing channels to boost the sensor's sensitivity. The polydimethylsiloxane (PDMS) coating on channel B's exterior allows it to detect changes in temperature, whereas the PDMS coating on channel A's exterior detects changes in refractive index. The sensor has a

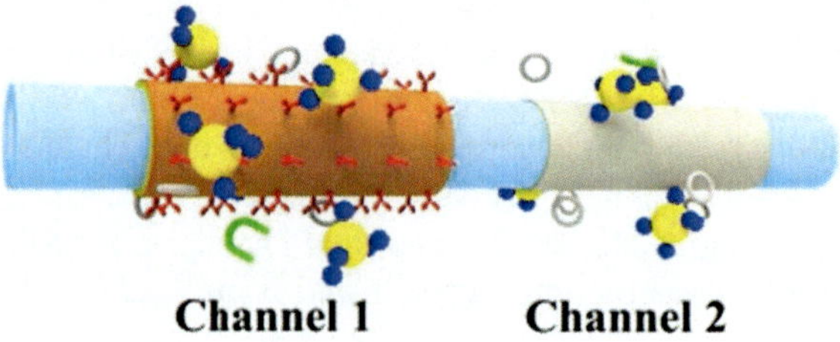

Fig. 4.4 Proposed dual-channel sensor. Reprinted with permission from Optics and Laser Technology. Copyright, 2020, Elsevier (Wang et al. 2019)

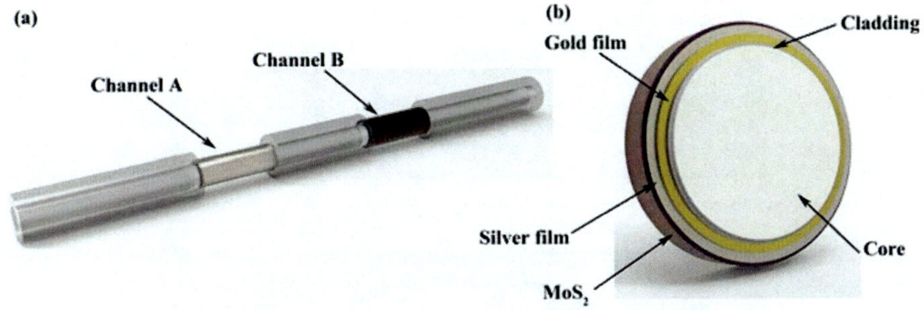

Fig. 4.5 **a** Cascade fiber optic SPR sensor; **b** Cross-section of sensor. Reprinted with permission from Optics Communications. Copyright, 2023, Elsevier (Tian et al. 2022)

refractive index sensitivity of 3820 nm/RIU in an environment with a RI between 1.34 and 1.38. Between 20 and 80 °C, the sensor's temperature sensitivity is −5.189 nm/ C. Correction for temperature allows for more precise measurements of the refractive index. With its precise measuring and detection abilities, this sensor is well suited for use in biosensing, food manufacturing, and medical diagnostics (Fig. 4.5).

4.2 Basic Theory and Sensor Structure

The mathematical model used for simulating fiber optic sensors and studying their response uses the transfer matrix method for optical layers arranged in stratification (Mishra et al. 2015). This method is widely employed for measuring thin films' reflectance and transmittance.

SPR-based fiber optic sensors are typically analyzed using the attenuated total reflection technique and the Krestchmann layout (Yuan et al. 2012). Yuan et al. (2012) suggested a four-layer model with a fiber core, metal, a sensing layer, and a sample. Using the proposed model, the SPR sensor's wavelength interrogation mode has been simulated. According to the proposed model, the first layer is a metal layer with a thickness of d_2 and a dielectric function of ε_m, while the second layer is a fiber core made of fused silica with a fixed refractive index of n_p (1.457). The third layer, which has a thickness of d_3 and a dielectric function ε_{sen}, is the sensing layer. The sample, whose refractive index is indicated by n_{sam} and whose dielectric constant is indicated by ε_{sam}, is the fourth layer (Fig. 4.6).

The power, dP, between the incident angles θ and $\theta + d\theta$, arriving at the opposite end of the fiber, can be written as (Yuan et al. 2012)

$$dP \propto d\theta \tag{4.1}$$

where the incident angle's corresponding modal power is (Yuan et al. 2012)

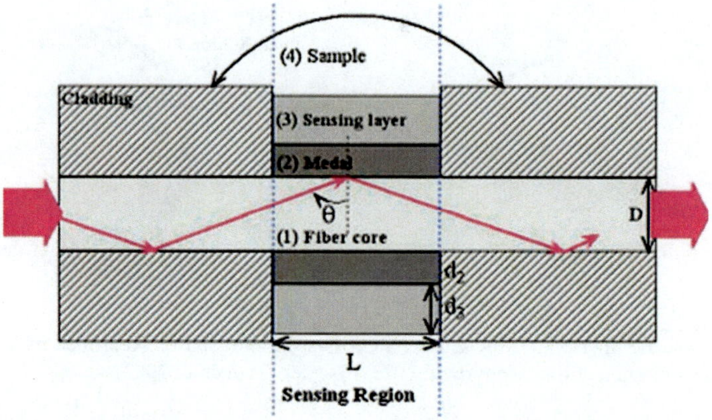

Fig. 4.6 A four-layer setup for fiber optic sensing based on SPR. Reprinted with permission from Sensors and Actuators B: Chemical. Copyright, 2012, Elsevier (Yuan et al. 2012)

$$P(\theta) = \frac{n_1^2 \sin\theta \, \cos\theta}{\left(1 - n_1^2 \, \cos\theta^2\right)^2} \tag{4.2}$$

It is possible to determine the normalized transmitted power of polarized light by measuring the reflectance after a single reflection at the core/metal interface (Yuan et al. 2012):

$$P_{trans} = \frac{\int_{\theta_{cr}}^{\frac{\pi}{2}} R_p^{N_{ref(\theta)}} P(\theta) d\theta}{\int_{\theta_{cr}}^{\frac{\pi}{2}} P(\theta) d\theta} \tag{4.3}$$

Here,

$$R_p = \left|r_p\right|^2 \tag{4.4}$$

$$\theta_{cr} = \sin^{-1}\left(\frac{n_{cl}}{n_1}\right) \tag{4.5}$$

In the above equations,

R_p reflection intensity
r_p amplitude reflection coefficient
θ_{cr} critical angle
n_{cl} refractive index of the fiber cladding.

Two cascaded SPR sensors with different structural configurations can be put together along an optical fiber to measure two different chemical substances simultaneously, as shown in Fig. 4.7. For p-polarized light, the total normalized transmitted power is expressed as (Yuan et al. 2012)

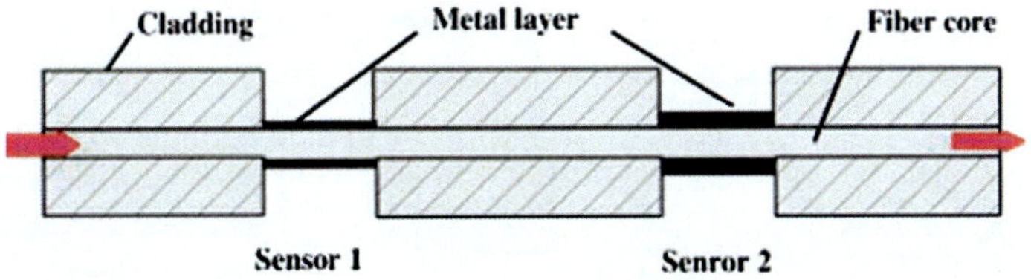

Fig. 4.7 Two SPR sensors in cascade. Reprinted with permission from Sensors and Actuators B: Chemical. Copyright, 2012, Elsevier (Yuan et al. 2012)

$$P_{total} = P^1_{trans} \cdot P^2_{tans} \tag{4.6}$$

where

$$P^{(i)}_{trans} = \frac{\int_{\theta_{cr}}^{\frac{\pi}{2}} \left[R^i_p\right]^{N^{(i)}_{ref}(\theta)} P(\theta)d\theta}{\int_{\theta_{cr}}^{\frac{\pi}{2}} P(\theta)d\theta} \tag{4.7}$$

Here, $i = 1$ or 2, and $R^{(i)}_p$ and $R^{(i)}_{ref}$ are the reflection intensity and the total number of light reflections in region i, respectively. An N-layer structure with an arbitrary layer k has been presented by Tabassum and Kant (2020b). Each layer k has a thickness d_k, RI n_k, and a dielectric constant ε_k. According to Fig. 4.8b, if each of an N-layered device's layers consists of an optical fiber core, a plasmonic metal, a high-RI material, and an analyte, then the device behaves similar to an SPR-based multi-layered fiber optic sensor. Comparing the multi-layered fiber optic SPR sensor in the Kretschmann configuration to the N-layered structure illustrated in Fig. 4.8a allows the transfer matrix method to be utilized to model an SPR-based fiber optic sensor.

Abelès matrix formalism is used for the mathematical evaluation of reflectance and transmittance of the strategically arranged multi-layer structure of thin films (Sun et al. 2018). The light directed through the core of the SPR-based optical fiber sensor generated evanescent waves and, hence, surface plasmon waves. At the output end, this transmitted light is analyzed using the Abelès formula for the intensity of the electric field. Using Abelès matrices, the tangential electric and magnetic field components at the first boundary (U_1 and V_1) are connected to counterparts in the last boundary (U_{n-1} and V_{n-1}) (Tabassum and Kant 2020b):

$$\begin{bmatrix} \frac{U_1}{V_1} \end{bmatrix} = M \begin{bmatrix} \frac{U_{n-1}}{V_{n-1}} \end{bmatrix} \tag{4.8}$$

$M =$ for an N-layered system; the total interference matrix, M, can also be denoted by

$$M = \prod_{K=1}^{N} M_K = \begin{pmatrix} M11 & M12 \\ M21 & M22 \end{pmatrix} = \begin{pmatrix} \cos\beta_K & (-i\sin\beta_k)/q_k \\ -iq_k\sin\beta_k & \cos\beta_k \end{pmatrix} \tag{4.9}$$

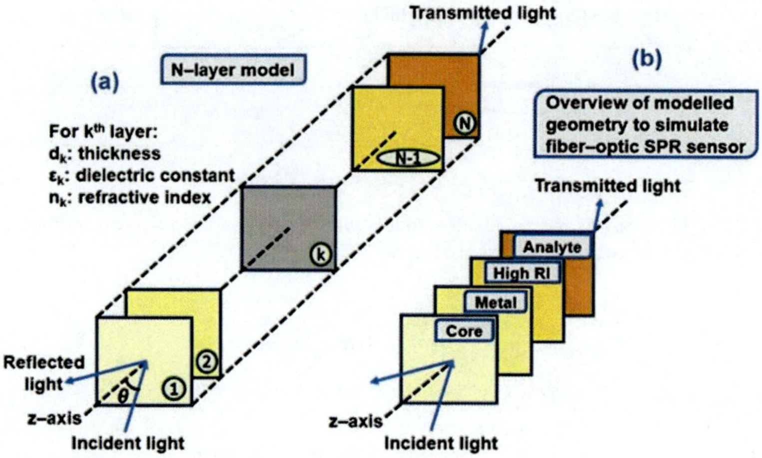

Fig. 4.8 **a** A schematic demonstrating the N-layer structure; **b** transfer matrix modeling of the geometry of an SPR-based multi-layered fiber optic sensor. Reprinted with permission from Journal of Applied Physics. Copyright, 2020, AIP Publishing (Tabassum and Kant 2020b)

$$\beta_k = \frac{2\pi d_k}{\lambda} n_k \cos\theta \tag{4.10}$$

$$\cos\theta_k = \frac{2\pi d_k}{\lambda\left(\varepsilon_k - n_1^2 \sin^2\theta_1\right)^{\frac{1}{2}}} \tag{4.11}$$

$$q_k = \left(\frac{\mu_k}{\varepsilon_k}\right)^{\frac{1}{2}} \tag{4.12}$$

$$\cos\Theta_k = \frac{\epsilon_k - n_1^2 \sin^2\theta_k}{\varepsilon_k} \tag{4.13}$$

θ_k is the angle at the kth interface, and $\mu_k = 1$. The total interference matrix can be calculated using the continuation of transverse electric and magnetic field components at every interface.

As a result, when a p-polarized incident wave travels through a multi-layered structure, the amplitude reflection coefficient (r_p) is calculated as (Tabassum and Kant 2020b)

$$R_p(\lambda, \theta) = \left|r_p(\lambda, \theta)\right|^2 \tag{4.14}$$

where R is the net reflectance. Cascaded channel fiber optic SPR is designed by creating many sensing zones on a single fiber optic probe. Claddings from areas, having a length of 1 cm, of fibers are removed, and these regions are then separated by a length of 1 cm. Layers of other plasmonic metals are then applied to the unclad

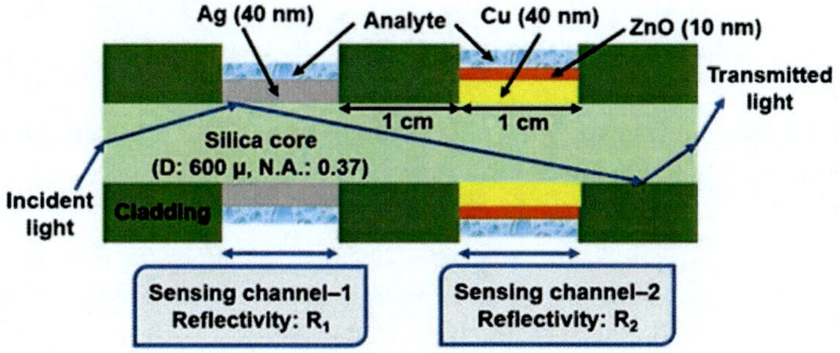

Fig. 4.9 Cascaded SPR fiber optic RI sensors with two channels. Reprinted with permission from Journal of Applied Physics. Copyright, 2020, AIP Publishing (Tabassum and Kant 2020b)

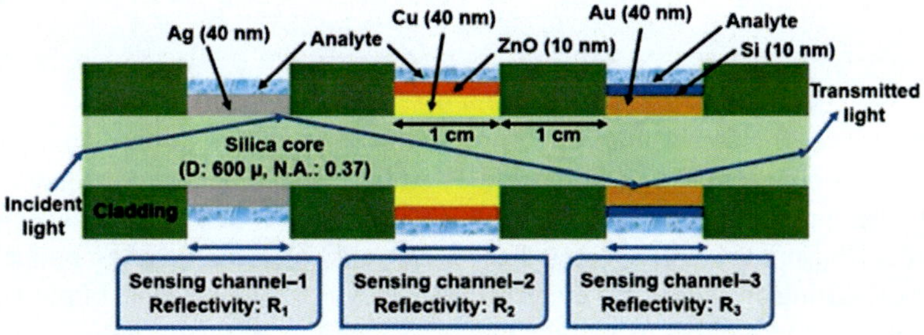

Fig. 4.10 Diagram of a cascaded triple-channel SPR fiber optic RI sensor. Reprinted with permission from Journal of Applied Physics. Copyright, 2020, AIP Publishing (Tabassum and Kant 2020b)

portions. The schematic for a cascaded dual- and triple-channel fiber optic sensor created in the manner described above, is shown in the Figs. 4.9 and 4.10.

To calculate the overall reflectance (R_p) of a cascaded channel of the fiber optic sensor, the reflectance (r_p) of each individual sensing channel is multiplied (Yuan et al. 2012). Total reflectance from all of the sensor's sensing channels is indicated by (Tabassum and Kant 2020b)

$$R_p = \left| r_{p,1} * r_{p,2} \right|^2 \tag{4.15}$$

Similarly, for triple-channel sensor net reflectance is given by (Tabassum and Kant 2020b)

$$R_p = \left| r_{p,1} * r_{p,2} * r_{p,3} \right|^2 \tag{4.16}$$

Only light rays that enter the fiber optic probe at an angle less than the critical angle (θ_{cr}) and that are perpendicular to the core–clad contact will be sent through the fiber (Tabassum and Kant 2020b):

$$\theta_{cr} = \left(\frac{n_{cladding}}{n_1}\right) \tag{4.17}$$

θ_1 is the fiber core's refractive index. The probe's output end produces power that is given by (Tabassum and Kant 2020b)

$$P_{trans} = \frac{\int_{\theta_{cr}}^{\frac{\pi}{2}} \cdot R_p^{N_{ref}(\theta)} \frac{n_1^2 \sin\Theta\cos\Theta}{\left(1-n_1^2\theta\right)^2} d\theta}{\int_{\theta_{cr}}^{\frac{\pi}{2}} \cdot \frac{n_1^2 \sin\Theta\cos\Theta}{\left(1-n_1^2\theta\right)^2} d\theta} \tag{4.18}$$

$N_{ref}(\theta)$ = number of ray reflections that are present within the sensing zone and extend through length (L). D is fiber's diameter, and $N_{ref}(\theta)$ is given by (Tabassum and Kant 2020b)

$$N_{ref}(\theta) = L/D(\tan\theta) \tag{4.19}$$

For the cascaded channel fiber optic probe, this approach is utilized to excite the SPR spectra connected to the different RI values of the analyte surrounding the probe. The figure of merit (FOM), sensitivity, and various other performance metrics are deduced using the SPR spectra. The author took the example of an optical fiber sensor consisting of a silica (Si) core surrounded by cladding made up of plastic. The fiber has a diameter of 600 μm and a numerical aperture of 0.24. These geometrical parameters enable the fiber to be multimode and robust, so that on a single probe, it can support up to three sensing regions (Diameter: 600 μm; Numerical Aperture: 0.24). The use of this optical fiber was done broadly for designing fiber optic SPR sensors used for both experimental and theoretical studies.

The core of an optical fiber is made of silica (SiO_2), and its refractive index (RI) varies with wavelength as shown by Mishra et al. (2015)

$$n_{SiO_2}(\lambda) = \sqrt{1 + \frac{0.6961663\lambda^2}{\lambda^2 - (0.0684043)^2} + \frac{0.4079426\lambda^2}{\lambda^2 - (0.1162414)^2} + \frac{0.8974794\lambda^2}{\lambda^2 - (9.896161)^2}} \tag{4.20}$$

The Drude dispersive model is used to derive the wavelength-dependent dielectric functions of silver (Ag), gold (Au), and copper (Cu). For an assembly of harmonic oscillators, this model assumes a conducting material. This assumption is based on the fact that electrons do not encounter restoration forces and hence can move freely around the metal lattice. The relationship between a metal's dielectric property and wavelength, as per the Drude model, is written as (Tabassum and Kant 2020b)

$$\epsilon(\lambda) = 1 - \frac{\lambda_c\lambda^2}{\lambda_p^2(\lambda_c + i\lambda)} \tag{4.21}$$

λ_c collision wavelength

λ_p the wave-length corresponding to bilk plasma frequency of the metal.

The Drude free electron theory considers certain approximations that do not take into account the loss in energy to a sizeable accuracy. The model provides precise estimates of the common plasmonic metals' dielectric constants in the visible and near-infrared regions. So, it is frequently used to investigate the transmission characteristics of fiber optic sensors based on SPR, as the region of cascaded fiber optic sensor operation is also in the visible and near-infrared regions of the electromagnetic spectrum.

4.3 Fabrication Process of the Cascaded Fiber Optic SPR Sensor

Wei et al. (2016) propose and show how to make a multi-channel SPR sensor out of cascading single-mode and multimode optical fiber. The proposed fiber uses both gold-coated multimode fiber and single-mode fiber in its construction (cascaded). The wavelength of the resonance changes with the grinding angle and the thickness of the gold layer. It has been shown by the author that the resonance wavelength grows with both the thickness of the gold layer and the grinding angle (Fig. 4.11).

Single-mode optical fiber was used to modify the grinding angle, while multimode optical fiber was used to modify the gold film thickness in this two-level SPR sensor. The single-mode and multimode optical fibers are fused and spliced together using industry-standard techniques. Hence, this multi-channel sensor maintains sensitivity while adjusting the dynamic range. Dual-channel refractive index measurements are possible with the multi-channel SPR sensor. For the detection of several analytes and the elimination of interference from the refractive index and temperature differential of the background, this is important (Fig. 4.12).

The fiber end was coated with gold using plasma sputtering. The probe, made up of a grinded and butt-joined eccentric core single-mode fiber and multimode fiber, was placed in the plasma cleaner for 8 min to remove any debris from the ground surface. Vacuum plasma sputtering is used to deposit the gold film on the fiber probe tip. Next, the probe was transferred into the vacuum environment of the plasma sputtering apparatus. Sputtering current and time were accurately controlled to create a 300 nm-thick gold film for reflecting light on the multimode fiber and a 50 nm-thick gold film for sensing light on the eccentric core single-mode fiber. A 3D morphology analyzer is used to calculate the thickness of the gold sheet. The detecting gold film on the multimode fiber is 35 nm thick, and this coated multimode fiber SPR sensor employs the aforementioned method for coating and film thickness control. The optical fiber probe needs to be fixed firmly to the specimen's clamping apparatus, which rotates horizontally, in order to keep the specimen's thickness constant (Fig. 4.13).

A dual-channel RI sensor has been suggested for fabrication by Zhang et al. (2019). It is accomplished by depositing Ag in various thicknesses. A sharp blade was used to remove 1 cm of the cladding from the RI sensor before coating it in

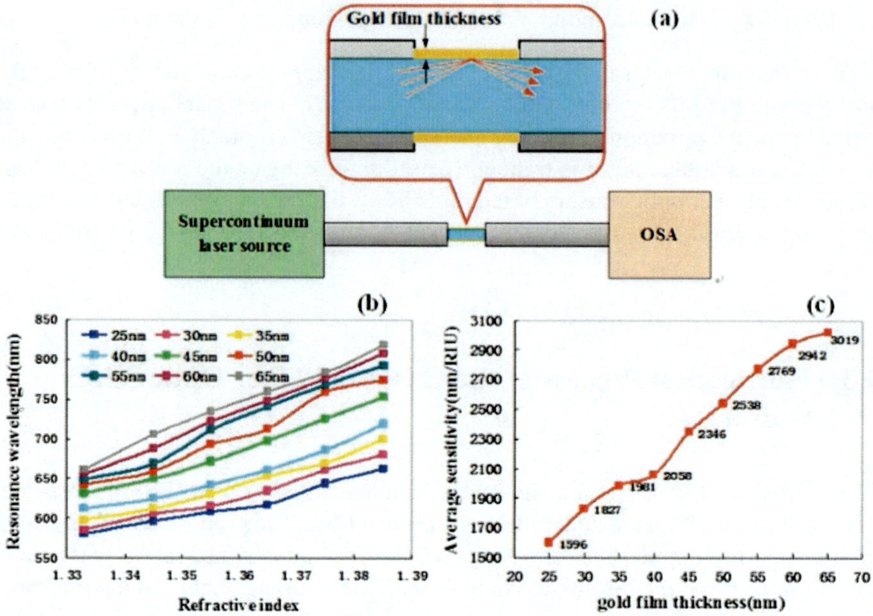

Fig. 4.11 **a** Gold-coated multimode fiber schematic; **b** results of experiments on the resonance wavelength and thickness of a gold film with a refractive index between 1.333 and 1.385; **c** average sensitivity and gold film thickness tests. Reprinted with permission from Optics Communications. Copyright, 2017, Elsevier (Wei et al. 2016)

ultrathin chromium and then coating it in Ag. The following procedures originally used chromium to stop Ag exfoliation. The coating process has been carried out using a sputter coater machine. It can adjust the sputtering current and time to control the thickness of the coated metal. The thickness data of sputtering metal can be referenced using a matched thickness monitor that is accessible at the same time. A metal layer measuring approximately 5 nm was created after 10 s of chromium sputtering. The positions of their resonance dips in the spectrum were then calculated for Ag layers with thicknesses spanning from 20 to 50 nm on the fiber core. Finally, two channels with different Ag thicknesses could produce two resonance dips. Below is a picture of the proposed sensor design (Fig. 4.14).

Wang et al. (2019) proposed a design where two cascaded photonic crystal fiber (PCF) are spliced together using a dual-channel fiber sensor. For the sensing area, PCF is coated with a coating of a noble metal. Each PCF measures 10 mm in length and 125 μm in outer diameter. The air hole measures 4.8 mm in width and 7.7 mm in pitch. The fusion splicer (Fitel, S178) was used to join the two PCFs. The air holes' collapse length during the fusion splice process is 225 μm. Thin layers of gold (Au) and silver (Ag) are coated on the surfaces of the sensing and reference channels. The detection accuracy suffers if the two channels are plated with the same metal film, which is why the two channels have different metal films on their

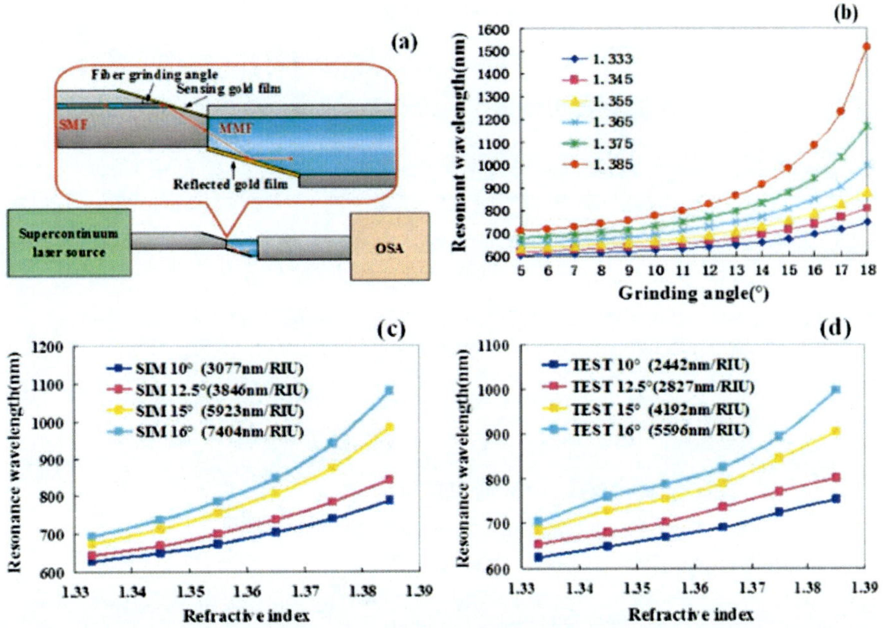

Fig. 4.12 **a** Schematic of single-mode fiber with grinding angle; **b** Simulations of the resonance wavelength-fiber grinding angle relationship; **c** Simulations of the refractive index–resonance wavelength relationship at 10°, 12.5°, 15°, and 16° grinding angles; **d** Results of tests on the relationship between the refractive index and the resonance wavelength using grinding angles of 10°, 12.5°, 15°, and 16°. Reprinted with permission from Optics Communications. Copyright, 2017, Elsevier (Wei et al. 2016)

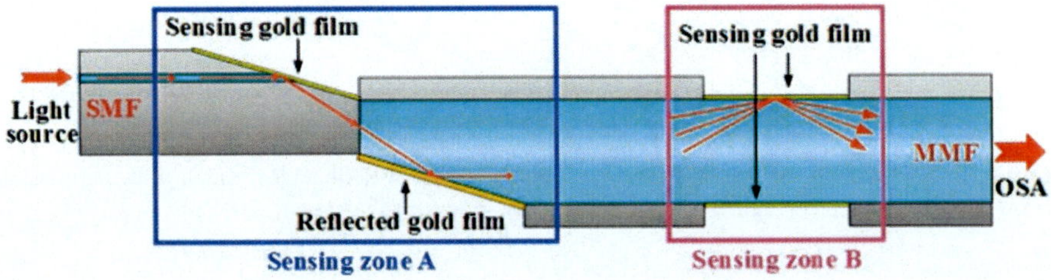

Fig. 4.13 Multi-channel SPR sensor architecture with sensing zones A (for altering the SPR resonance angle) and B (modifying the thickness of the gold layer). Reprinted with permission from Optics Communications. Copyright, 2017, Elsevier (Wei et al. 2016)

outside surfaces. Au material was sputtered onto the PCF surface using a magnetron sputtering apparatus for the sensing channel (channel 1). Au measured around 50 nm in thickness. For channel 2, Tollen's regent coats PCF with an Ag film with a thickness of approximately 50 nm (Fig. 4.15).

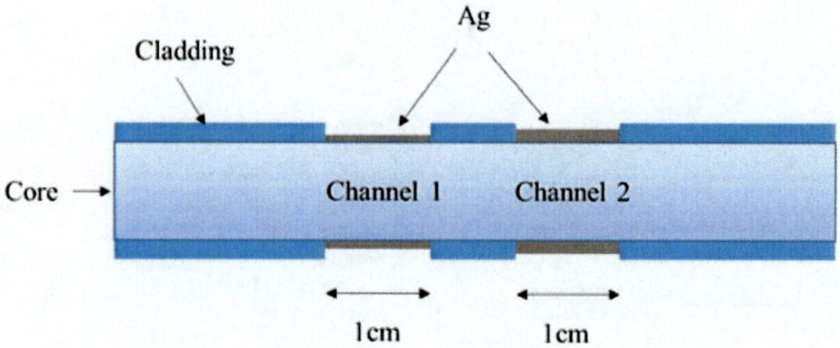

Fig. 4.14 The dual-channel optical fiber SPR sensor's structural profile. Reprinted with permission from Optics Communications. Copyright, 2020, Elsevier (Zhang et al. 2019)

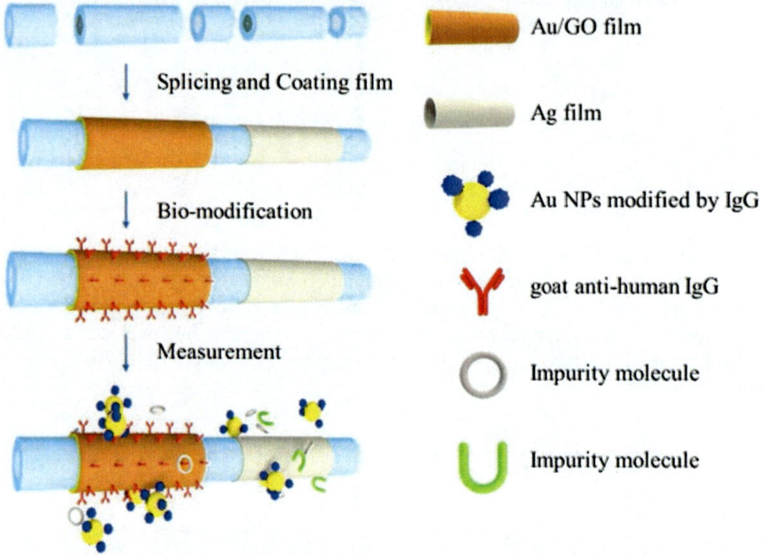

Fig. 4.15 An example of the fabrication of dual-channel SPR fiber. Reprinted with permission from Optics & Laser Technology. Copyright, 2020, Elsevier (Wang et al. 2019)

4.4 Principle and Method

It is found that SPR can be activated when the transverse component of the wave coincides with the real part of the surface plasmon wave's propagation constant (k_{sp}) (Zhang et al. 2022):

$$\frac{2\pi}{\lambda}n_1 \sin\Theta = Re(k_{sp}) \tag{4.22}$$

The transverse component is denoted by the left side of the equation. Right side of the equation denotes Re (propagation constant). The dielectric constant of the metal layer is related to the propagation constant (k_{sp}) of the surface plasmon wave by Zhang et al. (2022)

$$k_{sp} = \frac{2\pi}{\lambda}\sqrt{\frac{\varepsilon_m n_s^2}{\varepsilon_m + n_s^2}} \tag{4.23}$$

ε_m dielectric constant of the metal film
n_s refractive index of the sample
n_i refractive index of the fiber core

As the refractive index of the sample liquid changes (grows) or as the dielectric constant of the metal increases, the resonance dips will turn red.

Liu et al. (2017) built a modified hetero-core structure for multiplex analyte detection, as shown in Fig. 4.16b. The fiber probe is constructed from a commercial single-mode fiber (SMF-28) with a core diameter of 8.2 μm and a cladding diameter of 125 μm and a commercial multimode fiber (GIF-105) with a core diameter of 105 μm and a cladding diameter of 125 μm. A single MMF was combined with SMF at a polishing angle to create a spherical truncated cone. The SMF's cladding surface is then coated with a gold film 50 nm thick to create sensing channel I, and the polishing surface is coated with a gold film 50 nm thick to create sensing channel II. The cladding of the SMF would not block light rays from a source that were discharged into the lead-in MMF. The majority of the incident light would be deflected by total internal reflection at the cladding surface and stimulated by SPR in channel I. After that, channel II SPR would be reactivated, and the light wave would continue to transmit while it is clad; some of the light would undergo total internal reflection at the polishing surface. At last, OSA is able to gather light waves via the MMF lead-out. Channel I is 10 mm in length, and there is a 30-mm chasm between it and channel II. The incident angle range 1 of a light wave in channel I is given by (Liu et al. 2017)

$$\arcsin\left(\frac{n_1}{n_0}\right) \le \theta_1 \le 90^{\circ} \tag{4.24}$$

Here, n_0 is the cladding RI of the SMF ($n_0 = 1.4658$) and n_1 is the RI of the analyte. Since n_1 is the only variable, channel I's resonance wavelength range is fixed. The incident angle range θ_2 of a light wave in channel II, which is connected to the polishing angle α, can be expressed as follows:

$$\arcsin\left(\frac{n_1}{n_o}\right) - \alpha \le \theta_2 \le 90^{\circ} - \alpha \tag{4.25}$$

Comparing Eqs. (4.22) and (4.23), it can be assumed that θ_2 is smaller than θ_1. Given that the SMF is brief and straight, it is reasonable to assume that the light wave

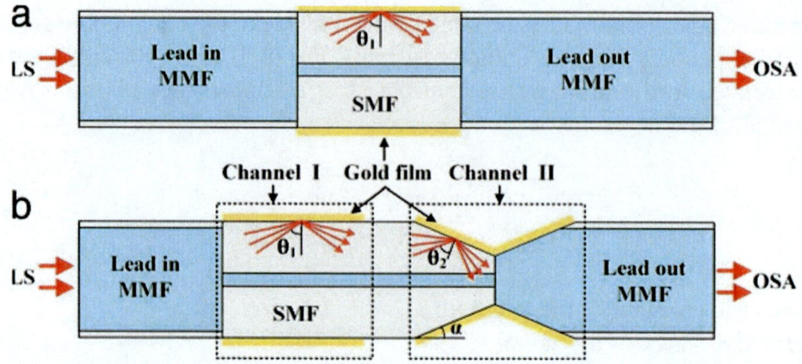

Fig. 4.16 Schematic for two different hetero-core structure SPR probes: **a** the hetero-core conventional SPR sensor with a single sensing channel; and **b** dual-channel sensing (I and II) with a refined hetero-core design for the SPR probe. Reprinted with permission from Optics Communications. Copyright, 2017, Elsevier (Liu et al. 2017)

path is nearly parallel to the SMF, which simplifies the range of incident angles. The incident angle can be written as $\theta_1 \approx 90°$, and $\theta_2 \approx 90° - \alpha$. As the incident angle θ decreases, the resonance wavelength range moves toward the red. The incident angle is getting closer to the critical angle θ_c as the polishing angle α rises, which can make sensors more sensitive (Liu et al. 2017; Zhang et al. 2022). It can be assumed that the resonance wavelength range of channel II may be considerably different from that of channel I. By using wavelength division multiplexing technology, dual-channel sensing can be accomplished sequentially.

4.5 SPR Enhancement with Cascaded Structure

To increase the effectiveness of SPR, a cascaded side-polished multimode fiber sensor is proposed by Lin et al. (2008). The cascaded SPR side-polished fiber sensing system offers a clear SPR reaction and a double variation for the sensitivity of intensity measurement without the use of large components or a complex and difficult signal processing scheme.

To improve the SPR effect, Woo-Hu et al. (2010) proposed a multi-step structure for a surface plasmon resonance-based side-polished optical fiber sensor. In comparison to other fiber optic sensors, the multi-stage SPR side-polished fiber sensor just needs one polishing step, making it well-suited for mass manufacturing.

Three different designs for SPR-based side-polished multimode fiber sensors are shown in Fig. 4.17a–c. The transverse-magnetic (TM) mode has decayed by the time the guided wave reaches the first SPR-detecting zone in the fiber. The transverse-electric (TE) field modes couple with the TM modes in the coupling region as a result of the mode random coupling phenomenon. To further enhance the SPR effect,

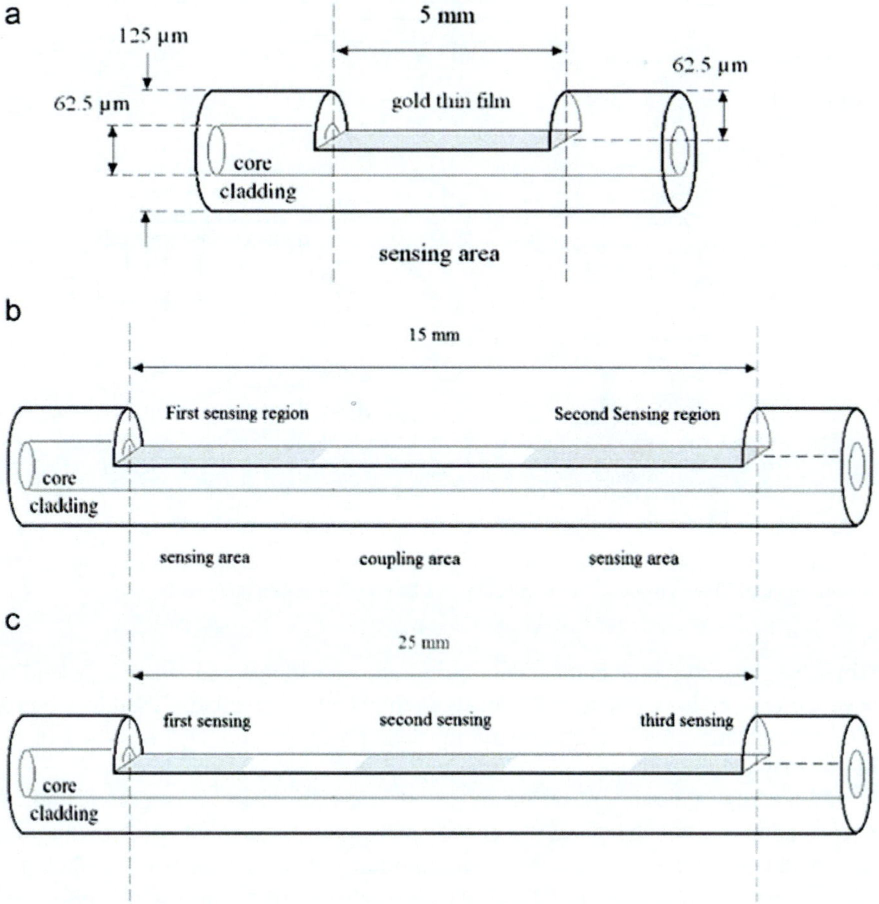

Fig. 4.17 **a** Single-step SPR, **b** two-step SPR, and **c** three-step SPR fiber sensor structures. Reprinted with permission from Optics and Laser Technology. Copyright, 2010, Elsevier (Woo-Hu et al. 2010).

additional TM modes are supplied into the second SPR-sensing region, where they are once again depleted.

4.6 Application of Cascaded Fiber Optic SPR Sensor

One application of cascaded fiber optic SPR sensors is the simultaneous detection of vitamin K1 and heparin. One approach for the simultaneous detection of heparin and vitamin K has been developed by Tabassum and Gupta (2016). One of the most important processes in the human body is blood coagulation, the incapability of

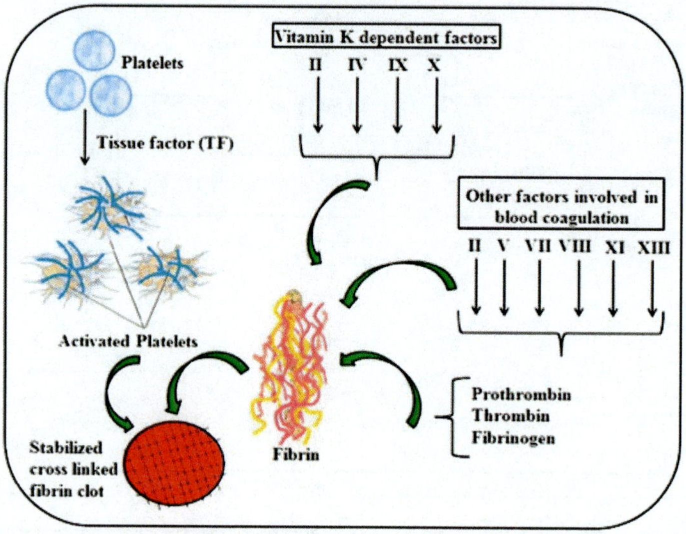

Fig. 4.18 Figure representing the blood coagulation pathway. Reprinted with permission from Biosensors and Bioelectronics. Copyright, 2016, Elsevier (Tabassum and Gupta 2016)

which might result in loss of excess blood after an injury and hence can lead to severe complications. Vitamin K1 levels in the blood are typically used to monitor blood coagulation, as it is one of the most dominant constituents involved in the blood coagulation pathway.

Chemically, vitamin K is made up of derivatives of 2-methyl-1 and 4-naphthoquinone, which belong to a class of structurally related fat-soluble vitamins. Among the derivatives, vitamin K1 is an important constituent for blood coagulation protein production. Figure 4.18 represents the pathway of blood coagulation.

A deficiency of VK1 in the human body results in bone weakening and artery calcification, whereas an excess of it may lead to blood clots that are mobile and circulate within the body. An excess of VK1 is controlled using heparin, an anticoagulant that prevents clot formation in veins and arteries.

Thus, VK1 and heparin are correlated to each other for clotting and de-clotting the blood, and it is required to measure them simultaneously. The use of SPR sensors for measuring VK1 and heparin simultaneously is one of the most widely used methods. It is advantageous to use SPR sensors for biomolecules as they provide selective detection of biomolecules with a minimum response time as compared to all the other sensors.

The prism-based sensors are bulky and costly. Meanwhile, there are several advantages to fiber-based sensors over prism-based sensors, as stated by Homola et al. (1999). For the sensing probe, multimode step-index fiber is used with plastic cladding. The core is 600 μm in diameter and the numerical aperture is in the range of 0.35–0.39. The fabrication of the fiber involves various steps.

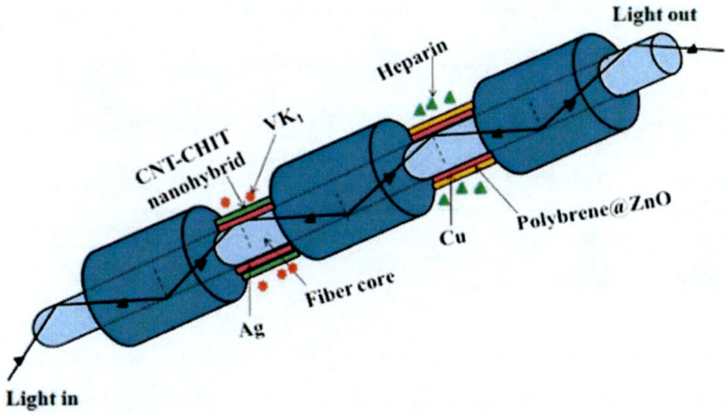

Fig. 4.19 Fiber optic sensor probe. Reprinted with permission from Biosensors and Bioelectronics. Copyright, 2016, Elsevier (Tabassum and Gupta 2016)

Step I: A fiber of length 20 cm is taken, and two separated regions of length 1 cm each are uncladded by removing the cladding with a sharp blade.

Step II: Plasmonic metals are deposited, using thermal evaporation coating units, over the uncladded areas of the fiber. A 40 nm-thick silver layer was deposited over the first uncladded region and was named channel I. Similarly, for the second region, a copper layer of 40 nm thickness was used, and it was termed channel II.

Step III: Additionally, MWCHNT-CHIT and polybrene@ZnO, which were prepared separately, were used to deposit both channels. Channel I was deposited as MWCHNT-CHIT, and channel II was deposited as polybrene@ZnO. This deposition was done using the dip-coating method.

The sensing probe built using the aforementioned techniques and based on a cascaded-channel fiber optic sensor is depicted in Fig. 4.19.

The figure below (Fig. 4.20) illustrates the experimental setup used to identify vitamin K1 and heparin. A W cutter was used to cut the ends of the fiber optic probe. This was done in order to enhance light coupling in the fiber. In the flow cell, the probe was fixed, and there was a facility for the inlet and outlet of the analyte solution. For the source, a tungsten lamp was used, and polychromatic light was launched into the input end. A spectrometer was connected to the output end, and experiments were conducted using a combination of solutions as well as different concentrations of vitamin K1 and heparin.

In the sensor probe, MWCNT-CHIT is being used for detecting VK1. The principle for this detection uses oxidation and reduction processes. A two-electron two-proton transfer reaction takes place between VK1 and MWCNT-CHIT when VK1 gets merged into MNCNT-CHIT. Due to a change in MWCNT-dielectric CHIT's constant, the SPR spectrum shows a shift in resonance wavelength. For the sensing of heparin, polybrene has been used as it neutralizes heparin, involving chemical reactions, according to the paper by Montalescot et al. (1990) and Tientadakul et al. (2013).

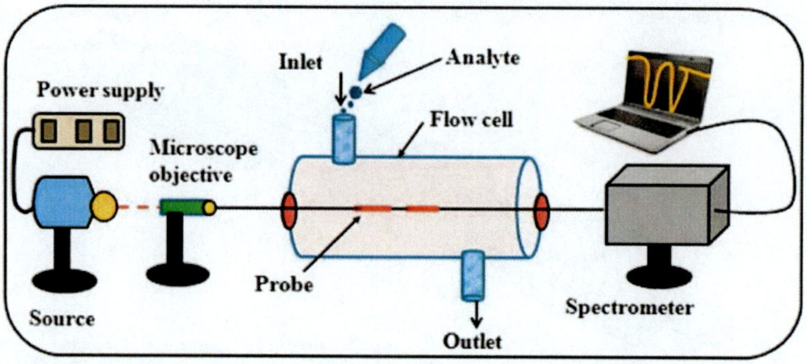

Fig. 4.20 Experimental setup for detecting VK1 and heparin. Reprinted with permission from Biosensors and Bioelectronics. Copyright, 2016, Elsevier (Tabassum and Gupta 2016)

Additionally, ZnO has been added as a central core in the form of a shell because it improves the sensor's performance by increasing the evanescent field.

Figure 4.20 in the plot above displays SPR spectra with vitamin K1 in aqueous solution representing the analyte. The solution interacts with both channels. There is a shift in the resonance wavelength in channel I with an increase in vitamin K1 concentration, but no shift is seen in channel II. It has been established that, for VK1, channel I is active while channel II is inactive. When heparin solution is introduced and the concentration is increased, a shift in resonance wavelength is noticed in channel II, whereas channel I remains inactive for heparin (Figs. 4.21, 4.22 and 4.23).

The spectra for a mixture of VK1 and heparin are shown in Fig. 4.22. It was observed that an individual analyte was sensed by its respective channel while the other channel was inactive. So the sensor can be used to measure VK1 and heparin

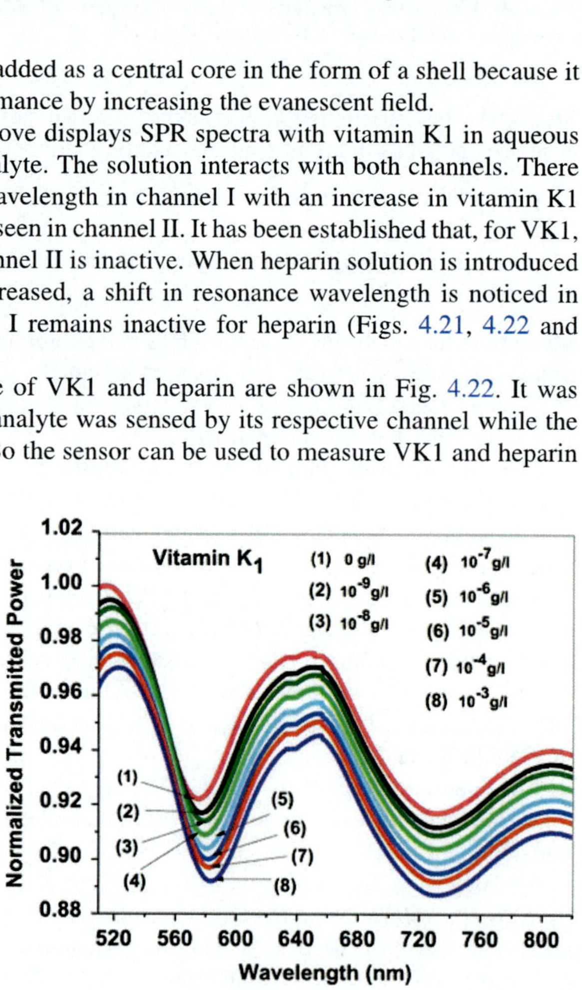

Fig. 4.21 Normalized transmitted power versus wavelength for VK1. Reprinted with permission from Biosensors and Bioelectronics. Copyright, 2016, Elsevier (Tabassum and Gupta 2016)

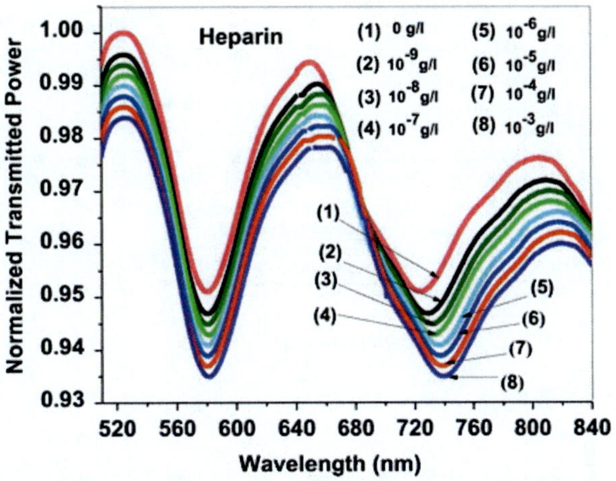

Fig. 4.22 Normalized transmitted power versus wavelength for heparin. Reprinted with permission from Biosensors and Bioelectronics. Copyright, 2016, Elsevier (Tabassum and Gupta 2016)

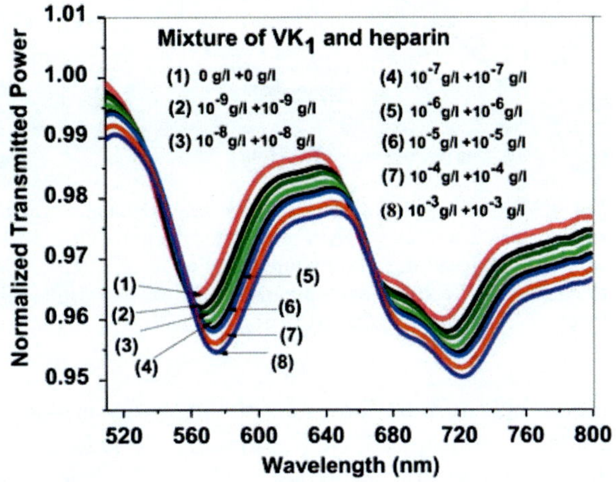

Fig. 4.23 Normalized transmitted power versus wavelength for VK1 and heparin mixture. Reprinted with permission from Biosensors and Bioelectronics. Copyright, 2016, Elsevier (Tabassum and Gupta 2016)

simultaneously. The most important parameter on which a sensor is judged is its sensitivity. The sensitivity of this sensor is defined as (Tabassum and Gupta 2016)

$$\text{Sensitivity} = \frac{\text{resonance wavelength}}{\text{concentration of analyte}} \tag{4.26}$$

From the resonance wavelength versus concentration curve, the slope gives us sensitivity. In Fig. 4.24a–d, we have the graph of resonance wavelength versus concentration of the analyte, but on a logarithmic scale. So the slope of this graph cannot be used for calculating the sensitivity of our sensor.

Now, for calculating the sensor's sensitivity, the data points from the above plots were fitted by second-order polynomials. The resonance wavelength values were determined by fitting the data points using second-order polynomials. Figure 4.25a

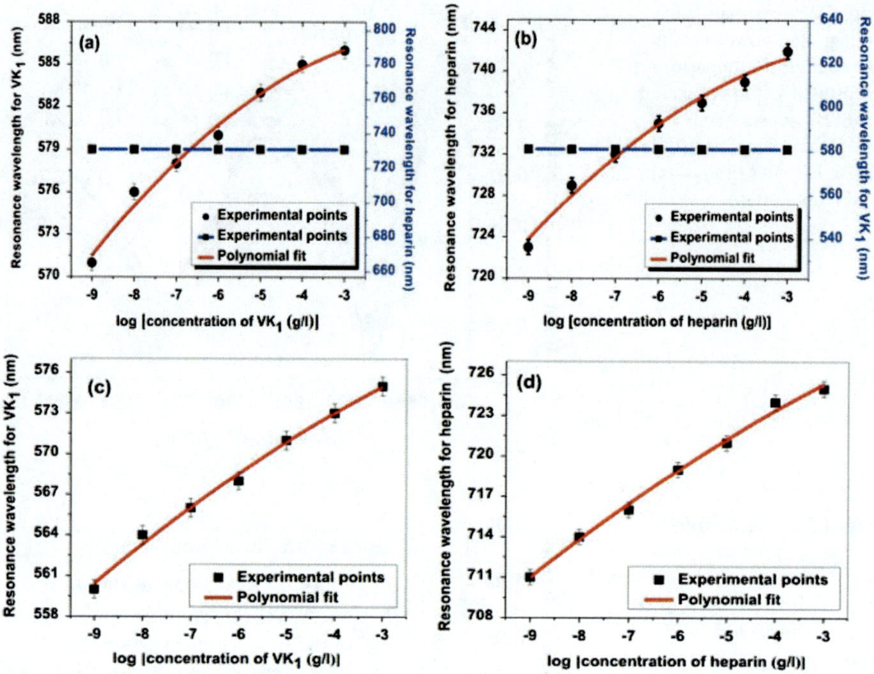

Fig. 4.24 Plots representing the change in resonance wavelength versus concentration [in logarithmic scale]: **a** VK1 supplied to both channels I and II; **b** Heparin supplied to both channels I and II; **c** VK1 and heparin mixture supplied to channel I; **d** VK1 and heparin mixture supplied to channel II. Reprinted with permission from Biosensors and Bioelectronics. Copyright, 2016, Elsevier (Tabassum and Gupta 2016)

represents the variation of sensitivity with varying concentrations of vitamin K1 supplied in channels I and II. Similarly, (b) is for sensing heparin, (c) for mixture in channel I, and (d) for mixture in channel II. The trend observed in the plots above stands true for low concentrations of the analyte. The quantity of NWCNT-CHIT or polybrene at ZnO per molecule of vitamin K1 or heparin is high. But when the concentration is increased, the ratio decreases significantly, and hence the sensitivity of the sensor drops. It follows that the current sensor can be utilized for medical purposes at low concentrations of analyte molecules.

4.7 Future Prospects

The future prospects of the chapter on cascaded fiber optic SPR sensors include the following: (1) Development of Multi-Analyte Sensing: The use of cascaded fiber optic SPR sensors offers great potential for multi-analyte sensing. Future research can focus on expanding the capabilities of these sensors to detect and analyze

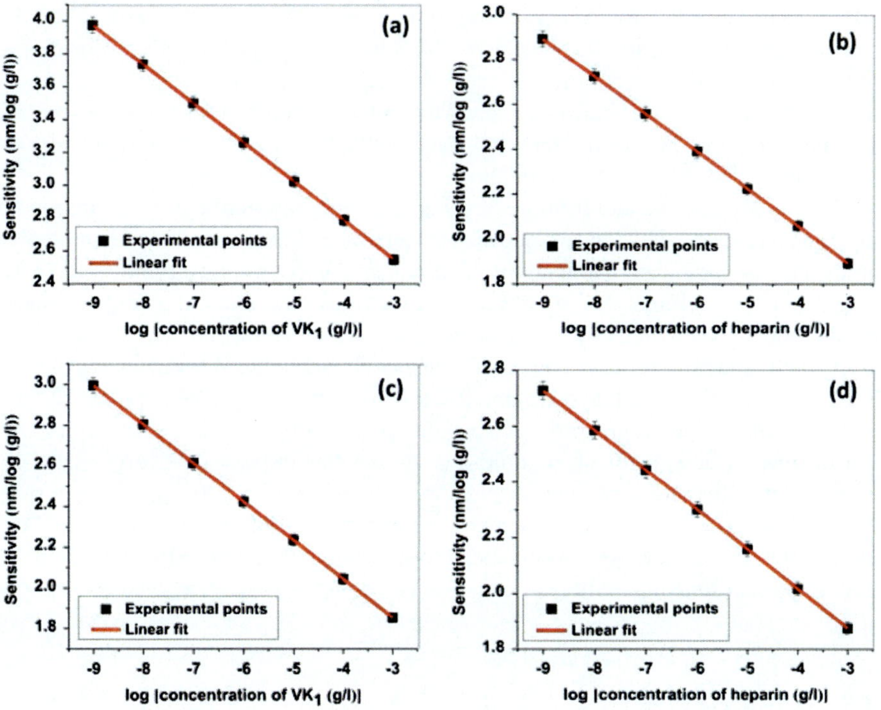

Fig. 4.25 Plots representing the change in sensitivity versus concentration [on a logarithmic scale]: **a** VK1 supplied to both channels I and II; **b** Heparin supplied to both channels I and II; **c** VK1 and heparin mixture supplied to channel I; **d** VK1 and heparin mixture supplied to channel II. Reprinted with permission from Biosensors and Bioelectronics. Copyright, 2016, Elsevier (Tabassum and Gupta 2016)

multiple analytes simultaneously. This could involve exploring new configurations and designs that enable the integration of additional sensing channels or analyte-specific receptors, enhancing the versatility and functionality of the sensors. (2) Improved Sensing Performance: Ongoing efforts can be directed toward improving the sensing performance of cascaded SPR sensors. This can involve optimizing the design parameters, such as the shape and dimensions of the modified hetero-core structure fiber, to enhance sensitivity, selectivity, and detection limits (Kumar et al. 2022a, b; Kumari et al. 2022; Li et al. 2022). Additionally, advancements in signal processing techniques and data analysis algorithms can further enhance the accuracy and reliability of the sensor measurements. (3) Integration with Other Technologies: The future prospects of cascaded fiber optic SPR sensors lie in their integration with other emerging technologies. This could include combining SPR sensing with microfluidics, lab-on-a-chip systems, or nanomaterials to create advanced sensing platforms. By integrating these technologies, researchers can develop highly sensitive, miniaturized, and portable SPR sensors for various applications, including point-of-care diagnostics and environmental monitoring (Liu et al. 2023; Pandey et al.

2022a, b). (4) Application-Specific Configurations: Future research can explore the customization of cascaded SPR sensors for specific applications. By tailoring the sensor design and configuration to meet the requirements of different analytes or target substances, researchers can expand the range of applications for these sensors. This may involve adapting the sensor surface chemistry, optimizing the sensor geometry, or exploring new sensing modalities to address specific sensing challenges. (5) Real-Time Monitoring and Remote Sensing: The integration of cascaded fiber optic SPR sensors with real-time monitoring systems and remote sensing technologies holds promise for future applications. This can enable continuous, remote, and real-time monitoring of analytes or substances in various environments. Such advancements would have implications in fields such as environmental monitoring, food safety, and industrial process control (Pandey et al. 2022c; Raghuwanshi et al. 2021; Shadab et al. 2022; Singh and Kumar 2020; Uniyal et al. 2022; Wang et al. 2021).

In summary, the future prospects of the chapter on cascaded fiber optic SPR sensors involve advancements in multi-analyte sensing, improved performance, integration with other technologies, application-specific configurations, and real-time monitoring capabilities. These prospects reflect the ongoing efforts to enhance the capabilities and explore new applications for cascaded fiber optic SPR sensors in the field of optical sensing.

4.8 Summary

Surface plasmon resonance sensors using cascaded fiber optics have been described in this chapter. Cascaded SPR sensors have a lot of potential for multi-analyte and multi-channel sensing. This chapter covered a modified hetero-core structure fiber-based dual-channel SPR refractive index sensor. The resonance wavelength range can be flexibly adjusted by shaping the traditional hetero-core structure fiber into a circular truncated cone with a different polishing angle. Then, it was discussed how two cascaded fiber optic surface plasmon resonance sensors may be used to test two separate chemicals at once, utilizing wavelength division multiplexing technology. A practical application of a cascaded SPR sensor is discussed further down, where the concentrations of coagulants and anticoagulants are contrasted for disease detection using the cascaded multi-analyte SPR sensor.

References

Homola J, Yee SS, Gauglitz G (1999) Surface plasmon resonance sensors: review. Sens Actuators B Chem 54(1):3–15. https://doi.org/10.1016/S0925-4005(98)00321-9

Homola J, Lu HB, Nenninger GG, Dostálek J, Yee SS (2001) A novel multichannel surface plasmon resonance biosensor. Sens Actuators B Chem 76(1–3):403–410. https://doi.org/10.1016/S0925-4005(01)00648-7

Kumar V, Raghuwanshi SK, Kumar S (2022a) Recent advances in carbon nanomaterials based SPR sensor for biomolecules and gas detection—a review. IEEE Sens J 22(16):15661–15672. https://doi.org/10.1109/JSEN.2022.3191042

Kumar V, Raghuwanshi SK, Kumar S (2022b) Advances in nanocomposite thin-film-based optical fiber sensors for environmental health monitoring—a review. IEEE Sens J 22(15):14696–14707. https://doi.org/10.1109/JSEN.2022.3185004

Kumari A, Vyas V, Kumar S (2022) Advances in electrochemical and optical sensing techniques for vitamins detection: a review. ISSS J Micro Smart Syst 11(1):329–341. https://doi.org/10.1007/s41683-021-00084-3

Li M, Singh R, Wang Y, Marques C, Zhang B, Kumar S (2022) Advances in novel nanomaterial-based optical fiber biosensors—a review. Biosensors 12(10):843. [Online]. https://www.mdpi.com/2079-6374/12/10/843

Lin YC, Tsai WH, Tsao YC, Tai JK (2008) An enhanced optical multimode fiber sensor based on surface plasmon resonance with cascaded structure. IEEE Photonics Technol Lett 20(15):1287–1289. https://doi.org/10.1109/LPT.2008.926874

Liu Z et al (2017) Dual-channel surface plasmon resonance refractive index sensor based on modified hetero-core structure fiber. Opt Commun 403:290–295. https://doi.org/10.1016/j.optcom.2017.07.064

Liu F et al (2023) Development of taper-in-taper-based optical fiber sensors for chemical and biological sensing. Photonics 10(5). https://doi.org/10.3390/photonics10050567

Mishra AK, Mishra SK, Gupta BD (2015) SPR based fiber optic sensor for refractive index sensing with enhanced detection accuracy and figure of merit in visible region. Opt Commun 344:86–91. https://doi.org/10.1016/j.optcom.2015.01.043

Montalescot G, Zapol WM, Carvalho A, Robinson DR, Torres A, Lowenstein E (1990) Neutralization of low molecular weight heparin by polybrene prevents thromboxane release and severe pulmonary hypertension in awake sheep. Circulation 82(5):1754–1764. https://doi.org/10.1161/01.CIR.82.5.1754

Pandey PS, Raghuwanshi SK, Kumar S (2022a) Recent advances in two-dimensional materials-based Kretschmann configuration for SPR sensors: a review. IEEE Sens J 22(2):1069–1080. https://doi.org/10.1109/JSEN.2021.3133007

Pandey PS, Raghuwanshi SK, Shadab A, Ansari MTI, Tiwari UK, Kumar S (2022b) SPR based biosensing chip for COVID-19 diagnosis—a review. IEEE Sens J 22(14):13800–13810. https://doi.org/10.1109/JSEN.2022.3181423

Pandey R, Singhal N, Kumar P, Choudhary K, Kumar S (2022) Graphene and its immense contribution in defense and security: a review. Anal Chem Lett 12(1):1–12. https://doi.org/10.1080/22297928.2021.1980099

Raghuwanshi SK, Kumar S, Singh Y (2021) 2D materials for surface plasmon resonance-based sensors. CRC Press

Shadab A, Raghuwanshi SK, Kumar S (2022) Advances in micro-fabricated fiber bragg grating for detection of physical, chemical, and biological parameters—a review. IEEE Sens J 22(16):15650–15660. https://doi.org/10.1109/JSEN.2022.3188813

Singh R, Kumar S (2020) Chapter Nine - Applications of graphene-based sensors for biomedical industries. In: Hussain CM (ed) Comprehensive analytical chemistry, vol 91. Elsevier, pp 201–233

Sun L, Qin J, Tong Z, Zhang W, Gong M (2018) Simultaneous measurement of refractive index and temperature based on down-taper and thin-core fiber. Opt Commun 426:506–510. https://doi.org/10.1016/j.optcom.2018.06.004

Tabassum R, Kant R (2020) Recent trends in surface plasmon resonance based fiber–optic gas sensors utilizing metal oxides and carbon nanomaterials as functional entities. Sens Actuators B Chem 310:127813. https://doi.org/10.1016/j.snb.2020.127813

Tabassum R, Kant R (2020) Cascaded wavelength multiplexed refractive index sensors in optical fibers based on surface plasmon resonances. J Appl Phys 128(7). https://doi.org/10.1063/5.0017256

Tabassum R, Gupta BD (2016) Simultaneous estimation of vitamin K1 and heparin with low limit of detection using cascaded channels fiber optic surface plasmon resonance. Biosens Bioelectron 86:48–55. https://doi.org/10.1016/j.bios.2016.06.030

Tian S et al (2023) Highly sensitive cascaded fiber SPR sensor with temperature compensation. Opt Commun 533:129277. https://doi.org/10.1016/j.optcom.2023.129277

Tientadakul P, Kongkan C, Chinswangwatanakul W (2013) Use of an automated coagulation analyzer to perform heparin neutralization with polybrene in blood samples for routine coagulation testing. Arch Pathol Lab Med 137(11):1641–1647. https://doi.org/10.5858/arpa.2012-0599-OA

Uniyal S, Choudhary K, Sachdev S, Kumar S (2022) Recent advances in K-SPR sensors for the detection of biomolecules and microorganisms: a review. IEEE Sens J 22(12):11415–11426. https://doi.org/10.1109/JSEN.2022.3172115

Wang Q, Wang XZ, Song H, Zhao WM, Jing JY (2020) A dual channel self-compensation optical fiber biosensor based on coupling of surface plasmon polariton. Opt Laser Technol 124:106002. https://doi.org/10.1016/j.optlastec.2019.106002

Wang Y et al (2021) Biocompatible and biodegradable polymer optical fiber for biomedical application: a review. Biosensors 11(12):472. [Online]. https://www.mdpi.com/2079-6374/11/12/472

Wang R et al (2022) Research and application of multi-channel SPR sensor cascaded with fiber U-shaped structure. Optik (Stuttg) 266:169603. https://doi.org/10.1016/j.ijleo.2022.169603

Wei Y et al (2017) Multi-channel SPR sensor based on the cascade application of the single-mode and multimode optical fiber. Opt Commun 390:82–87. https://doi.org/10.1016/j.optcom.2016.12.069

Woo-Hu T, Yu-Cheng L, Jiu-Kai T, Yu-Chia T (2010) Multi-step structure of side-polished fiber sensor to enhance SPR effect. Opt Laser Technol 42(3):453–456. https://doi.org/10.1016/j.optlastec.2009.09.008

Xia L, Shuai B, Li W, Liu D (2012) Polarization detection analysis of dual-channel surface plasmon resonance sensing for silicone oils based on the D-shaped fiber with a central hole. Opt Commun 285(18):3730–3734. https://doi.org/10.1016/j.optcom.2012.05.001

Yuan Y, Wang L, Huang J (2012) Theoretical investigation for two cascaded SPR fiber optic sensors. Sens Actuators B Chem 161(1):269–273. https://doi.org/10.1016/j.snb.2011.10.029

Zhang Y et al (2017) Cascaded distributed multichannel fiber SPR sensor based on gold film thickness adjustment approach. Sens Actuators A Phys 267:526–531. https://doi.org/10.1016/j.sna.2017.09.038

Zhang Y et al (2022) Cascaded dual-channel fiber SPR sensor based on Ge2Sb2Te5. IEEE Sens J 22(5):4083–4089. https://doi.org/10.1109/JSEN.2022.3142085

Zhang M, Zhu G, Li T, Lou X, Zhu L (2020) A dual-channel optical fiber sensor based on surface plasmon resonance for heavy metal ions detection in contaminated water. Opt Commun 462:124750. https://doi.org/10.1016/j.optcom.2019.124750.

Chapter 5
Symmetric Versus Asymmetric Coated (Half Coated) Fiber Optic SPR Sensor

5.1 Symmetric Fiber Optics SPR Sensor

Fiber optic Surface Plasmon Resonance (SPR) sensors are advanced optical sensors that can detect and analyze changes in the refractive index (RI) of a sample by employing the principles of SPR. When light strikes the surface of a conducting material, it causes electrons to oscillate, creating what is known as "surface plasmons." A fiber optic SPR sensor works by depositing a thin metal coating on its surface, usually consisting of gold or silver. The fiber acts as a waveguide to carry light signals. Light gets "coupled" into the fiber and travels along it until it reaches the metal coating. Here, SPR occurs as a result of light's interaction with the metal film at the contact. The behavior of SPR is affected by the RI of the surrounding medium. A change in the reflected or transmitted light spectrum occurs when a sample is brought close to the metal film, and its RI affects the resonance state of the surface plasmons. Changes in the RI of the sample may be detected and quantified by the sensor by monitoring this spectral shift. There are many benefits to using fiber optic SPR sensors. As a first advantage, they can detect even the most minor changes in analyte concentration or biomolecular interactions by measuring the RI, as can be depicted in Fig. 5.1. And unlike fluorescent or radioactive indicators, they may be detected in real time. Because of their small size and adaptability, they may be included in a wide range of analytical systems, such as those used in biomedical research and environmental monitoring. The biochemical, pharmaceutical, environmental monitoring, and food security industries are just a few that benefit from the use of these sensors. They are often employed in real-time process monitoring, chemical and biological analyte detection, and biomolecular interaction research. The miniature ability, high sensitivity, and lack of need for labels in the fiber optic SPR sensor make it a desirable instrument in a wide range of scientific and industrial settings. High sensitivity can be achieved using either symmetric or non-symmetric fiber optic SPR (SPR) sensors; however, whatever design is used for the sensor will depend on the needs of the particular application in addition to the desired trade-offs. The sensing

S. K. Raghuwanshi et al., *Geometric Feature-Based Fiber Optic Surface Plasmon Resonance Sensors*, Springer Tracts in Electrical and Electronics Engineering,
https://doi.org/10.1007/978-981-99-7297-5_5

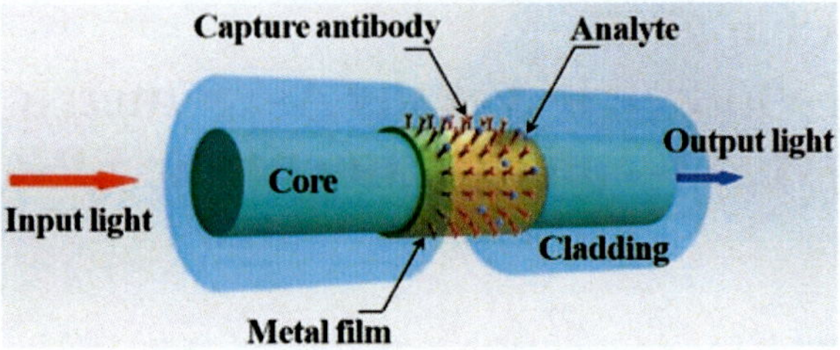

Fig. 5.1 Fiber optic coupling excites the SPR phenomenon. Reprinted with permission from Optics and Laser Technology. Copyright, 2022, Elsevier (Liu et al. 2022)

region of a symmetric fiber optic SPR sensor is often a homogeneous layer of the metal coating (like gold or silver) on the fiber's surface itself. The SPR phenomenon is affected by changes in the RI of the medium that surrounds the metal surface, which interacts with the incoming light and the metal surface itself. Due to the fact that light interacts directly with the metal layer, symmetric sensors have the potential to provide an excellent degree of sensitivity. However, there are a number of parameters, including the thickness of the metal layer and the mode field distribution of the fiber, that might restrict the sensitivity of these devices. However, asymmetric fiber optic SPR sensors incorporate asymmetrical features, such as grooves or gratings, to enhance the interaction between the sensing area and the coming light. By planning for multiple resonances or localized plasmons to be stimulated, the sensitivity of these structures can be improved. The asymmetric design allows for a customized sensor response that is optimized for a specific analyte or RI range. However, it could be more expensive to produce, involving a more complicated fabrication process and the risk of coupling losses. Ultimately, a number of factors, including the desired sensitivity, the range of interest for the RI, the availability of manufacturing processes, and cost concerns related to the specific application, determine whether symmetric or asymmetric fiber optic SPR sensors are used.

Although silica fibers are commonly used in fiber optic SPR sensors, the high cost of these fibers prevents their widespread use in price-sensitive applications. Plastic-clad silica (PCS) fibers are inexpensive yet hard to work with because they contain a large silica core (Dwivedi et al. 2008). Therefore, plastic optical fibers (POF) are an appealing option for economically motivated applications. Easy manipulation, a high numerical aperture and diameter, and outstanding adaptability are only a few of the benefits they offer in the SPR sensor's context (Cennamo et al. 2011). However, due to the high number of guided modes in POF-based SPR sensors, the spectral width of the resonance spectrum is larger than in silica fiber-based sensors. The majority of POF-based SPR sensors have been made by polishing the polymer cladding on one side. To create a D-shaped structure, Cennamo et al. (2011) adopted a traditional manufacturing method that involved polishing the side of POF's cladding around half its circumference. Polishing the plastic fiber, however, breaks the optical fiber's circular symmetry and leads to extra polarization-dependent losses (Krchnavek 1994). The circular asymmetry of these arrangements restricts the cross-section of the field interactions, which weakens the field interactions. Moreover, side-polishing is a mechanically demanding operation, and the resulting side-polished fibers are difficult to splice, resulting in additional losses (Kvavle et al. 2008).

Al-Qazwini et al. (2016) constructed a symmetrically etched POF-based SPR sensor. The sensor fabrication process was greatly optimized, and overall system costs were cut by utilizing an etching approach that maintained the POF's cylindrical shape during sensing area preparation. Figure 5.2 depicts the experimental setup. System components include a white light source (Ocean Optics HL-2000, wavelength range: 360–2000 nm), fiber optic connections, a spectrometer (Ocean Optics USB4000-VIS-NIR), and the proposed SPR sensor installed in a u-groove frame. The transmitted SPR spectra in the region of 350–1100 nm are calculated and shown on a computer using SpectraSuite software. They studied how the thickness of gold (Au), the thickness of residual fiber, and the sensing length affected the constructed sensor's performance. They discovered that resonance depth, cost, and adaptability had been compromised in favor of increased sensitivity and detection accuracy in sensors with thick Au layers and shorter sensing lengths. They also observed that sensors with smaller diameters were more sensitive to variations in the RI of their environment. Nonetheless, the residual thickness only has a small impact on the resonance's depth and width. Also, sensors with small diameters are especially susceptible to damage and breakage. Therefore, they have created a low-cost SPR sensor with exceptional performance, achieving a sensitivity of 1600 nm/RIU and an FWHM of roughly 154 nm when optimized for a residual thickness of 964 m, an Au thickness of 55 nm, and a sensing length of 10 mm.

As can be seen in Fig. 5.3, the hollow fiber was initially used in SPR sensing by Liu et al. (2013) to detect materials with a high RI. The hollow fiber's inner wall was symmetrically coated with silver, and its air core was filled with a liquid sensing medium having a high RI. The RI detection range was 1.509–1.763, and the Ag film thickness was 57 nm. The sensor's sensitivity ranged from 2185 nm/RIU to 6607 nm/RIU. With a resolution of 10^{-4} RIU, such a sensor would be able to detect the RI.

In order to further optimize the RI detection ranges, Chen et al. (2016) used a tunable nanoporous silica (NPS) coating, as illustrated in Fig. 5.4. In order to fulfill the complete reflection criterion and the SPR condition, variable RI (nNPS) NPS coatings can be utilized. The results demonstrate the effectiveness of the sensor in these two ranges when coated with nanoporous silica. They measured a sensitivity of 5840 nm/RIU (1.33–1.34) for the low RI range and 5120 nm/RIU (1.42–1.44) for the high RI range. Furthermore, the nanoporous silica coating's RI may be modified to tailor sensor performances and working wavelengths to a variety of applications. They also examine the impact of single-incident light modes on sensor performances by focusing on single-mode incidence scenarios.

Liu et al. presented a core fiber with a controllable RI in 2020 (Liu et al. 2020) using an ultraviolet-curable adhesion with a variable RI and a hollow fiber, as depicted in Fig. 5.5. In this structure, the ultraviolet lamp was used to cure the ultraviolet-curable adhesive in the porous fiber. The maximum sensitivity, 3467 nm/RIU, was obtained using a fiber SPR sensor with a core RI of 1.454.

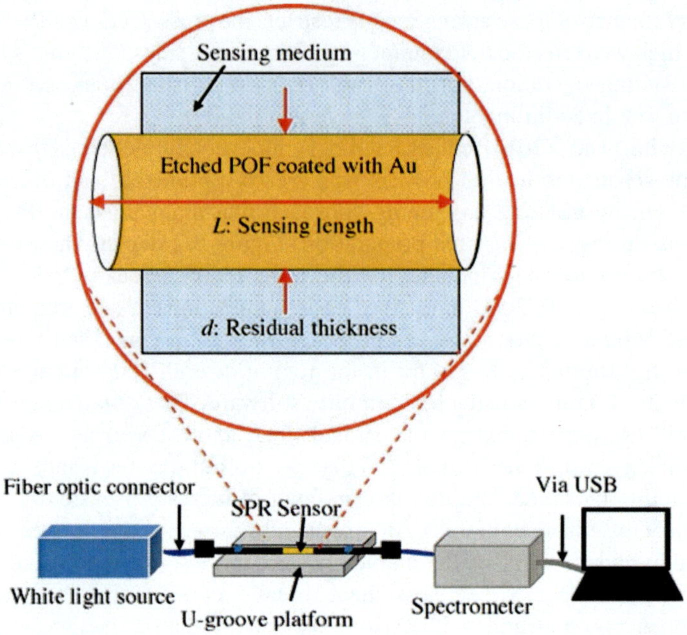

Fig. 5.2 A simplified illustration of the experimental setup. Reprinted with permission from Sensors and Actuators A: Physical. Copyright, 2016, Elsevier (Al-Qazwini et al. 2016)

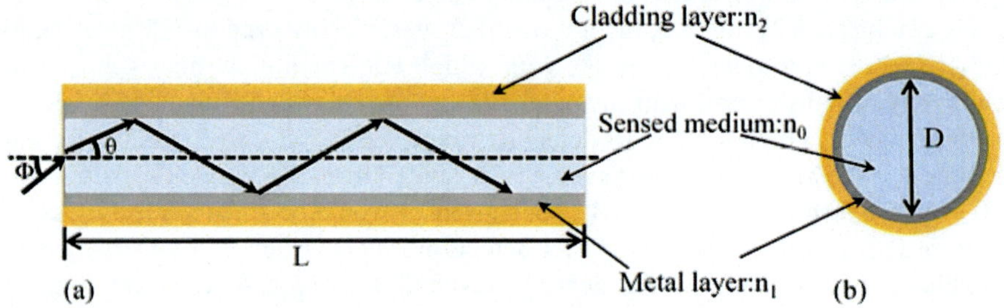

Fig. 5.3 The hollow fiber SPR sensor is depicted in a sketch and a ray model: **a** a segment that runs lengthwise; **b** a segment depicted horizontally. Reprinted with permission from Optics Express. Copyright, 2013, Optica Publishing Group (Liu et al. 2013)

As shown in Fig. 5.6, Lu et al. (2016) developed a novel SPR temperature sensor based on liquid crystal-filled hollow fiber to further extend the use of SPR in temperature sensing. Because liquid crystal has such a high thermo-optical coefficient, the liquid crystal-filled hollow fiber SPR sensor in this design is susceptible to changes in temperature and offers a broad dynamic RI detection range. The temperature sensitivity was 4.72 nm/°C from 20 to 34.5 °C, while the sensitivity to the RI was as high as -4.68×10^3 nm/RIU.

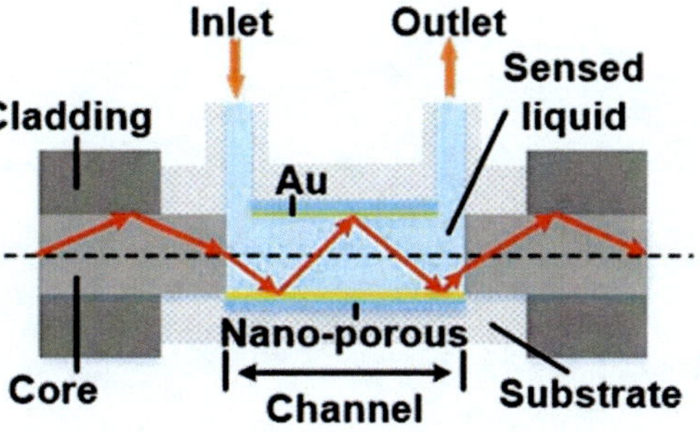

Fig. 5.4 A simplified representation of the SPR sensor based on nanoporous silica-coated hollow fiber. Reprinted with permission from Optics and Laser Technology. Copyright, 2022, Elsevier (Liu et al. 2022)

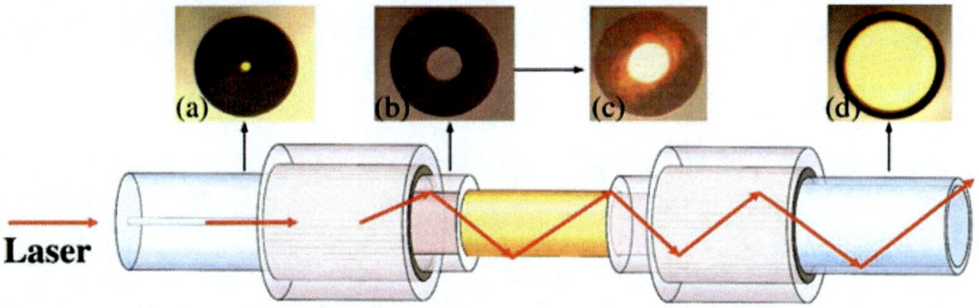

Fig. 5.5 Fiber SPR-RI sensing probe, sectional view of **a** single-mode fiber, **b** capillary fiber, **c** hollow capillary fiber filled with ultraviolet-curable adhesive, and **d** step-index multimode fiber. Reprinted with permission from Applied Optics. Copyright, 2020, Optica Publishing Group (Liu et al. 2020)

Wei et al. (2023) developed a fiber optic SPR-based salinity sensor. When the salinity of the surrounding saltwater is altered, the transmission sensing strategy's signature peaks and valleys will shift. Salinity detection is achieved by keeping an eye on the distinctive wavelength change. The sensor improves its detection sensitivity by using fusion taper innovation, which is widely regarded as having production repeatability and boosting the evanescent field. The ideal construction of a tapered optical fiber SPR sensor is developed using evaluation parameters like FOM. The sensor has a tapering area length of 27 mm, a thickness of gold film of 40 nm, and a waist area diameter of 25 μm. Figure 5.7 depicts the design and sensing mechanism of the tapered optical fiber used in the SPR salinity sensor. The experimentally optimized maximum sensor's sensitivity is 0.708 nm/%. As a further advantage, the structure's metal coating protects its mechanical strength and durability, and repeat measurements have an error of less than 3 nm. When the margin of error is taken into account, R^2 becomes 0.999, indicating high reliability and durability. The experimental findings demonstrate the structure's potential as a sensitive, long-lasting salinity sensing system.

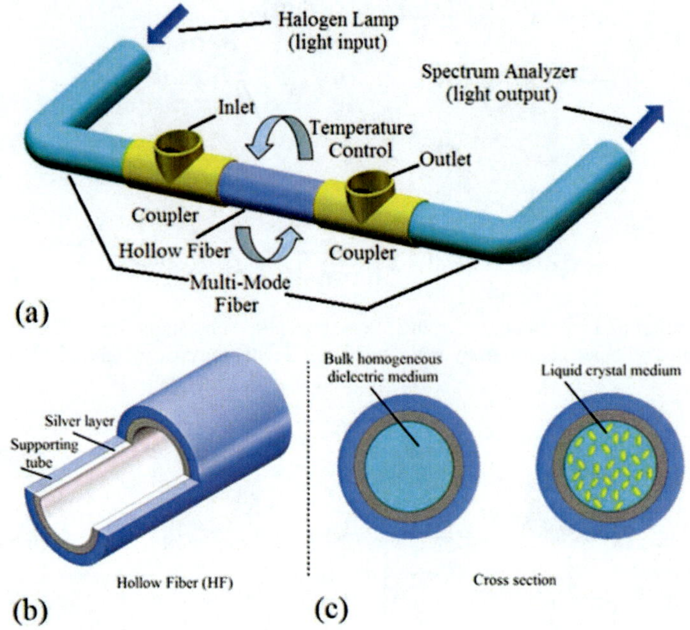

Fig. 5.6 SPR thermometer design. **a** Schematic of hollow fiber with liquid crystal medium and external heat source. A spectrophotometer measured the hollow fiber's temperature response to white light, **b** silver-coated hollow fiber design, and **c** hollow fiber filling cross-section with bulk homogenous dielectric or liquid crystal medium. Reprinted with permission from Optics Express. Copyright, 2016, Optica Publishing Group (Lu et al. 2016)

5.2 Asymmetric Coated (Half Coated) Fiber Optics SPR Sensor

Using a non-spectroscopic method based on SPR-induced changes in birefringence and intensity, Nguyen et al. (2014) produced Ag–Al coated multimode fiber optical sensors. Experimentally, a single-wavelength operation with an RI range of 0.05 yielded a minimum resolvable RI of 5.8×10^{-6}. Their detection approach is sensitive enough to pick up on the simultaneous impacts of SPR-induced birefringence shift and SPR-attenuated intensity, which contribute to the system's broad working range. The asymmetric shape of the fiber core's metal coating, as well as the broad range of incidence angles for multimode propagation, can further provide a large performing range. The Ag and Al coatings on the unprotected fiber core were applied using a thermal evaporator. The fibers were oriented so that they were perpendicular to the metal vapor deposition direction inside the vacuum chamber of the thermal evaporator, as illustrated in Fig. 5.8a. To achieve the profile of the asymmetric cross-section of coating thickness seen in Fig. 5.8b, they deposited each metal twice, first on one side of the fiber and again on the other side, with the fiber rotated along its axis by 180° between each deposition. Incomplete control over the evaporation direction,

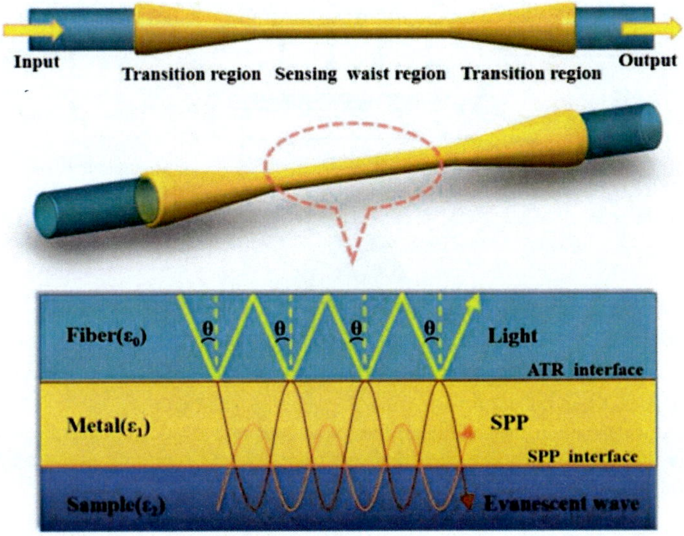

Fig. 5.7 The construction of the SPR salinity sensor, which is based on a tapered fiber optic, as well as the sensing mechanism behind it. Reprinted with permission from Optical Fiber Technology. Copyright, 2023, Elsevier (Wei et al. 2023)

such as experimental errors in the coating, would result in a non-circular symmetric profile of thickness different from that shown in Fig. 5.8b, but the asymmetry of the coated profile allows for varying penetration depth into the sensing region. Figure 5.9 is a schematic illustrating the fiber optical sensor apparatus used in the experiment. Polydimethylsiloxane (PDMS) was formed into a ring-shaped flow cell with inlet and exit valves for the analyte solution, and the metal-coated fiber was installed in a groove constructed within the ring-shaped interior of the flow cell. Since the fiber device in this system doesn't need a spectrograph, it could be possible to integrate and miniaturize it cost-effectively while still keeping the system sensitive enough over a wide range of wavelengths to be used in label-free biochemical and biomedical sensing applications.

Esteban et al. (2011) established that indium nitride (InN) is a suitable material for SPR-based sensor development. Since there are many available nitrogen atoms on the surface of this material, it can be employed to fix the recognizing components when transducers are functionalized for analyte identification. Figure 5.10 depicts a schematic of the manufactured transducers. The Optical Spectrum Analyzer (OSA) measures the spectral transmittance of the devices in response to changes in the RI of the surrounding liquid media. Since its relatively high RI makes the layer thinner than is needed when using oxides of metal like TiO_2 or Al_2O_3, a comprehensive analysis of the relationship between the transducers' behavior and InN thickness has been performed. It has also been demonstrated that the sensitivity improves with decreasing InN thickness, reaching a maximum of 11,800 nm/RIU at 20 nm thickness for an RI range of 1.415–1.429, a noteworthy result in comparison with the most

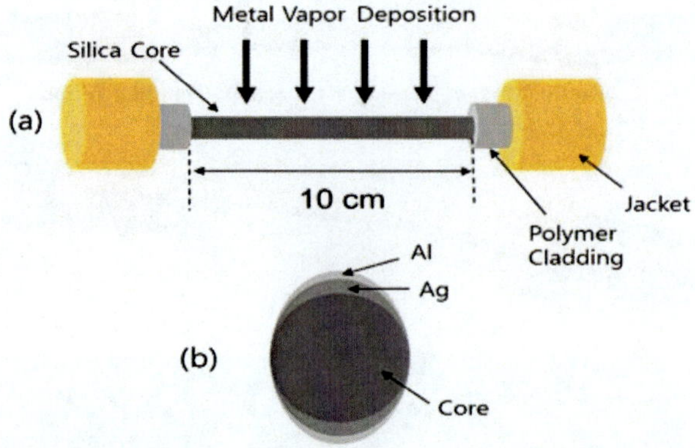

Fig. 5.8 **a** The clad-free optical fiber has metal vapor deposited onto its core's exposed surface, **b** after bimetallic deposition, the coating profile on the fiber core should look like this, as seen in the cross-section for SPR. Reprinted with permission from Optics Express. Copyright, 2014, Optica Publishing Group (Nguyen et al. 2014)

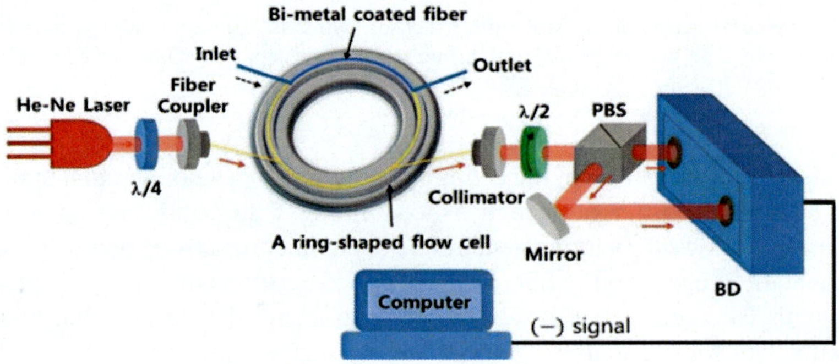

Fig. 5.9 Analyte solution sensing experimental setup diagram. Quarter-wave and half-wave plates, a polarizing beam splitter (PBS), and a balanced detector (BD) are all indicated by the notation λ/4 and λ/2, respectively. Reprinted with permission from Optics Express. Copyright, 2014, Optica Publishing Group (Nguyen et al. 2014)

sensitive devices shown to date. For a relatively low-resolution spectrometer of 1 nm, the resolution of the sensors varies between 2×10^{-4} and 8×10^{-5}, depending on the range. Long-term measurement series have also been used to assess the transducer's dependability, revealing that its behavior is consistent and reproducible even in the absence of any particular preservation efforts.

Bueno et al. (2004) have employed asymmetric devices and given equal weight to both total power and spectral transmittance as metrics for success. Surface plasmon excitation in the metal layer provides a simple explanation for the observed relationship in both circumstances. To create the asymmetrical device, a double layer is deposited on only one side of the tapered fiber. In Fig. 5.11, we can see a schematic diagram of the whole device. They have also investigated the effect of

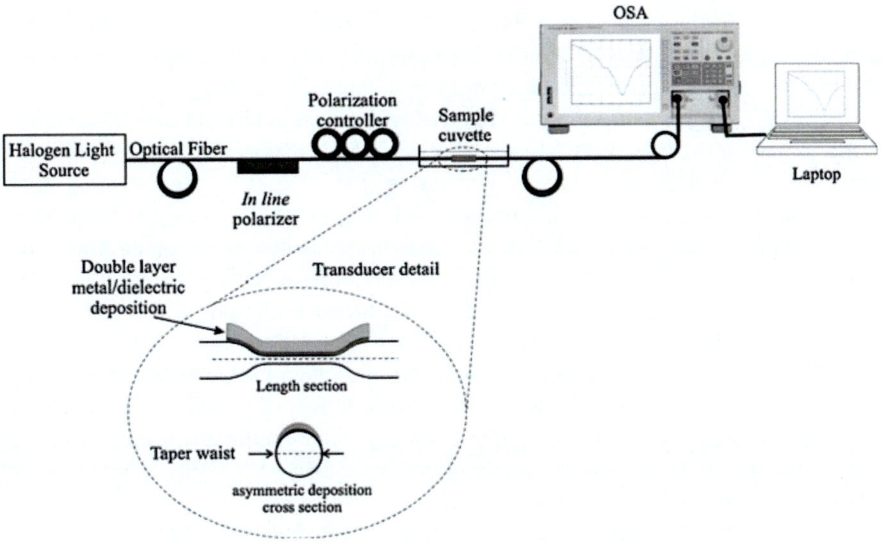

Fig. 5.10 Experimental design diagram and images of the final transducers. Reprinted with permission from Sensors and Actuators B: Chemical. Copyright, 2011, Elsevier (Esteban et al. 2011)

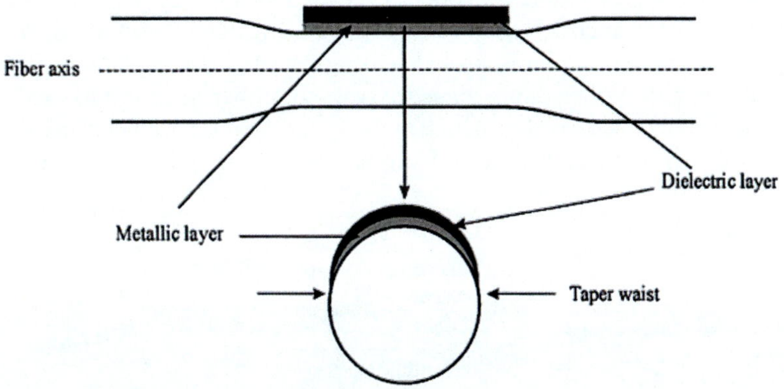

Fig. 5.11 The asymmetrical device, shown in cross-section and from the side. The in-line waist is superimposed with an image of a double-layer deposit. Reprinted with permission from Applied Optics. Copyright, 2004, Optica Publishing Group (Bueno et al. 2004)

the taper waist's diameter, which is a unique degree of freedom that may be utilized to choose the device's working points. It has been demonstrated that the sensors can function both with and without polarization control, and the impact of polarization on device behavior has been tested.

For the purpose of measuring both the temperature and RI of a liquid sample in real time, Velázquez-González et al. (2017) have designed and implemented a fiber optic sensor based on a dual-channel SPR platform. Three different metal layer geometries, shown in Fig. 5.12b as an asymmetric, (c) as two diametrically opposed, and (d) as a quasi-symmetric gold layer, were obtained by coating a length of 5 mm of single-mode fiber using three distinctive, yet simple and stationary procedures. While polarization is essential to the functioning of all three structures, its effect diminishes with increasing layer count due to the fiber perimeter being covered across a larger area. The high TOC polymer (PDMS) coated half of the SMF segment that had been coated with gold. The wavelength resonance of the bared fiber gold-coated segment is adjusted by the RI and temperature of the surrounding medium, causing one of the dips to be produced. The center wavelength of the 900 nm dip generated by the polymerized gold-coated fiber segment corresponds to the PDMS RI at 20 °C and is utilized solely for monitoring temperature changes in the liquid sample. The suggested device's high-temperature sensitivity and RI, −2.850 nm/°C and 2323.4 nm/RIU, as well as its compact size, simple manufacturing technique, and bio-compatibility, are all compelling features that make it well-suited for real-world bio-sensing applications.

Using a twin-core fiber, Liu et al. (2016) developed and investigated a reflective-distributed SPR (SPR) sensor. First, they looked at how changing the grinding angle of the fiber optic dual-tapered (DT) probe affected the dynamic range of SPR. The outcomes demonstrated that when angles of grinding rise, the resonant wavelength also increases, leading to greater testing sensitivity. This technique allows the sensor to function in a highly efficient frequency range. Second, they grounded the probe of DT into an asymmetric wedge form with two angles of grinding (6° and 17°) configured to carry out the distributed sensing based on the outcomes previously reported. The 125 μm cladding-diameter twin-core fiber used in this study has a 3.8 μm core diameter. The construction of the sensing probe is depicted in Fig. 5.13a, and the values for the angles of grinding are as follows: grinding angle (1) is α,

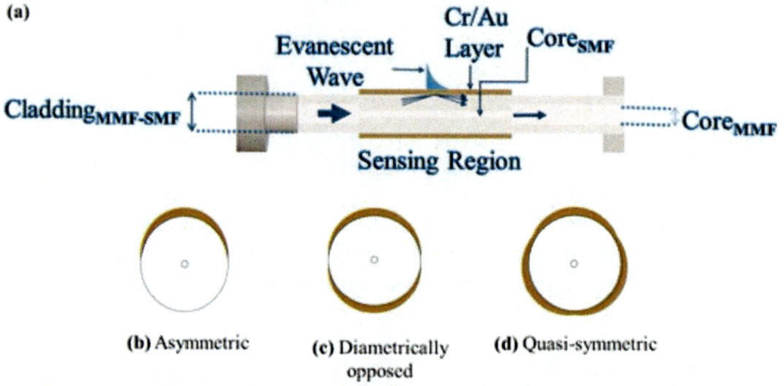

Fig. 5.12 **a** A MM-SM-MM structured fiber, which consists of a length L-long portion of single-mode fiber placed into a multimode fiber, **b** shows the cross-section of fiber after first evaporation, **c** shows the cross-section after second evaporations; and **d** shows the cross-section after third evaporations. Reprinted with permission from Sensors and Actuators B: Chemical. Copyright, 2017, Elsevier (Velázquez-González et al. 2017)

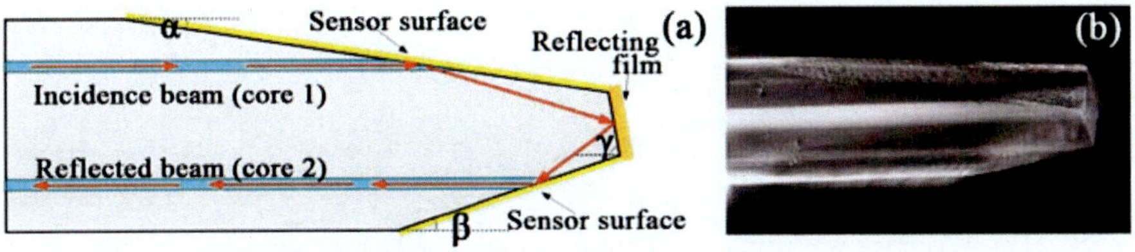

Fig. 5.13 Schematic image (**a**) and picture (**b**) DT probe of the twin-core fiber. Reprinted with permission from Optics Communications. Copyright, 2016, Elsevier (Liu et al. 2016)

grinding angle (2) is β, and the resonant angle is θ. The picture of the ground fiber tip is shown in Fig. 5.13b. The fiber tip has been ground, and then a gold layer 50 nm thick has been plated on the tilted surface. Figure 5.14 depicts the experimental setup. By means of a lens-based transform system (Fig. 5.14b), the authors focused a supercontinuum light source (SuperK compact, NKT Photonics) with a one core of a twin-core fiber that can transmit light between 450 and 2400 nm in which the collimating lens is placed in front of a pair of single-mode fiber and the twin-core fiber in front of the objective microscope lens, which has a 25× magnification. According to the results, the testing sensitivity is 4558 nm/RIU and 2385 nm/RIU for two distinct sensing zones, 729–966 nm and 591–715 nm, respectively, with an RI detecting range of 1.333–1.385. This method has allowed for the simultaneous detection of multiple analytes in the same sensing area, and it has effectively compensated for errors brought on by index interference of background and the RI change brought on by physical absorption or by non-specific binding, among other factors. Due to its compact size and ability to fit into tight testing spaces (μm scale), this reflective dispersed fiber optic-based sensor is ideal for biochemical sensing in real-world applications. The twin-core fiber's tiny 125 μm diameter makes it suitable for use in a microfluidic chip. To mimic real-world online monitoring of blood vessel function, the authors of this study included a fiber probe inside an infusion needle.

5.3 Sensing Probe Preparation

The symmetric multimode fiber utilized in this experiment has a core diameter of 105 μm and was produced by Wei et al. (2023). To get the desired length, the fiber is heated and stretched using a fiber fusion taper system. It is important to note that although the conventional nonadiabatic structure will obtain a strong evanescent field, the taper used in this experiment is a thick and long taper (the diameter of the waist area is several microns, and the length of the sensing area is several millimeters). By being so thick and lengthy, this design successfully accomplishes the goal of "adiabatic" by preventing the maximum possible amount of light loss that would otherwise be caused by the fiber core. Although the strength of the evanescent field is inadequate, it can still be reflected in the absorption waveform because

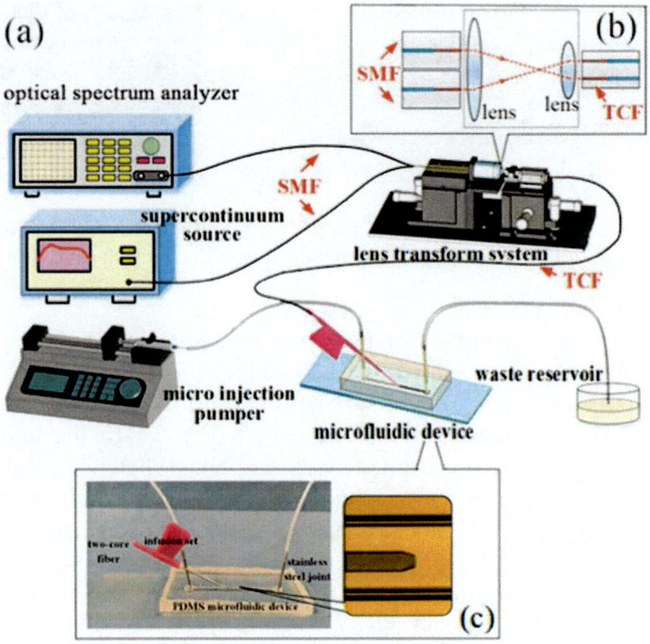

Fig. 5.14 **a** A sketch diagram of the reflecting distributed two-core fiber SPR sensor used in the experiment setting; **b** an outline schematic of the lens transform system; and **c** pictures of the microfluidic device, with an inset image displaying the microfluidic channel and fiber probe in greater detail. Reprinted with permission from Optics Communications. Copyright, 2016, Elsevier (Liu et al. 2016)

of the high initial signal. This is the same reason why multimode fiber was chosen over single mode. Furthermore, the strength will be substantially increased since the taper is quite thick, and the contact area will be larger, both of which are helpful in enhancing the quality of the coating. After that, magnetron sputter coating equipment is used to cover the sensor structure. Figure 5.15 depicts the sensor assembly procedure. Figure 5.16 illustrates the experimental setup. The equipment includes a computer, a sensor structure, an ocean spectrometer (QE Pro), and a halogen light source (HL-2000). First, the construction of the sensor is improved. Theoretically, the sensor's ability to detect the RI of seawater will directly correlate to its accuracy in measuring salinity. Therefore, a NaCl solution whose RI varies from 1.3327 to 1.3711 (configuration of solution based on references) (Tan and Huang 2015) is used to test structural performance. Salinity detection will take advantage of the optimized structure. The concentrations of seawater utilized in this study ranged from 0, 5, 20, 30, 35, and 40%, all of which were manufactured by the National Marine Standards and Metrology Centre.

For the subsequent immunoassay investigation, Mai et al. (2019) produced a disposable symmetric fiber optic SPR biosensor probe. Two multimode fibers (MMF, core diameter 62.5 μm and cladding diameter 125 μm) are spliced together to create

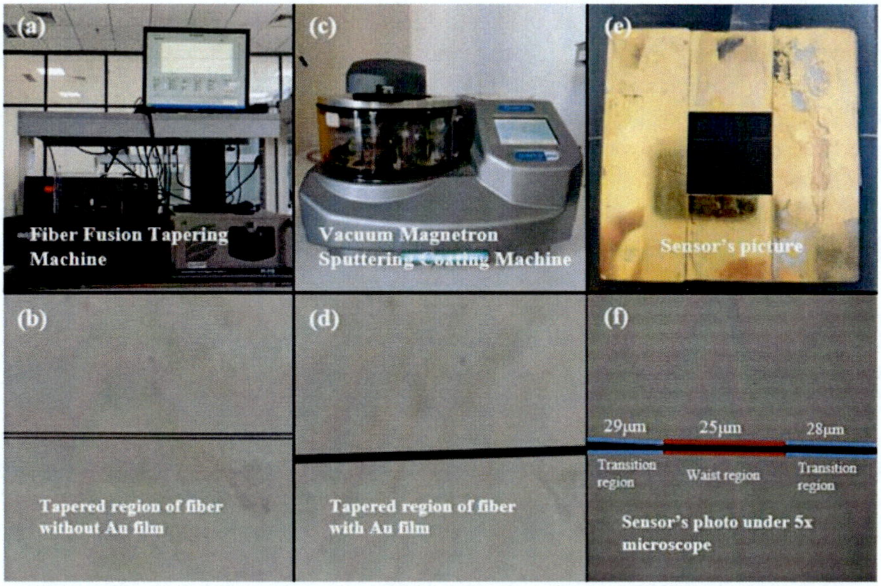

Fig. 5.15 Sensor's fabrication process images of **a** the optical fiber fusion taper system, **b** the microscope image of the tapered area of the optical fiber, **c** the magnetron sputtering coating machine, **d** the microscope image of the tapered area of the optical fiber after coating, **e** the sensor, and **f** the sensor under a 5× microscope. Reprinted with permission from Optical Fiber Technology. Copyright, 2023, Elsevier (Wei et al. 2023)

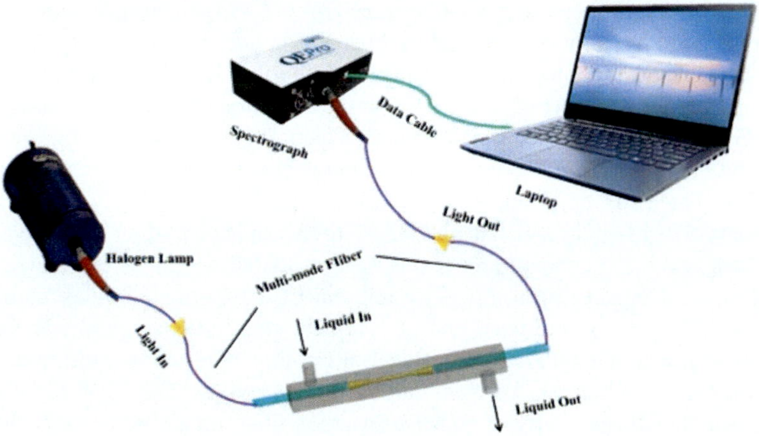

Fig. 5.16 Schematic diagram of the examination setup for measuring salinity. Reprinted with permission from Optical Fiber Technology. Copyright, 2023, Elsevier (Wei et al. 2023)

a single-mode fiber (SMF, core diameter 9 μm and cladding diameter 125 μm), as shown at the bottom of Fig. 5.17. The proposed biosensor with hetero-core fiber, i.e., M-S-M, was fabricated, enabling collision excitation of SMF cladding modes. Here, the SMF area was extended to 12 mm, and anhydrous ethanol and distilled water were used to carefully wipe away any remaining dust. Using the JGP450A magnetron sputtering fiber coating instruments (SKY Technology Development Inc.), they deposited a gold layer over the SMF area once they had cleaned its surface. As shown in Fig. 5.17, a sensing system was then prepared in order to test the proposed biosensor. This system consisted of the fiber optic SPR biosensor, a computer, a spectrometer, and a light source with a wide wavelength range (360–2600 nm, Thorlabs Inc. SLS201/L). Transmission fibers made of multimode fiber (MMF) were used to link the proposed biosensor, the light source, and the spectrometer. For the first four hours, they exposed pure antibody (5 mg/mL) to 50 mM 2-MEA in PBS containing 10 mM EDTA-2Na, pH 6.0. This created half-antibody fragments that could be used to furnish the surface of the sensor for the particular detection. By targeting the disulfide bonds in the hinge areas of IgG, this method generated two monovalent IgGs, or half-antibody fragments, each with one or two free sulfhydryl groups. The 2-MEA was then removed by ultrafiltration through Millipore (30 K) five times in anaerobic 1X PBS. Next, the concentrated IgGs were frozen at −20 °C to prevent further degradation before being used to furnish the proposed fiber probes. Finally, lowered IgGs (in a PBS buffer with a pH value of 7.4) were added to the proposed symmetric fiber probe sensing region for 48 h at 4 °C to allow free SHs (sulfhydryl groups) to self-assemble on the gold layer. The fiber probe of GAM-IgG-furnished was prepared for the detection of M-IgG after being thoroughly cleaned with PBS. The steps involved in creating the fiber probe of GAM-IgG-furnished are depicted in Fig. 5.18a. The authors also made a fiber probe adorned with M-IgG to detect GAM-IgG. The detection of M-IgG using a fiber probe GAM-IgG-furnished is one-half of an immunoassay (Fig. 5.18b), while the detection of GAM-IgG using a fiber probe GAM-IgG-furnished is the other half (Fig. 5.18c). The optimal concentrations for M-IgG and GAM-IgG furnishing were determined by determining the concentration at which the target molecule was most easily detected. The sensitivity of the fiber probes M-IgG-furnished and GAM-IgG-furnished to their respective target molecules at varying concentrations was then evaluated.

Velázquez-González et al. (2017) interconnected two graded-index multimode fibers (62.5/125 μm) onto each end of a typical step-index single-mode fiber (9/125 μm) segment of length L to create the multimode-single-mode-multimode (MM-SM-MM) optical fiber construction employed in the study (see Fig. 5.12). In order to increase the RI sensitivity of an MM-SM-MM fiber, it can be coated with a thin coating of gold to create SPR and boost the evanescent wave interaction with the external medium. Despite silver's superior adhesion to silica fiber for RI measurements, gold's resistance to the organic solvents often employed for cleaning the fiber makes it a better choice. Traditionally, intermediate bond materials like chromium were used to address the adhesion problem. The fiber coating procedure is complicated by the fiber's cylindrical shape. As shown in Fig. 5.12, the MM-SM-MM fibers were coated in three separate ways, each of which resulted in a different metal

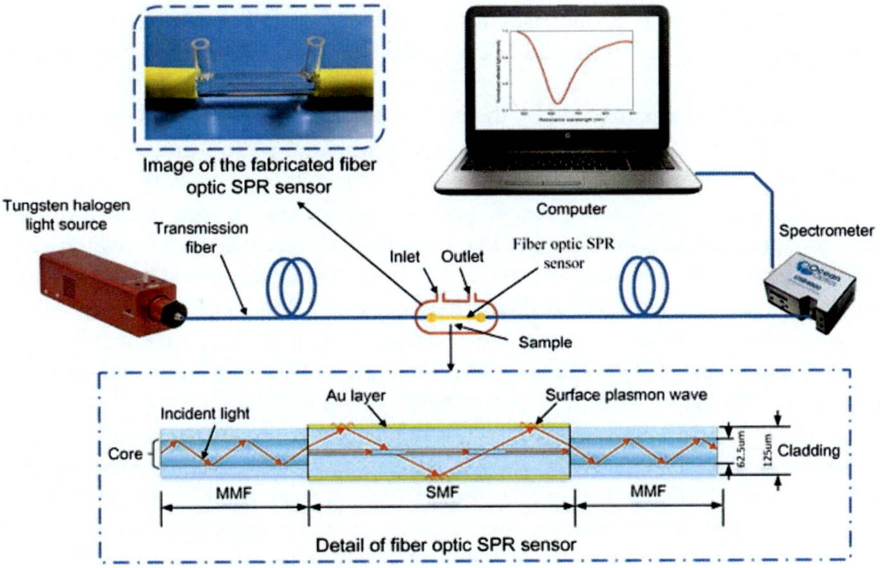

Fig. 5.17 A schematic illustration of the sensor system. Reprinted with permission from Biosensors and Bioelectronics. Copyright, 2019, Elsevier (Mai et al. 2019)

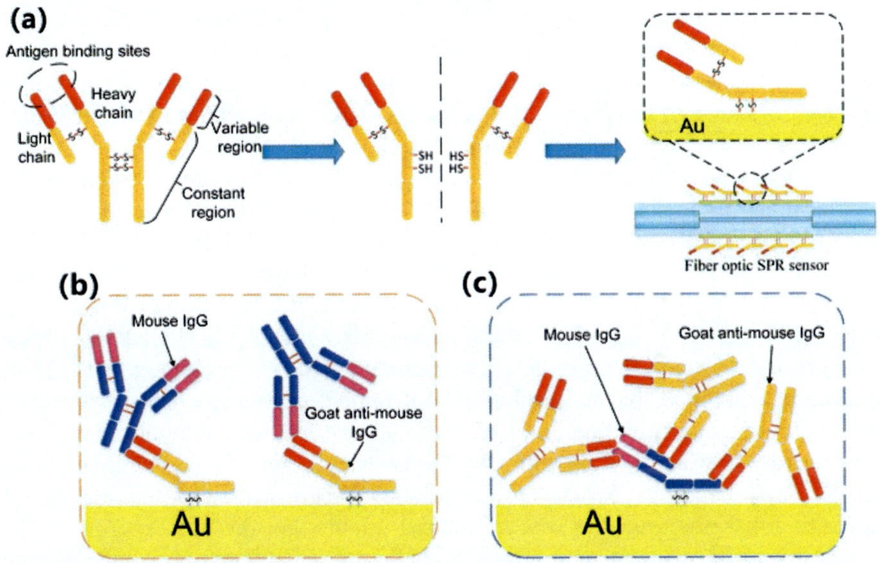

Fig. 5.18 Fiber optic SPR (SPR) biosensor surface decorating and selective detection: **a** fiber probe of GAM-IgG furnishing process; **b** detection of M-IgG with a furnished fiber probe of GAM-IgG; and **c** GAM-IgG detection with a furnished fiber probe of M-IgG. Reprinted with permission from Biosensors and Bioelectronics. Copyright, 2019, Elsevier (Mai et al. 2019)

layer geometry: (b) an asymmetric, (c) two diametrically opposed, and (d) quasi-symmetric gold layer. Polarization is essential to all three structures, but its effect diminishes with layer count since more and more of the fiber's perimeter is masked. The coating process performed by the authors to achieve such asymmetric geometries is detailed in Luna-Moreno and Monzón-Hernández (2007), and it enables us to produce many samples concurrently with high repeatability. In addition, the device can be thought of as a small sensor with a sensitive region long enough to ease manufacture. A simple summary of the steps taken to create the sensor would go as follows: the PDMS polymer was combined with the curing agent and left to rest until all air bubbles were removed. A Peltier plate was placed in an acrylic piece, and the fiber was then suspended above it; this setup places half of the SMF one millimeter above the acrylic. See Fig. 5.19a for the completion of the process, which involved pouring PDMS over an acrylic slice covering half the length of the gold-coated fiber segment. The PDMS did not drip down from the acrylic because of surface tension. After curing the PDMS at 80 °C for 6 h, the device was prepared for simultaneous RI and temperature monitoring due to the PDMS film embedding the gold-coated SMF half-section. Fiber transmission was measured during the PDMS coating and curing process. A resonant dip in the transmitted spectrum at 900 nm developed at the precise instant when the polymer had been spread over the fiber. A little change in the resonance dip was seen after curing, and the device was taken out of the mount. The fabricated device is seen in Fig. 5.19b; the polymer layer of their sensor is rectangular in form, making it more manageable, but a variety of PDMS shapes may be achieved depending on the application's needs. The configuration of the experimental components to characterize the response of the sensor to varying temperatures and values of the RI is shown in Fig. 5.19c. This sensor is extremely sensitive to bending, thus the device must remain stationary throughout the test.

Esteban et al. (2011) have developed a method to manufacture tapered asymmetric single-mode fibers from ordinary silica fibers. These fibers are optimized for transmission at 850 nm. The resulting tapers have a waist diameter of 40 μm, a waist length of 12 mm, a total taper length of about 50 mm, and losses that are less than 0.2 dB. They deposited an Al/InN bilayer on these tapers using RF sputtering with a 2″ confocal magnetron cathode (AJA International, ATC ORION-3-HV). This process was carried out at room temperature. This is the method that is most frequently employed for the development of optical coatings, which involves the deposition of nanocrystalline thin films onto a substrate. The aluminum layer was produced by heating an aluminum (Al) target using argon plasma while applying 75 W of DC electricity. When these conditions for growth were used, the rate of deposition that was produced was 2.8 nm per minute. For the InN experiment, the target was a disk made of pure In (4N5), while the reactive gas was made of pure N_2 (6N). Plasma at a radio frequency was used to deposit the nitride at a power level of 40 W. The deposition was carried out at a working pressure of 3.5 mTorr throughout, whereas the initial pressure for it was on the order of 10^{-5} mTorr. Using these characteristics, they were able to build a number of sensors, each of which had an Al layer that was 8 nm thick but had a varied InN thickness, namely 40, 30, and 20 nm. The procedure for depositing has been carried out in a consistent manner.

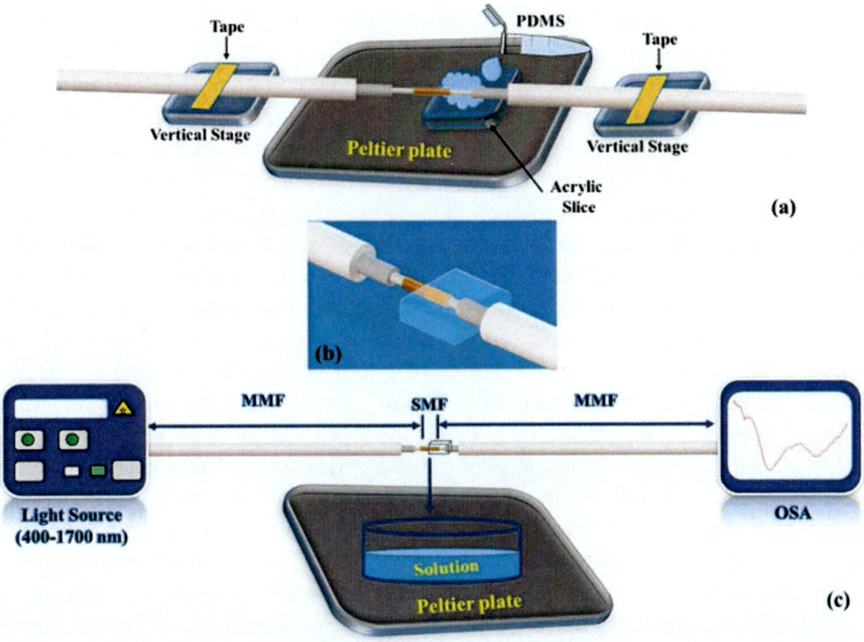

Fig. 5.19 Depicts **a** the sensor manufacturing process, **b** a proposed dual-channel SPR optical fiber sensor, and **c** an experimental setup for characterizing the sensor's response to changes in RI and temperature. Reprinted with permission from Sensors and Actuators B: Chemical. Copyright, 2017, Elsevier (Velázquez-González et al. 2017)

Therefore, the devices that were produced are what are referred to as "asymmetric ones". In these devices, only about half of the taper is covered by the bilayer structure, and the layer thicknesses vary across the device. A schematic representation of how the transducers were manufactured may be seen in Fig. 5.10. It is based on the measurement of the spectral transmittance of the devices through an OSA for various refractive indices of the outer media, and it is always done with the medium in liquid condition. As a result, the process begins with the measurement of the device's transmission with air functioning as the surrounding medium; this measurement serves as the reference for the complete characterization. Because the results are spectrally based, a halogen lamp was utilized as the light source for this experiment. A polarization-controlling device that is based on a linear polarizer and a set of Lefèvre loops has been incorporated into the setup since the SPR phenomena in asymmetric tapers depend on polarization. However, polarization is not an issue with double-layer uniform-waist tapered fibers (DLUWTs) because polarization is not a factor in the SPR phenomenon. Because the authors of this research altered the polarization at the beginning of the tests to create the highest contrast possible, there is nothing further that has to be done to prevent any additional effect on the data. When the polarization plane is aligned with the maximum thickness of the deposited layers, this system should exhibit a significant drop in the amount of light that it transmits.

The asymmetric device, the preparation of which can be broken down into two phases, was carried out by Bueno et al. (2004). The first option is to reduce the diameter of the fiber in order to get closer to the evanescent field of the mode that is being transmitted by the fiber that has not been disturbed. The authors have used the so-called "travelling burner" method created by Kenny et al. Many writers have utilized this method to create a wide range of devices with tapered optical fibers. They were able to generate consistent waists of tapered fiber that extend many millimeters in length with minimal loss by employing this method. The authors have employed a standard, single-mode, step-index optical fiber operating at a nominal wavelength of 820 nm. The waists of the various devices they have created are around 30–40 μm, and the tapers' losses are around 0.3 dB. The deposition of layers onto the substrates acquired in the first step described above is the second stage of manufacturing. The technique used for depositing the material is physical vapor deposition, as seen in the approach described by Alonso et al. The deposition of a double layer on one side of the tapered fiber yields the asymmetric device. In Fig. 5.11, we see a simplified diagram of the whole apparatus. Deposited materials range in thickness from zero at the borders to a maximum value at the top of the waist. They've settled on an 8 nm thick aluminum layer and a 60 nm thick titanium dioxide layer. Their ability to measure in the 780–880 nm wavelength range for outer medium refractive indices between 1.32 and 1.41 is a direct result of the author's previous papers, in which D-type fibers with a double metallic–dielectric layer were used as refractive index sensors.

5.4 Asymmetric Fiber Optic Chemical Detector

As a first step in determining the magnitude of the SPR-induced dimming (Nguyen et al. 2014) measured the relationship between optical power at the fiber output and glycerol content for both uncoated and metal-coated fibers. Increased glycerol content in the absence of a metal coating boosts optical power at the fiber's output by decreasing the loss of propagating mode energies over the clad-free sensing length. In contrast, the SPR-induced reduction of output power is clearly visible when the metal coating is applied to the fiber core and the glycerol concentration is increased. Figure 5.20a–d display the observed signal S for varying composition ratios of Ag and Al deposited thicknesses on the fiber core as the glycerol concentration increases. Signal (S) showed various sensitivity characteristics, notably at or around zero concentration (C), while decreasing with concentration for all composition ratios of Ag–Al coating thicknesses. Compared to $t_{Ag}/t_{Al} = 20$ nm/5 nm and $t_{Ag}/t_{Al} = 36$ nm/4 nm, the coating thickness ratios of $t_{Ag}/t_{Al} = 7$ nm/30 nm and $t_{Ag}/t_{Al} =$ 30 nm/10 nm resulted in more nonlinearly varying signals at zero concentration. The two distinct sensitivity (slope) regions, depicted in Fig. 5.20c, d, occurred at concentrations of around 1 and 0.05%. At near-zero concentration, a thickness ratio of $t_{Ag}/$

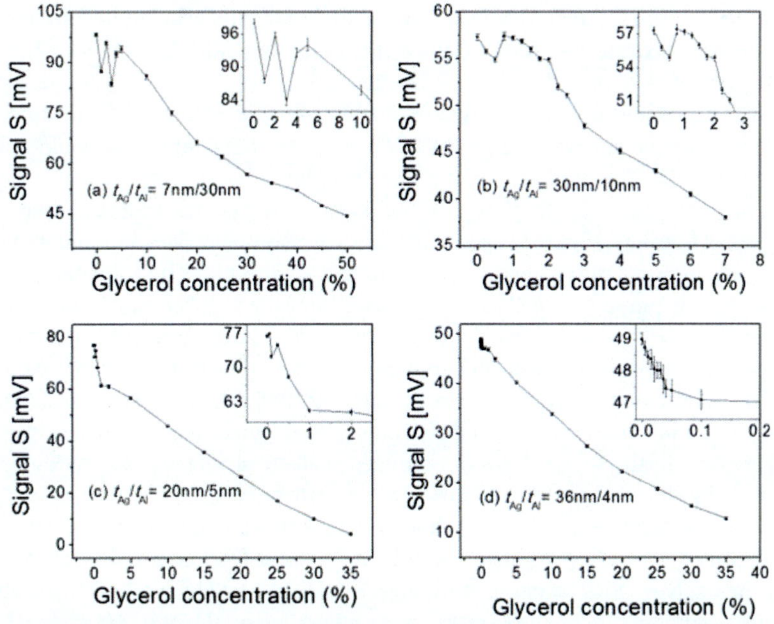

Fig. 5.20 Glycerol concentration sensing using volume-to-volume coating thickness ratios of Ag and Al. Reprinted with permission from Optics Express. Copyright, 2014, Optica Publishing Group (Nguyen et al. 2014)

t_{Al} = 36 nm/4 nm proved to be more sensitive than a ratio of t_{Ag}/t_{Al} = 20 nm/5 nm (15.2 mV/%), as measured by the signal-to-concentration sensitivity *S/C* = 32.4 mV/%. As shown in the inset of Fig. 5.20d, they used a fiber device with this bimetallic coating (t_{Ag}/t_{Al} = 36 nm/4 nm) to probe smaller concentrations near zero, and they achieved the minimum detectable concentration of 5×10^{-3}%, where the increase of concentration from zero caused the signal change to be comparable or slightly greater than a standard deviation of the signal near zero. With their present configuration, they have experimentally obtained a minimum resolvable RI (MRI) of 5.8×10^{-6} RIU, based on a signal standard deviation near zero employing the C/S. Despite only using a single wavelength, Fig. 5.20d demonstrates that the fiber device's proven working RI range permits concentrations from 0 to greater than 35%. In this case, the predicted linear connection between the RI and the concentration provides a range of >0.05RIU for possible RI operation. The asymmetric metal coating profile of the fiber core is thought to play a role in the large operating range by providing different SPR probe angles in addition to the wide range of incidence angles for multimode fiber propagation. The non-golden bimetallic coating was selected to prevent excessive SPR-induced attenuation while still guaranteeing sufficient device sensitivity, and this coating is observed to be able to accommodate a broad working RI range.

Another version, with an InN layer 30 nm thick, was constructed by Esteban et al. (2011) to demonstrate the devices' tunability. The transmittance for the characterized transducer is depicted in Fig. 5.21; the outer RI of 1.3950 yields the shortest resonance wavelength, while a value of 1.4089 yields the longest. Once again, the numbers for the center wavelength of the range, about 900 nm, have been approximated as the RI. The predicted resonance at higher RI values in the same spectral region was seen when the thickness of the dielectric layer was reduced, confirming the expected findings. Figure 5.22 shows that the response behaves in a linear fashion, with a very high average sensitivity of roughly 10,800 nm/RIU obtained from the slope of the displayed curve. This number is significantly greater than what has been reported before for similar devices. As before, no extra care was taken to preserve the manufactured transducers before the stability test was conducted after a period of time. For the same values of outer RI, the transmittance was consistent across many transducers made using the same process, and the resonance wavelength hardly drifted at all. When the RI is reduced instead of increased, no changes are seen, proving that the measurements may be performed in either direction. Minor changes in the measured RI values due to temperature dependence are below the spectrometer's resolution limit, as are the very small fluctuations in the measurements themselves. SPR sensing utilizing DLUWTs has always had no problems with measurement stability, repeatability, or reversibility; this will not alter with the introduction of InN.

Velázquez-González et al. (2017) saw a second dip down at a lower wavelength than that of PMDS after submerging the device in water–glycerol mixtures. Both dips have wavelengths that change according to the solutions' refractive indices and temperatures. RI solutions of 1.346 (black line), 1.365 (red line), and 1.388 (blue line) at 20 °C were used to provide transmitted spectra for comparison in Fig. 5.23a. Abbe

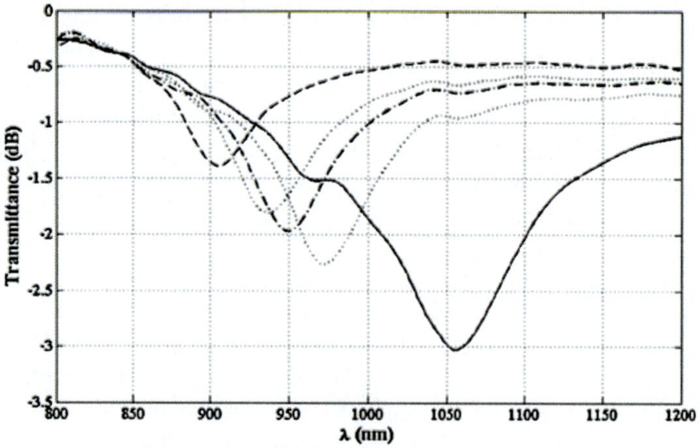

Fig. 5.21 Device's transmission at 900 nm computed for an Al/InN bilayer with thicknesses of 8 nm and 30 nm, respectively, with outer refractive indices of 1.3950–1.4089. Reprinted with permission from Sensors and Actuators B: Chemical. Copyright, 2011, Elsevier (Esteban et al. 2011)

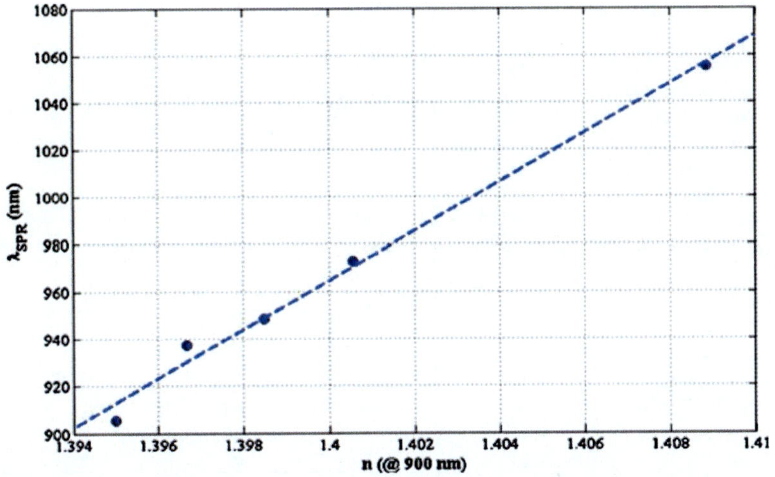

Fig. 5.22 Variation in the outer RI causes a shift in the resonant wavelength for the device seen in Fig. 5.21 (bilayer Al/InN, 8 nm/30 nm thick, respectively). Based on the curve's slope, it is estimated a median sensitivity of about 10,800 nm/RIU. Reprinted with permission from Sensors and Actuators B: Chemical. Copyright, 2011, Elsevier (Esteban et al. 2011)

refractometers were used to determine the RI of the water–glycerol solutions. The gap between dips and the form of the dips is substantially affected when the solution's RI increases, as in the case of a solution with an RI of 1.388, when bare shifts toward longer wavelengths. The upper bound of the dynamic range can be established as this RI. It is crucial to note that the polymerized fiber section is insensitive to the RI change, which is seen in Fig. 5.23b as a shift in the resonance dips owing to RI variations. The sensitivity of the bare fiber portion to changes in the RI was 2323.4 nm/RIU, while the polymerized gold-coated part was insensitive at 0 nm/RIU. Temperatures in these solutions were raised from 20 to 60 °C while the fiber sensor was submerged there. The spectrum of light that was transmitted through the fiber when it was submerged in a 90% water–10% glycerol solution (RI: 1.346) at 25, 35, 45, and 55 °C. There was a blueshift in both resonance dips, but the λ_{PMDS} displacement was 10 times higher than the λ_{bare}. For a 90% water/10% glycerol solution, the temperature sensitivities of the first and second peaks were −0.28 and −2.85 nm/°C, respectively. Both the estimated and measured temperature sensitivity of the dual-signal SPR sensor in ambient air are exceptionally close to one another.

The following set of equations quantitatively expresses the relationship between changes in temperature (ΔT) and RI (Δn) and the resulting changes in the wavelength of the resonance dips (λ_{bare} and λ_{PDMS}):

$$\begin{bmatrix} \Delta\lambda_{bare} \\ \Delta\lambda_{PDMS} \end{bmatrix} = \begin{bmatrix} -0.280 & 2323.4 \\ -2.85 & 0 \end{bmatrix} \begin{bmatrix} \Delta T \\ \Delta n \end{bmatrix}. \tag{5.1}$$

The sensing matrix can finally be expressed as:

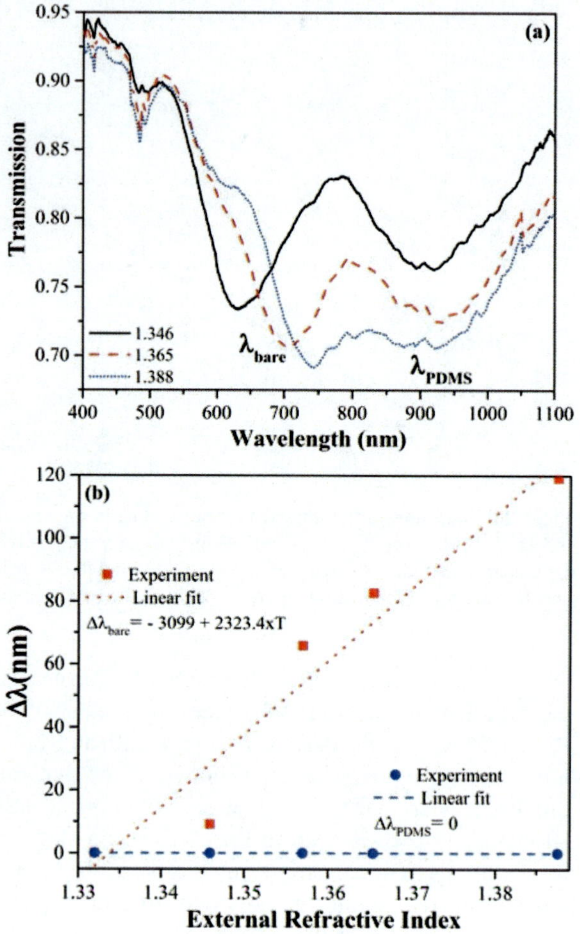

Fig. 5.23 **a** Transmission spectra measured experimentally for a fiber sensor submerged in three distinct water-glycerol solutions at 20 °C: 1.346 (black line), 1.365 (red line), and 1.388 (blue line). **b** A response curve characterizing the sensor's sensitivity to experimentation with varying RI. Reprinted with permission from Sensors and Actuators B: Chemical. Copyright, 2017, Elsevier (Velázquez-González et al. 2017)

$$\begin{bmatrix} \Delta T \\ \Delta n \end{bmatrix} = \frac{1}{(6{,}621.69)} \begin{bmatrix} 0 & -2323.4 \\ 2.85 & -0.280 \end{bmatrix} \begin{bmatrix} \Delta\lambda_{bare} \\ \Delta\lambda_{PDMS} \end{bmatrix}, \tag{5.2}$$

where λ_{bare} and λ_{PDMS} are measured in nm, ΔT in °C, and Δn in RIU. The complete RI change can now be calculated using the measured temperature change ΔT and the λ_{PDMS} characterized as:

$$\Delta n = 4.3 \times 10^{-4} \times \Delta\lambda_{bare} + 1.2 \times 10^{-4} \times \Delta T. \tag{5.3}$$

It's important to note that the experiments were conducted using a hot plate that was kept at a constant temperature of 1-grade centigrade above room temperature. With a sensitivity of about 2.9 nm/°C, their technology was able to detect this shift. Assuming an OSA resolution of 0.1 nm, they estimated a temperature detection limit of roughly 0.05 °C. The same reasoning could potentially be applied to their device's

RI detection limit; they estimated it to be somewhere around 5×10^{-5} RIU. Taking into account this RI detection limit, they determined that a temperature change of 0.5 °C is required to cause a detectable RI change in a fiber sensor submerged in water. It's worth noting that the PDMS polymer is hydrophobic. Therefore, it retains its characteristics even when immersed in aqueous solutions.

The distributed DT probe was made by Liu et al. (2016) using a twin-core fiber. The sensor probe construction is shown in Fig. 5.13a, and the value for grinding angle 1 is α and for grinding angle 2 is β. The picture of the ground asymmetric fiber tip is shown in Fig. 5.13b. In this case, they used the grinding angle γ to manipulate the reflected light. The fiber tip was similarly prepared by grinding and plating a thin gold layer (50 nm thick) on its angled surface. Authors adjusted the α to 6°, β to 17°, and γ to 79° based on the above simulation and experimental outcomes. Here, they examined the location of the resonance dip, that coincides with the predicted outcomes even if the transmitted amplitude attenuating spectrum does not. When the grinding angle α is 6°, and β is 17°, respectively, the resonance wavelength is shown in Fig. 5.24, along with simulated and experimental findings. Changing the fiber grinding angle allowed them to modify the sensor's operational range, as shown in Fig. 5.24. The redshift of the resonance wavelength, the expansion of the resonance range, and the enhancement of the testing sensitivity occur with an increase in the fiber grinding angle. The experimental results reveal that in the RI detecting range of 1.333–1.385, the sensitivity of average testing rises from 2308 nm/RIU to 5096 nm/RIU when the grinding angle of the DT probe is increased from 10° to 16°. This sensor allows for the simultaneous detection of numerous analytes in the same sensing region, as well as the effective compensation of errors brought on by such things as background interference and changes in RI due to non-specific binding or physical absorption, among other capabilities. This reflective-distributed fiber-based sensor has a tiny form factor and can fit into tight testing locations, making it ideal for use in real-world biochemical sensing applications. The twin-core fiber has a very small diameter of 125 μm, making it easy to include in a microfluidic chip; in this part, they incorporated the fiber probe into an infusion needle to mimic blood vessels for real-time monitoring.

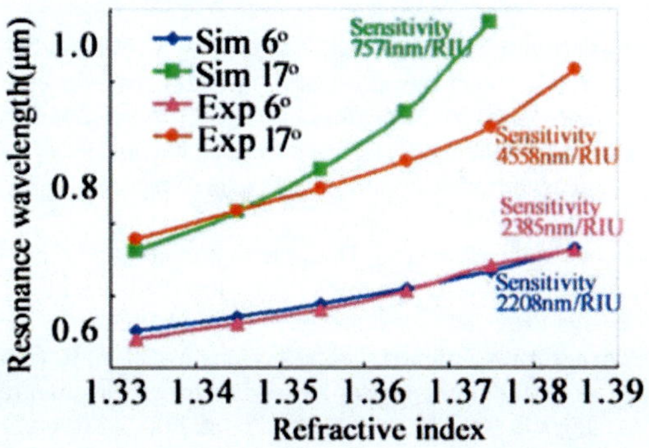

Fig. 5.24 **a** Simulation and **b** experiment outcomes for sensitivity testing using a 6° and 17° grinding angle, respectively. Reprinted with permission from Optics Communications. Copyright, 2016, Elsevier (Liu et al. 2016)

5.5 Summary

In order to detect analytes with high sensitivity without using a label, fiber optic SPR sensors are frequently employed. SPR, which occurs when light interacts with a thin metal layer and its environment, is used in these devices. One optical fiber is coated with a metal (usually gold or silver) and a dielectric substance to create a symmetric fiber optic SPR sensor. In order to generate surface plasmon waves, the metal coating is patterned into a specific form, such as a grating or a periodic array of nanostructures. Surface plasmons at the metal–dielectric contact interact with the light that has been linked to the fiber. To ascertain the analyte concentration or binding events occurring on the sensor surface, the resonant wavelength of the surface plasmons may be detected as it varies in response to variations in the RI of the surrounding medium. The core of an asymmetric fiber optic SPR sensor is exposed by polishing away a part of the cladding on one side of the fiber, a process also known as side-polished or tilted-fiber SPR sensors. A thin coating of metal, usually gold or silver, is subsequently applied to the exposed core. Total internal reflection occurs when light is linked to the fiber at a certain angle. The angle of total internal reflection changes because the analyte's binding to the metal-coated area changes the RI at the core-analyte interface. This shift is measurable and can be linked to binding events or analyte concentration. Fiber optic SPR sensors, whether symmetric or asymmetric, have benefits in sensitivity, adaptability, and real-time monitoring. If your application doesn't need a very precise spatial location, a symmetric sensor may be the way to go. Asymmetric sensors, on the other hand, allow for localized sensing at certain places along the fiber and provide more sensor design versatility. In the end, symmetric and asymmetric fiber optic SPR sensors provide adequate resources for label-free sensing and detection in fields as diverse as biomedical research, environmental monitoring, and chemical analysis. Because of their great sensitivity and capacity to provide readings in real time, they find use in a wide variety of disciplines.

References

Al-Qazwini Y, Noor ASM, Al-Qazwini Z, Yaacob MH, Harun SW, Mahdi MA (2016) Refractive index sensor based on SPR in symmetrically etched plastic optical fibers. Sens Actuators A 246:163–169. https://doi.org/10.1016/j.sna.2016.04.064

Bueno FJ, Esteban O, Diaz-Herrera N, Navarrete MC, González-Cano A (2004) Sensing properties of asymmetric double-layer-covered tapered fibers. Appl Opt 43(8):1615–1620. https://doi.org/10.1364/AO.43.001615

Cennamo N, Massarotti D, Conte L, Zeni L (2011) Low cost sensors based on SPR in a plastic optical fiber for biosensor implementation. Sensors 11(12):11752–11760. https://doi.org/10.3390/s111211752

Chen Y, Li X, Zhou H, Hong X, Geng Y (2016) Refractive index detection range adjustable liquid-core fiber optic sensor based on surface plasmon resonance and a nano-porous silica coating. J Phys D Appl Phys 49(35):355102. https://doi.org/10.1088/0022-3727/49/35/355102

Dwivedi YS, Sharma AK, Gupta BD (2008) Influence of design parameters on the performance of a surface plasmon sensor based fiber optic sensor. Plasmonics 3(2–3):79–86. https://doi.org/10.1007/s11468-008-9057-z

Esteban Ó, Naranjo FB, Díaz-Herrera N, Valdueza-Felip S, Navarrete MC, González-Cano A (2011) High-sensitive SPR sensing with indium nitride as a dielectric overlay of optical fibers. Sens Actuators B Chem 158(1):372–376. https://doi.org/10.1016/j.snb.2011.06.038

Krchnavek RR (1994) Precision fabrication of D-shaped single-mode optical fibers by in situ monitoring. J Lightwave Technol 12(9):1524–1531

Kvavle JM, Schultz SM, Selfridge RH (2008) Low loss elliptical core D-fiber to PANDA fiber fusion splicing. Opt Express 16(18):13552. https://doi.org/10.1364/oe.16.013552

Liu B-H, Jiang Y-X, Zhu X-S, Tang X-L, Shi Y-W (2013) Hollow fiber surface plasmon resonance sensor for the detection of liquid with high refractive index. Opt Express 21(26):32349. https://doi.org/10.1364/oe.21.032349

Liu C, Zhang X, Gao Y, Wei Y, Wu P, Su Y, Wu P (2020) Fiber SPR refractive index sensor with the variable core refractive index. Appl Opt 59(5):1323. https://doi.org/10.1364/ao.380665

Liu W, Liu Z, Zhang Y, Li S, Zhang Y, Yang X, Zhang J, Yuan L (2022) Specialty optical fibers and 2D materials for sensitivity enhancement of fiber optic spr sensors: a review. Opt Laser Technol 152(December 2021):108167. https://doi.org/10.1016/j.optlastec.2022.108167

Liu Z, Wei Y, Zhang Y, Zhu Z, Zhao E, Zhang Y, Yang J, Liu C, Yuan L (2016) Reflective-distributed SPR sensor based on twin-core fiber. Optics Commun 366:107–111. https://doi.org/10.1016/j.optcom.2015.12.018

Lu M, Zhang X, Liang Y, Li L, Masson J-F, Peng W (2016) Liquid crystal filled surface plasmon resonance thermometer. Opt Express 24(10):10904. https://doi.org/10.1364/oe.24.010904

Luna-Moreno D, Monzón-Hernández D (2007) Effect of the Pd-Au thin film thickness uniformity on the performance of an optical fiber hydrogen sensor. Appl Surf Sci 253(21):8615–8619. https://doi.org/10.1016/j.apsusc.2007.04.059

Mai Z, Zhang J, Chen Y, Wang J, Hong X, Su Q, Li X (2019) A disposable fiber optic SPR probe for immunoassay. Biosens Bioelectron 144(June):111621. https://doi.org/10.1016/j.bios.2019.111621

Nguyen TT, Lee E-C, Ju H (2014) Bimetal coated optical fiber sensors based on surface plasmon resonance induced change in birefringence and intensity. Opt Express 22(5):5590. https://doi.org/10.1364/oe.22.005590

Tan CY, Huang YX (2015) Dependence of refractive index on concentration and temperature in electrolyte solution, polar solution, nonpolar solution, and protein solution. J Chem Eng Data 60(10):2827–2833. https://doi.org/10.1021/acs.jced.5b00018

Velázquez-González JS, Monzón-Hernández D, Moreno-Hernández D, Martínez-Piñón F, Hernández-Romano I (2017) Simultaneous measurement of refractive index and temperature using a SPR-based fiber optic sensor. Sens Actuators B Chem 242:912–920. https://doi.org/10.1016/j.snb.2016.09.164

Wei X, Peng Y, Chen X, Zhang S, Zhao Y (2023) Optimization of tapered optical fiber sensor based on SPR for high sensitivity salinity measurement. Opt Fiber Technol 78(January):103309. https://doi.org/10.1016/j.yofte.2023.103309

Chapter 6
D-shape Fiber Structure-Based SPR Sensor

6.1 Introduction

Research into optical fiber sensors has been prevalent because of their desirable sensing and physical properties. The cladding of conventional optical fiber is etched for use in sensing applications, making the fiber's properties (refractive index and deformation) more sensitive to environmental changes. However, if the cladding's corrosion depth is excessive, it will have diminished mechanical strength and be more susceptible to damage. In order to address these drawbacks and enhance the sensing properties, a wide variety of specialty optical fibers have been developed. These fibers include photonic crystal fiber, Bragg fiber, W-shaped fiber, and microstructure fiber. An expanded frequency range and higher measurement sensitivity are two of the many enhancements. However, it is difficult to monitor the detection signal strength with sensors based on transmitted wave absorption when materials with a high optical absorption coefficient are employed. It has been proposed to use a D-shaped optical fiber for assessing environmental conditions. The absorption of evanescent waves was used to make this a reality. In addition to its other benefits, it may be used to effectively track even hazy substances. Optical fiber sensors based on the D-shaped structure have found widespread use in a variety of sensing applications in recent years. These applications include monitoring changes in refractive index, temperature, magnetic field, pressure, biological characteristics, and more (Ying et al. 2016).

D-shaped optical fibers have several potential uses in sensing applications due to their adaptability in terms of effective index and energy distribution during transmission. The optical wavelength shifts as the effective index is changed, and these shifts can stand in for other environmental factors. Observing changes in light intensity can reveal a shift in energy distribution. While there has been progress in both approaches, their sensitivities are still rather low. This calls for research into methods of boosting the sensitivity of D-shaped optical fibers. Surface plasmon resonance (SPR) and a D-shaped optical fiber were combined in a new sensing method first published in 2006

S. K. Raghuwanshi et al., *Geometric Feature-Based Fiber Optic Surface Plasmon Resonance Sensors*, Springer Tracts in Electrical and Electronics Engineering,
https://doi.org/10.1007/978-981-99-7297-5_6

by Wang et al. (2006). It was found that the fiber's sensitivity to changes in refractive index may be greatly improved by depositing a gold coating on it in this configuration. In 2011, Lanza et al. (2011) introduced a D-shaped optical fiber Bragg grating (FBG) sensor. He looked into the impact magnets have on the structure's efficiency. A sensitivity of 1.4403.0 pm/G was reported. In 2014, Ubeid and Shabat (2015) introduced a D-shaped optical fiber loaded with graphene and investigated its reflection properties. It was demonstrated that the sensitivity to changes in the refractive index was much improved by the addition of a graphene layer. The excellent sensitivity and exceptional versatility of D-shaped optical fiber sensors suggest that they could find many applications.

The configuration of the D-shaped optical fiber sensor must be analyzed since it measures the evanescent wave's interaction with the medium around it. Figure 6.1a depicts the design process for the fundamental structure, which involves side polishing a single-mode fiber. This type of fiber consists of a cladding, a core, and a metal film. The optical fiber's cross-section is depicted in Fig. 6.1b. The cladding's flat side is given a metal film deposition. Both ends of the single-mode fiber are linked to the sensing portion, as shown in Fig. 6.1c.

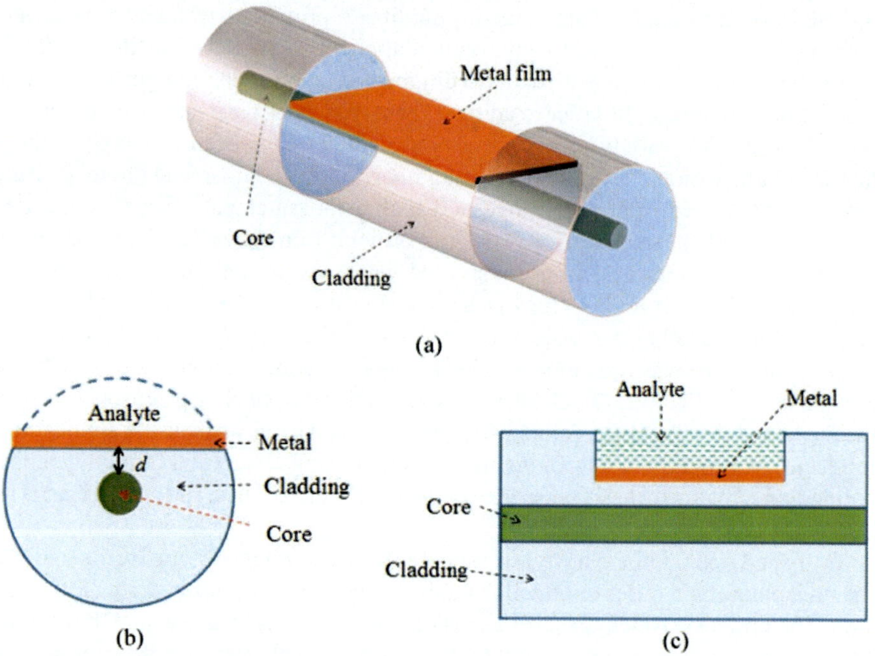

Fig. 6.1 Shows **a** a structure stereogram, **b** a cross-section view, and **c** a detailed schematic of a D-shaped optical fiber. Reprinted with permission from Optics & Laser Technology. Copyright, 2017, Elsevier (Ying et al. 2016)

6.2 SPR Sensor Based on Nanophotonics and Plasmonics

As a rapidly expanding field of study, plasmonics has brought together experts from several disciplines, including chemistry, physics, and medicine. At the nanoscale, the interaction of photons with surface electrons is studied in plasmonics. Surface plasmons are oscillations of a group of surface electrons induced by incoming photons from a light source. This can result in the transmission of surface waves between dielectric nanofilms and metal, surface plasmon polarization, or, in nanoparticles and nanostructures, a strong localized oscillation of surface electrons (Stockman 2011; Liu et al. 2017). Nanoparticles and nanostructures have a strong surface electron oscillation that makes them move quickly. At the nanoscale, since LSPR, localized surface plasmon resonance, and SPP are both capable of causing energy transfer, strong light absorption and scattering occur between light and matter. Sensors based on the Kretschmann configuration for surface plasmon resonance (SPR) are the most popular type of plasmonic biosensors. In biological and chemical applications, SPR sensors based on prisms are frequently used to detect unknown analytes in real time as extremely accurate biosensors (Luo et al. 2019). A major drawback to using prism-based SPR sensors is the setup's bulkiness, which makes it less useful for on-site sensing, especially for sensing in difficult-to-reach places (Liu et al. 2017; Esfahani Monfared 2020a). Scientists in the domains of biology, chemistry, and medicine, on the other hand, have an exciting new potential to miniaturize SPR sensors thanks to the junction of plasmonics (SPR sensors) with fiber optics (Liu et al. 2017; Esfahani Monfared 2020a). In addition to allowing for significant reductions in sensor size, these tiny architectures also provide for device design, material, and sensing performance flexibility. Fiber optic plasmonic sensors can overcome some of the fundamental problems of traditional SPR sensors while also adding potential new qualities such as the ability to withstand harsh environments, remote sensing, and distributed sensing (Liu et al. 2017). Several fiber optic plasmonic biosensing setups have been presented in recent years in order to detect biological samples and chemicals. In recent years, optical fibers with various plasmonic materials have also been researched due to the fact that different metals can generate varied plasmonic properties. In this research, the most widely employed plasmonic materials are gold (Au) and silver (Ag), which are noble metals. In this research, however, the field is currently seeing a rise in other materials, such as graphene (Rifat et al. 2016), transition metal oxide (Jabin et al. 2019), and transition metal nitride (Esfahani Monfared 2020a) in the design of biosensors. For instance, plasmonic graphene nanostructures can be powered by surface plasmon resonances (SPR). For sensors in the THz range, graphene-plasmonic couples have excellent optical confinement, extremely low losses, and EM (electromagnetic) tenability. While metal–dielectric nanostructures are used in plasmonic nanostructures, in graphene, carbon atoms are organized in two-dimensional lattices (Sharma and Gupta 2018). The interaction of free electrons in a metal with an electromagnetic light wave known as surface plasmon resonance (SPR) can overcome the constraints of light imposed by nanoscale diffraction at wavelengths lower than the wave lengths that are propagating. Thus, graphene-plasmonic structures are useful as absorbers (Chau et al. 2018a), temperature and refractive index sensors (Chou Chao et al. 2020), or biosensors for the detection of essential constituents such as hemoglobin (Sharma and Gupta 2018) and glucose (Panda et al. 2020). There are a variety of ways to fabricate graphene-plasmonic-based nanostructures. Sensors based on plasmonic graphene can be developed in a variety of ways. To construct biosensors with high sensitivity and accuracy, we should look to RI-based architectures (Saifur Rahman et al. 2020). Various configurations of plasmonic-graphene nanostructures have been investigated. It has been shown that plasmonic materials, such as gold nanorods, can be studied using the finite element method (FEM). Plasmon-mode wave functions caused rhythmic patterns to appear on the surface (Chau et al. 2009). Another study used the FEM method to conduct computational examinations of adjustable SPRs and dimple cavity plasmon modes of scattering cross-section (SCS) on various nanorod configurations (Chou Chau et al. 2016). Nanorods with diverse properties can be tweaked and regulated to produce specific wave lengths based on the properties of these structures (Chou Chau et al. 2016). In another work, sensors with variable dielectric cores for plasmonic-dielectric combinations have been proposed and studied. With the combination of lattice resonance, localized SPR, and cavity plasmon resonant modes, the suggested structure performs extremely well in terms of function (Chau et al. 2018b). This biosensor can detect blood glucose concentrations between 25 and 175 mg/dl with a RI range of one to one thousand seven (Sengupta and Hussain 2021). In terms of sensitivity, the proposed sensor measured 275.15/RIU. Researchers have suggested graphene-based biosensors for virus detection since they are more affordable, accurate, and quick in light of the COVID-19 outbreak (Sengupta and Hussain 2021). In a graphene-plasmonic biosensor that was proposed and examined, nanoparticles of silver with graphene film layers were used to construct the structure. A 9 nm graphene layer was used to achieve a 304.6% increase in sensitivity over a non-graphene structure (El barghouti et al. 2020). For hemoglobin concentration sensing, plasmonic nanostructures of various forms (spherical, cubic, and cylindrical) have also been proposed (Heidarzadeh et al. 2020). Resonance frequencies for each of the three nanoparticle morphologies were found to be distinct. Graphene-based nanostructures have been shown to be capable of being employed as biosensors in a variety of ways by controlling and manipulating their adjustable properties (Vermisoglou et al. 2020). Another study takes into account the refractive index of blood, glucose, and other biological elements, as well as biomolecules and biological cells (Sadeghi et al. 2020).

It has been an issue to directly coat the outside of the fibers to increase sensor performance because of the difficulty of the aforementioned operation to make sensor performance better. SPR sensors with Au films and $BaTiO_3$ thin layers were developed by Wang et al. (2020b). The transmission of charge between Au and $BaTiO_3$ improves the sensor's sensitivity compared to the typical Au film SPR sensor. Compared to a standard Au film SPR sensor, the fiber/Au film/$BaTiO_3$ film SPR sensor has a refractive index range of 1.3332–1.3710 and a sensitivity of 2543 nm/RIU. According to Niu et al. (2019), a D-type large-core fiber sensor was proposed, utilizing an Au film and Au nanoparticles coupled together in a single fiber. According to the data, simulated electric field intensities in the gap between Au nanoparticles and Au film are 4–5 times greater than those of Au film, and the electric field intensity on the surface of Au nanoparticles is double that of Au film. Compounds previously unknown to chemists can be detected by changes in the analyte's core refractive index. Hossen et al. (2018) reported that gold-coated PCF-SPR biosensors might be used to detect between 1.35 and 1.4 mV with a maximum amplitude sensitivity of 442.11 RIU^{-1}. Additionally, An et al. (2017) proposed a D-shaped plasmonic fiber optic sensor using a gold layer as the plasmonic material, with a maximum sensitivity of 10,493 nm/RIU at 1.38 and a detection range of 1.33–1.38. In the same way, a PCF that is only filled with the channels that you want can do this. Biosensors with high-index PCF-SPR biosensors, such as those developed by Chu et al. (2019) can detect analytes with a RI range of 1.44–1.57. Another biosensor developed by Chu et al. (2019) has a detection range of 1.46–1.52 nm, and its highest sensitivities are 6100 nm/RIU. A D-shaped high-index PCF-SPR sensor with a maximum amplitude sensitivity of 85 RIU^{-1} at 1143 nm and 90 RIU^{-1} at 1190 nm has recently been proposed by Luan and colleagues (2018). They have a spectral sensitivity of 7200 nm/RIU for their sensor (Luan et al. 2018). The sensor detection range and sensitivity should be improved even further, notwithstanding the high potential shown by these numbers. D-shaped PCFs with the same air hole diameters can also be used to make plasmonic sensors that are easier to make than high-index PCF sensors that were difficult to make before.

6.2.1 Nano Film D-shaped Plasmonic Fiber Sensor

Plasmonic fiber optic biosensors can also be created by side polishing the cladding section of a normal optical fiber. Plasmon nanofilms or nanoparticles are first placed on the fiber's flat surface using a vacuum evaporation method, and then the fiber's cladding portion is polished to a high gloss (Dong et al. 2019). Depending on the biological sample refractive index (RI) on the surface of nanofilms or nanoparticles, the interaction between fiber and plasmonic modes can generate a transmission dip (absorption peak) in light transmission spectra. With this technique, an unknown chemical can be detected and the changes in the analyte monitored (Fig. 6.2).

In the year 2019, Dong et al. (2019) investigated a gold thin film-based SPR biosensor using a few-mode fiber (FMF) that has been side-polished for the first time. The biosensing effectiveness of their sensor was proved by evaluating bovine serum albumin (BSA) solution, which they found to have a detection range for the RI of 1.333–1.404. BSA concentrations with average sensitivities of 1.17 nm/mg of BSA were reported by the researchers (Dong et al. 2019).

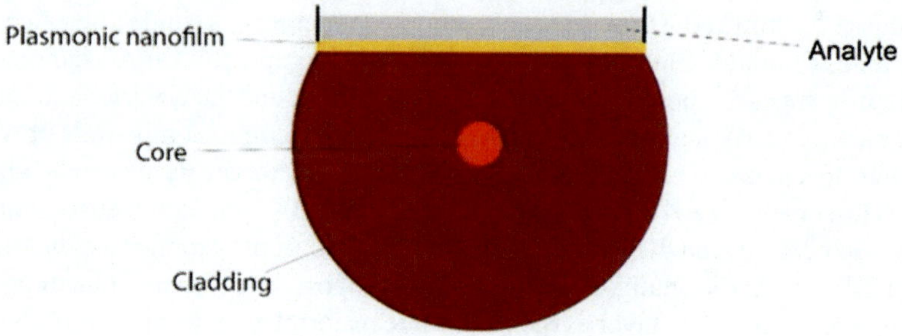

Fig. 6.2 A plasmonic fiber optic biosensor for liquid analyte detection with a D-shaped cross-section. Adapted from Esfahani Monfared (2020b)

6.2.1.1 Plasmonic Sensors Using Photonic Crystal Fibers

A flexible and customizable optical fiber type, photonic crystal fibers (PCFs), open out new optical sensing, nonlinear optics, and imaging capabilities. As Fig. 6.3 depicts the situation, PCFs feature a core that is ringed by a regular pattern of air holes. For example, PCFs may achieve ultra-flat dispersion, operate in a single mode forever, have low propagation losses, and have adjustable nonlinearity (Monfared and Ponomarenko 2019; Monfared et al. 2021; Unterhuber et al. 2019). Based on their construction, PCFs can be divided into two primary categories: PCFs with a hollow core and a solid core (Monfared and Ponomarenko 2019; Monfared 2018). The TIR mechanism is used by solid-core PCFs, just as it is by traditional optical fiber. HCPCF guiding mechanisms are heavily reliant on the core material in hollow-core PCFs. High-index materials, such as high-index liquids, can be used to guide HCPCFs through their cores (Monfared and Ponomarenko 2019). However, if we put a substance with a low index of refraction, like a gas, into the entire internal reflection, it is impossible (Monfared 2018). Even though PCF has a low-index material-filled hollow core and a photonic bandgap mechanism, light steering can still be achieved over a restricted bandwidth (Monfared 2018).

PCF includes a number of design characteristics that can be tweaked to achieve the desired optical qualities. Core size, air hole dimension, material for the backdrop, the distance between air holes from center to center, and the number of air holes are all design characteristics. Both solid-core and hollow-core PCFs can be used as host mediums for plasmonic materials and analytes in plasmonic biosensing applications. For high refractive index biosensing applications, liquid analytes can be injected into the hollow core of PCFs. D-shaped PCF biosensors are the most important PCF-based plasmonic biosensors (Monfared and Ponomarenko 2019).

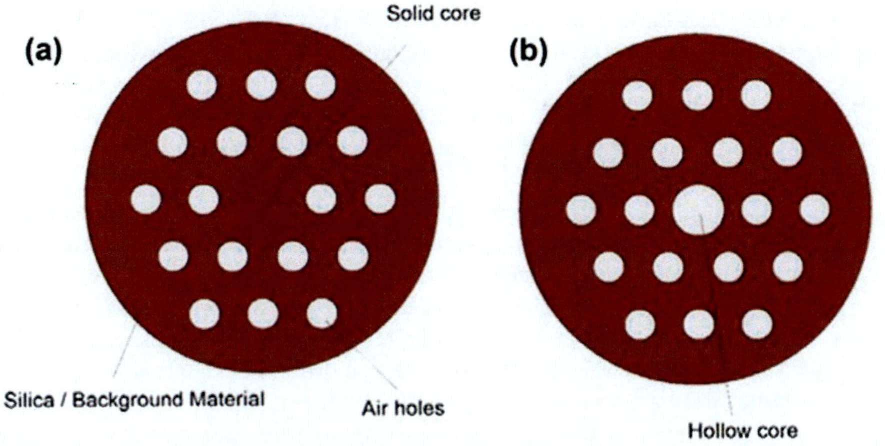

Fig. 6.3 Shows a photonic crystal fiber (PCF) with a solid core and a hollow core in cross-section. Adapted from Esfahani Monfared (2020b)

6.2.2 Nano Film with a D-shaped PCF Plasmonic Sensor

Biosensors that are based on an unknown analyte or a change in the physical or chemical characteristics of the analyte can be detected using SPR-based biosensors or D-shaped PCF Plasmonic by interacting with the core fiber mode and surface plasmonic modes of nanofilms and nanoparticles. For fiber mode and plasmonic mode phase matching (index matching), the PCF's air holes can help (Monfared et al. 2021; Lu et al. 2018). When it comes to employing plasmonic materials, PCF biosensors with a D shape are one of the most common designs since they may be utilized to detect analytes with low or high RI analytes (Monfared et al. 2021; Liu et al. 2018). It is shown in Fig. 6.4 what it looks like when you cut through the biosensor in the D shape made of PCF plasmonics.

Fig. 6.4 Low and high refractive indices analytes can be detected using a surface plasmon resonance (SPR) biosensor with a D-shaped PCF. Adapted from Esfahani Monfared (2020b)

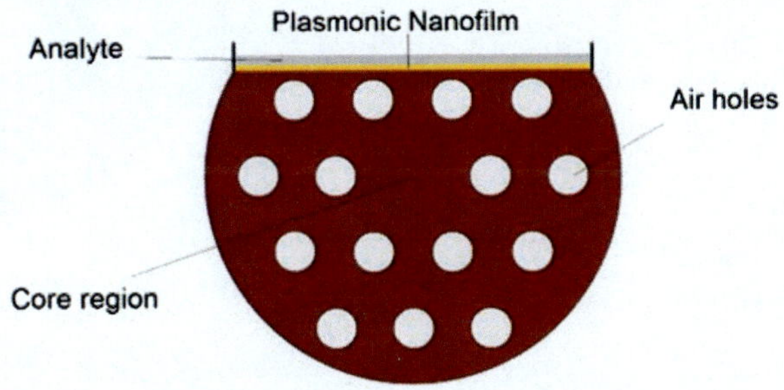

A gold-grating-based high-resolution D-shaped PCF-SPR biosensor was presented in 2018 by Lu et al. In the RI range of 1.36–1.38, a maximum resolution of 5.98106 RIU and a spectral sensitivity of 3340 nm/RIU were reported (Lu et al. 2018). Liu et al. (2018) discovered a symmetrical dual D-shaped structure.

By injecting the PCF core with an analyte, the fiber optic SPR sensor can be made to work. Also, there is a small layer of gold that serves as the plasmonic material, allowing for an ultra-high sensitivity to be achieved by interactions between fiber and plasmonic modes. Several research studies have evaluated the feasibility of fabricating D-shaped metal-coated PCF using straightforward techniques such as chemical vapor deposition (CVD) (Wu et al. 2017; Wang et al. 2017). When fiber optics and plasmonics are coupled, the resulting structure is likely to exhibit both fiber and plasmonic modes. It's possible that different excitation wavelengths will result in varied amounts of energy loss as the fundamental fiber mode and the plasmonic mode interact. Resonance occurs when the plasmonic mode's effective mode index (or n_{eff}) matches the fundamental modes. Changes in the analyte's core refractive index can produce distinct resonance frequencies, allowing the detection of previously unidentified compounds.

We present a simple, miniaturized fiber optic-based SPR sensor using a PCF with a quasi-D-shaped hollow core and analyte to address issues. The setup's bulkiness, manufacturing challenges, and RI detection restrictions also included is a small layer of gold that serves as the plasmonic material, allowing for ultra-high sensitivity to be achieved by interplay between fiber and plasmonic modes (Fig. 6.5).

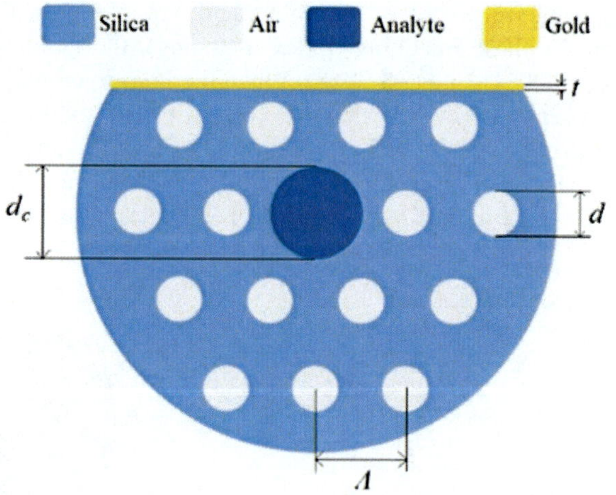

Fig. 6.5 Air hole diameter (d), analyte-core diameter (d_c), hole pitch, and gold layer thickness (t) of the proposed quasi-D-shaped photonic crystal fiber plasmonic biosensor. Adapted from Esfahani Monfared (2020b)

6.2.3 Experimental Setup of D-shape Plasmonic Fiber Optic Sensor

A silver nano-film-based SPR-based optical fiber biosensor with a side-polished refractive index was theoretically, mathematically, and empirically proposed by Zainuddin et al. (2019). An example of a wheel polishing schematic diagram is shown in Fig. 6.14a, which shows the fabrication of the side-polished optical fiber device. Two centimeters of the SMF-28 optical fibers from Corning coating were removed in order to reveal the fiber's glass core and cladding, which were then polished (Monfared et al. 2021). In order to lower the fiber's thickness, a sandpaper grinding wheel was used to control the wheel's rotational speed using a computer program. On top of the fiber core, a 40 nm silver nanofilm was deposited on the D-shaped fiber probe (Monfared et al. 2021). The experimental setup used in this investigation includes a spectrometer for light detection, a side-polished optical fiber, and a broadband light source with a wavelength range of 400–1000 nm. The sensitivity to water ($n = 1.333$) and alcohol ($n = 1.345$), nm/RIU of 2166 and 208.333, were found, respectively, using the previously described side-polished fiber designs. Figure 6.6a illustrates that it is first necessary to remove 2 cm of coating from a Corning SMF-28 optical fiber to reveal the core and cladding of the fiber. As shown in Fig. 6.15a, Grinding wheels with sandpaper on them were used to lower the thickness of the fibers by varying the wheel's rotational speed using an on-board computer program. In the end, a 40 nm silver nanofilm was applied to the D-shaped fiber probe (Zainuddin et al. 2019). An optical fiber with a side-polished surface was used in their study, as well as a spectrometer, as shown in Fig. 6.15b, to take a reading of the light (Zainuddin et al. 2019). They found distilled water has a sensitivity of 2166 nm/RIU and a sensitivity of 208.333 nm/RIU ($n = 1.333$) for alcohol ($n = 1.345$) using the previously mentioned side-polished fiber designs.

It was also proposed numerically by Melo and colleagues (2018) in the form of plastic and graphene/silver-coated D-shaped fiber optic biosensors with peak sensitivity of 5161 nm/RIU. For the biosensor, they looked at whether the polishing depth and sensing area length had an impact. It turns out, based on these findings, that polishing depth is more important than graphene layer count for obtaining narrower curves for normalized transmitted power, and smaller sensing is a length (Melo et al. 2018).

6.3 Metamaterials-Based D-type of Fiber Optics SPR Sensor

The new era of research for high-sensitivity optical sensing has undergone a revolution in the last decade owing to how surface plasmons and light interact resonantly in metal sheets. It is now possible to create optical metamaterials and other innovative artificial materials, leading to improved fabrication techniques for metal-dielectric

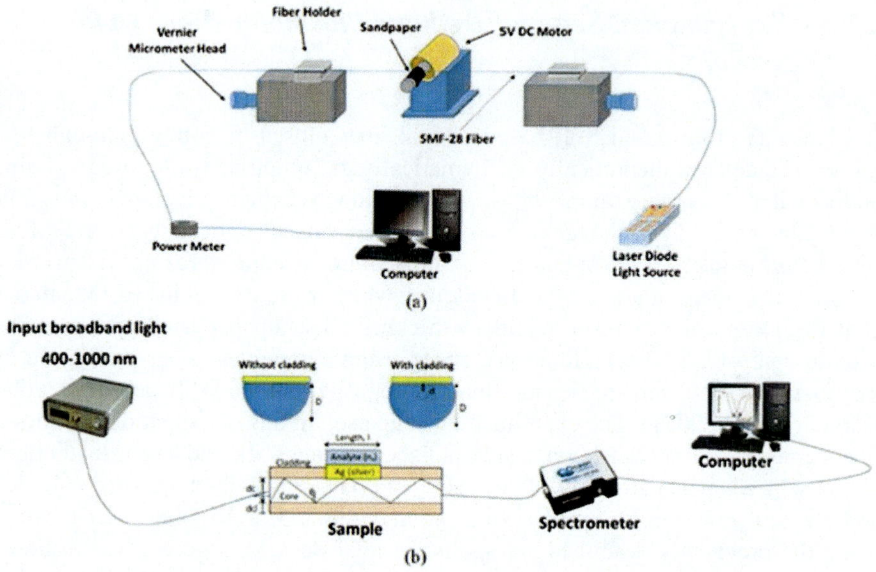

Fig. 6.6 **a** Schematic of the side-polished optical fiber fabrication process; **b** D shaped fiber optic sensor with plasmonics, experimental setup with broadband light source and spectrometer. Adapted from Esfahani Monfared (2020b)

nanostructures. New approaches to analyzing the interaction of light with materials, particularly the interaction of light with surface plasmons, can be found in these materials. Optical metamaterials have recently been investigated in more modern SPR sensor development, which incorporates many materials. The optical features of these materials, which cannot be found in naturally occurring ones, are derived via the combination of two or more different materials, resulting in a new material with unique optical properties (Chen et al. 2005). Their constituent materials' optical characteristics, as well as their relative quantity and the shape of the nanostructures employed to assemble them, all play a role in these qualities. For example, metamaterials with nearly customized optical properties can be developed using the large latitude available during manufacturing to improve sensor performance (for example, loss, light wavelength range detection, and refractive index range detection) above traditional SPR sensors (Leite et al. 2014). A D-type PCF fiber is combined with an Al_2O_3–Ag metamaterial in order to create a sensor based on SPR. To determine the optical properties of metamaterials, sensor analysis focuses on the fraction of each component and layer thickness. Because of the time and money that can be saved by using a computer to simulate different sensor ideas and combinations, only the ones that work best can be made. Sensor analysis focuses on the proportion of each component and layer thickness to identify the optical properties of metamaterials. Because of the efficiency gained by utilizing a computer to simulate numerous sensor concepts and combinations, constructing only the ones that perform best can save a significant amount of time and money.

To improve the behavior of optical fiber sensors, several optical fiber topologies have been investigated, including straight strip cladding (Cennamo et al. 2011), D-shaped cladding (Wang et al. 2020a), tapered cladding (Cennamo et al. 2014), and U-bent fiber cladding (Zhang et al. 2017a). The D-POF has a number of advantages, including easy evanescent field access, molecular fixing assistance, hyper-machinability, and low cost (Zeni et al. 2018; Liu et al. 2019). Strengthening the materials used as sensing layers, as well as selecting distinct fiber structure shapes, is another way to create a stronger SPR signal. In recent years, as sensing layers, metals (such as Au and Ag) and composite metal fabrication have been proposed (Zhang et al. 2017a, b; Xi et al. 2020). A sensitivity of 1227 nm/RIU for a G/Au/D-POF SPR sensor was proposed by Liu et al. (2019). However, with the rapid advancement of biological detection, the SPP stimulated by the pure metal structure is unable to meet the criteria for detection accuracy. As a result, the new sensing layer structure must be further developed and investigated.

A plasmonic sensor made of silver and titanium dioxide (Ag/TiO_2) bilayers is formed on an SPF covered with HMM. SPF and HMM benefits are combined in our design, resulting in a high-performance sensor with a small all-fiber form factor. With the use of the sensor, the sensor is more sensitive in the long wavelength range due to the effective medium theory (EMT) and resonant coupling situations. The amount of metal filling and the number of bilayers were also optimized using a finite element approach. The best settings were picked when FOM $= 0.7$ and $N_{bi} = 3$, according to the maximum FOM. The greatest average S and FOM values were 5114.3 nm/RIU and 182.0 RIU^{-1}, respectively, which may be reached with the optimized parameters in the range of 1.33–1.40 RIU with the optimum parameters. The enhancement in performance is examined in detail from both the dispersion relationship and the increased electric field viewpoints. In this way, the fiber plasmonic sensor could be made with a layered HMM as its foundation. This way, it could be made with high-performance fiber plasmonic sensors with dispersion that can be easily tuned.

Experimental demonstrations of optical metamaterials synthesized artificially with better magnetic characteristics were possible with micro/nano processing technologies (Poddubny et al. 2013). When it comes to light scatterers and emitters, hyperbolic metamaterials (HMMs) have quickly taken on a central role because of their capacity to access and alter the near field, which is directly derived from the activation of coupled surface plasmons (CPS) in HMMs (Maier and Atwater 2005). HMMs are unusual artificial electromagnetic metamaterials with hyperbolic dispersion and anisotropic permittivity in bulk 3D sub-wavelength structures. HMMs have recently been used to achieve exceptional sensitivity (Hu et al. 2020; Santos et al. 2017). In 2009, Kabashin et al. proposed the use of a prism-coupling technique to create plasmonic nanorod metamaterials for use in highly sensitive plasmonic biosensors sensitive to RI changes of over 30,000 nm/RIU (Kabashin et al. 2009). Miniaturized plasmonic biosensors using hyperbolic metamaterials of three-dimensional multilayer metals ($AuAl_2O_3$) and unusual 2D gold diffraction gratings were created in 2016 by Sreekanth et al. (2016a). These sensors can detect biomarkers with an unprecedented level of precision. Angular scans of 7000° per RIU were also obtained in the wavelength range of the near-infrared spectrum (Sreekanth et al.

2016b). However, the sophisticated configurations of the system-state biosensors have several drawbacks, such as the inability to address high prices and the difficulty of miniaturizing.

A 3D Au/Al_2O_3 HMM, graphene sheet, and D shape plastic optical fiber are used in this design. An SPR optical fiber biosensor has been developed. Numerical simulation and experimental demonstration were used to evaluate the system's performance (Li et al. 2021). For example, the multilayer HMM was found to significantly improve the appliance's sensitivity over the standard fiber optic SPR sensor used. Graphene has been demonstrated to enhance the sensitivity, chemical affinity, and stability of a biosensor in three different ways (Li et al. 2020; Xu et al. 2017; Sun et al. 2019b). Resonance location can be successfully tuned between visible and near-infrared wavelengths using D-type fibers and HMM structures. With a simple fabrication and inexpensive cost, the G/HMM/D-POF sensor suggested in this study has great sensitivity (4461 nm/RIU), stability, linearity, and repeatability. Additionally, the SPR biosensor was successfully used to evaluate DNA hybridization kinetics that were time- and concentration-dependent, demonstrating outstanding performance in the field of biological detection.

Hyperbolic metamaterial (HMM) is described as a type of artificial material that displays hyperbolic dispersion due to one of its primary permittivity components (Poddubny et al. 2013). The other two have the opposite sign. Recent studies have shown that plasmonic sensors with HMMs are extremely sensitive (Hu et al. 2020; Sreekanth et al. 2016a; Yang et al. 2021). Our research has focused on the type II HMM, which is a multilayer of metal and dielectric bilayers that may transmit and propagate surface waves along a metal surface. One of the qualities of the BPP wave is its ability to propagate through the majority of the material while retaining its characteristics as an external propagating wave (Liu et al. 2021). Experiments have shown that BPPs are particularly sensitive to changes in the evanescent field dielectric constant. By adding graphene to the sensors, it is possible to further improve their performance. According to the following advantages, graphene is worth considering: Graphene's atomic-scale thickness has little effect on sensing performance, but the carbon atoms arranged in a honeycomb pattern interact with aromatic rings to generate a stacking interaction, which aids in the fixation of biomolecules (Song et al. 2010). In addition, the SPP electric field can be strengthened by a graphene layer, which promotes interactions between biomolecules and the evanescent field (Zhang et al. 2017b). The sensitivity of 4461 nm/RIU was achieved by Li et al. by combining HMM and several graphene layers (Li et al. 2021). According to Zhang et al. (2017b), SPP intensity can be increased by roughly 30.2% in monolayer graphene, while the electron energy loss in numerous graphene layers reduces the SPP intensity. However, there are still certain issues with graphene preparation on D-type optical fiber. Two examples of this are: first, the monolayer of graphene that can be made by chemical means is not guaranteed to be monolayer; and second, traditional wet-transfer graphene can't remove all of the polymethyl methacrylate molecules from the graphene. The most efficient way to transport monolayer graphene to optical fiber must therefore be proposed. A highly sensitive G/HMM/D-POF SPR sensor is created by combining monolayer graphene (Au/Al_2O_3) composite HMM with D-POF. PMMA residue causes transmission loss while simultaneously strengthening the electric field of SPP. The method of salvaging monolayer graphene and transferring it to an optical fiber via a thin water coating on glass is proposed. According to a study published in ACS Nano, the G/HMM/D-POF sensor has been found to offer various advantages, including low cost, simple manufacturing, high sensitivity, and linear response. Highly sensitive D-POF SPR sensors will help find chemical and biological molecules (Gao et al. 2021). A silver and titanium dioxide (Ag/TiO_2) bilayer is used to form a plasmonic sensor on an SPF coated with HMM. SPF and HMM benefits are combined in our design, resulting in a high-performance sensor with a small all-fiber form factor. When the effective medium theory (EMT) and resonant coupling conditions are utilized, the sensor becomes more sensitive in the long wavelength region. Its metal filling percentage and the number of bilayers were also optimized using a finite element approach. The best settings were picked when $N_{bi} = 0.7$ and $N_{bi} = 3$, according to the maximum FOM. The maximum average value was 5114.3 nm/RIU, with a FOM of 182.0 RIU^{-1} that may be reached with the optimized parameters in the range of 1.33–1.40 RIU with the optimum parameters (Hu et al. 2020). The improvement in performance is examined in detail from both the dispersion relationship and the electrical field enhancement viewpoints. In this way, the fiber plasmonic sensor could be made with a layered HMM as its foundation. This way, it could be made with high-performance fiber plasmonic sensors with dispersion that can be easily tuned (Hu et al. 2020).

6.3.1 A Metamaterial-Based Optical Fiber Sensor's Design and Implementation

6.3.1.1 Design of a 3D Fiber Sensor with the Use of Metamaterial Film

Consider how a refractive index sensor might be constructed based on the PCF fiber shown in Fig. 6.16 with a D-type profile. An array of dielectric structures (matching the holes in the PCF fiber) surrounds the glass core, giving the fiber its refractive index of n_g. The Sellmeier equation is used to compute the refractive index of a material. An external refractive index and an analytic medium are assumed to fill the fiber's outer space. Metamaterial layer thickness (d_m), hole diameter (d_{hole}), and hole spacing (K) are all represented by their respective letters. The gap between the fiber and the metamaterial layer is referred to as "residual cladding." In contrast to PCF, the D-type profile of the fiber encourages core modes to interact with their surroundings. The metal layer that supports surface plasmons enhances the sensor's sensitivity by mediating the electromagnetic modes' interactions with external media. A metamaterial coating of silver and alumina (sub-wavelength) is used to substitute the metal in this study (see the inset of Fig. 6.7). The thickness of the silver and alumina layers

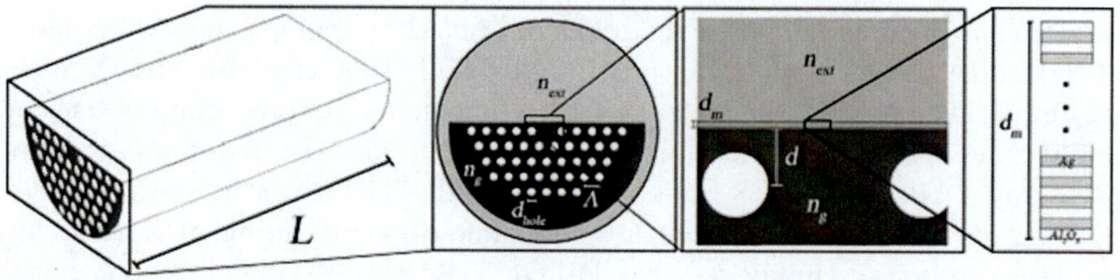

Fig. 6.7 This is a cross-section of the fiber sensor in three dimensions. The fiber sensor surface structure shows the metamaterial film's location, and an illustration of the thin metal-dielectric layers that make up this metamaterial film is shown from left to right in the images. Reprinted with permission from Optical Fiber Technology. Copyright, 2017, Elsevier (Santos et al. 2017)

affects the metamaterial's optical characteristics. In order to compute the metamaterial film's effective refractive index, one can utilize an effective medium technique, which relies on the Bruggeman Effective Medium Theory and the Maxwell–Garnett Theory (Cai and Shalaev 2010).

Nanostructures of Al_2O_3–Ag, such as silver nanowires or nanorods embedded in an alumina matrix, can also be employed. Using the same homogenization model, the form factor g can be substituted with the appropriate value based on the nanostructure geometry in these circumstances as well. Our focus is on the simplest form to build because it has similar general optical properties to those found in other metamaterials (Leite et al. 2014). Its metal and dielectric components combine to form the metamaterial complex dielectric. It is possible to think of the metamaterial layer as an artificial metal layer that can support plasmonic modes that work well in real life.

6.3.2 *Hyperbolic Metamaterial Used in Side-Polished Few-Mode Fiber Sensor*

Using side-polished few-mode fiber and HMM layers, the proposed sensor is built (HMM-SP-FMF). The RFT of the few-mode fiber is 72 nm. Multiple periodic bilayers of silver and titanium dioxide are used to coat the HMM, with a thickness of 30 nm for each bilayer. The commercial few-mode step-index fiber (FMF, OFS) has a core/cladding diameter of 19–125 nm, which is consistent with the parameters used in the simulation. Figure 6.8a and b show the HMM-SP-FMF sensor schematic diagram and cross-sectional view, respectively. To illustrate, the *x*-axis is perpendicular to the fiber direction in Fig. 6.8a, where the *y*- and *z*-axes are perpendicular to the fiber direction. Ignoring optical fiber dispersion, the cladding and core refractive indices are 1.449 and 1.444, respectively. The EMT can be used to demonstrate the hyperbolic properties of multilayer metal-dielectric films. In this case, the HMM equivalent dielectric tensor can be judged using EMT criteria because the Ag/TiO_2

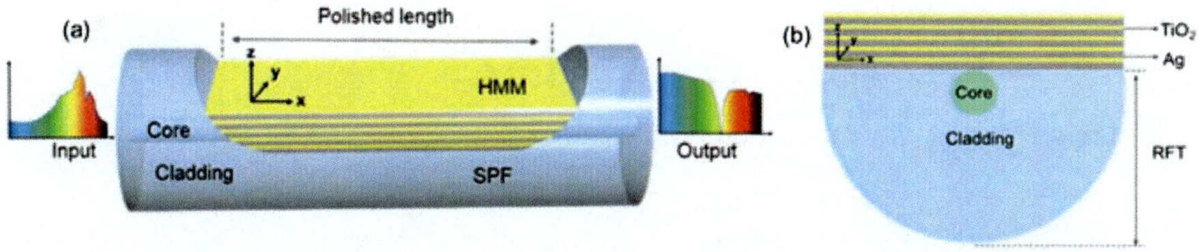

Fig. 6.8 **a**, **b** Show a schematic diagram and a cross-sectional view of the proposed side-polished fiber-based plasmonic sensor coated with a layered hyperbolic metamaterial (HMM-SP-FMF). SPF is an abbreviation for side-polished fiber; HMM is an abbreviation for hyperbolic metamaterial; and RFT is an abbreviation for residual fiber thickness. Reprinted with permission from Optics Express. Copyright, 2020, Optica (Hu et al. 2020)

bilayer is only 30 nm thick, which is less than a few hundred nanometers (Cortes et al. 2012).

6.3.3 Design of a D-shaped Plastic Optical Fiber Sensor Based on Graphene and Hyperbolic Metamaterial

The 3D Au/Al_2O_3 composite HMM, a graphene film, and a D-shaped plastic optical fiber (D-POF) (G/HMM/D-POF). In Fig. 6.9, the synthesis of G/HMM/D-POF is depicted. Sequential gold and Al_2O_3 layer deposition resulted in a series of composite HMM where the Al_2O_3 layers split the gold film into a number of distinct layers (n) Au/Al_2O_3, with n ranging from 2 to 5. The total gold layer thickness in the HMM was selected at 50 nm since the functional layer thickness in the traditional single-layer gold SPR sensor ranges between 30 and 80 nm, with an optimal thickness of 50 nm (Neff et al. 2006; Tabassum and Gupta 2016; Tadepalli et al. 2015). The functional layer thickness was reduced to less than 80 nm. There were two stages to this process: First, gold pellets were heated and evaporated at a rate of $0.7\,\text{Å}\,s^{-1}$ ($1\,\text{Å} = 0.1$ nm), and then the gold layer was applied to the top of the POF. A 6 nm thick layer of Al_2O_3 was applied to the gold film by the oxidation of aluminum and other processes (Jeurgens et al. 2002). The gold layer on top of the HMM was used to increase the surface plasmon resonance effect of graphene. Metal-to-dielectric thickness per unit period is crucial in HMM. The best candidate (HMM/POF) for graphene development and biological alteration was chosen from the list of structures. An SPR sensor based on a 50 nm thick gold film (Au/D-POF) was constructed as a comparison.

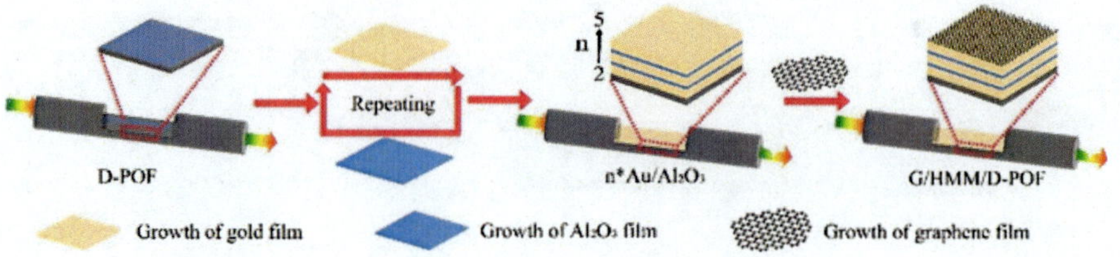

Fig. 6.9 Preparation process: graphene, hyperbolic metamaterial, D-shaped plastic optical fiber sensor. Reprinted with permission from Photonics Research. Copyright, 2021, Optica (Li et al. 2021)

6.3.4 Metamaterial Based D-shape Plastic Optical Fiber Sensor for DNA Detection

To alter the G/HMM/D-POF for DNA detection, the sensor was immersed in a 1 M (1 M = 1 mol/L) PBASE solution for 4 h (Xu et al. 2017; Sun et al. 2019a). Pyrene and graphene might form an aptamer conjugate containing an aptamer probe with an NH2 group if PBASE was used as the scaffold. Triple washing with DMSO and deionized water was used to eliminate the unmodified PBASE from the sample. It was then fixed by placing the sensor into the solution (1 M) for four hours on the G/HMM/D-POF surface. Once the unreacted aptamer had been removed, the probe was washed with a PBS solution and deionized water. Finally, it is possible to detect DNA hybridization using a modified G/HMM/D-POF sensor. Various amounts of tDNA and misDNA were used to test the sensor described in this work. Before each detection, the sensor was washed using a PBS solution. Pure PBS buffer (0.01 mL) was used to extract target DNAs from the relevant solution for the DNA dissociation kinetics. After a 60-s immersion in 10 mm of aqueous sodium hydroxide solution and washing with PBS solution, the dissociation was completed. In addition, the transmission spectrum was used to track the complete reaction in real time.

6.3.5 Experimental Setup of Graphene and Hyperbolic Metamaterial-Based D-shape Plastic Optical Fiber Sensor

Figure 6.10 depicts an experimental setup for evaluating the performance of a custom-built G/HMM/D-POF sensor. The sensor was mounted in a reaction cell made of polyethylene (PE) with a diameter of 5 mm and a length of 2 cm, specifically for the detection of probe solutions. As an excitation light source, an Ocean Optics HL-2000 tungsten lamp producing light with a wavelength range of 360–2000 nm was used to record the SPR peak shifts. This was done with a Zeiss Gemini Ultra-55 scanning electron microscope. It was used to look at the surface morphology of 3D nanostructures.

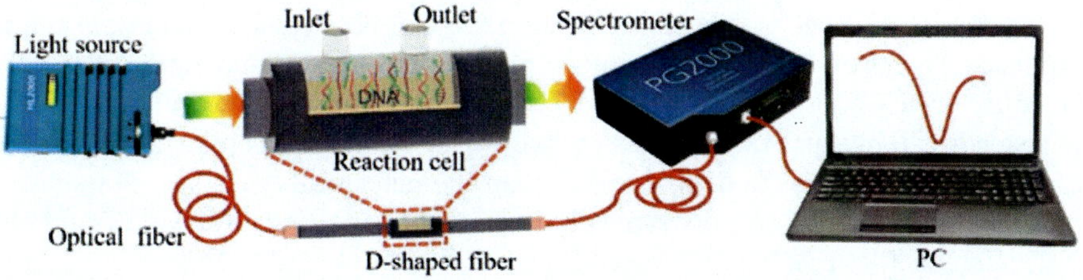

Fig. 6.10 Experimental setup of the G/HMM/D-POF sensor. Reprinted with permission from Photonics Research. Copyright, 2021, Optica (Li et al. 2021)

6.4 Graphene Injected into the D Type of Fiber Optics SPR Sensor

Numerous SPR sensors based on graphene have been invented in recent years. The carbon atoms in graphene are organized into a honeycomb-like lattice in a two-dimensional plane. It possesses optical conduction from visible to infrared wavelengths due to its interband transition. In addition, the material's high surface-to-volume ratio and wide band optical and plasmonic capabilities make it a good candidate for usage as a coating substance that can be used to coat existing plasmonic devices (Wu et al. 2013). Biomolecules' adsorption in a large region is also possible due to the high surface-to-volume ratio. A dense cloud of graphene atom orbitals fills the gap between its atomic rings, making it impossible to see through. Comparing graphene-coated SPR sensors coated in gold and silver, we found a wider SPR curve in the latter group (Wu et al. 2010; Byun et al. 2011). It is possible to employ the Kretschmann configuration of graphene on silver for distant sensing and refractive index monitoring, but the bulky structure prevents this. There have been a number of fiber-based SPR sensors reported in order to achieve miniaturization (SPR Based Fiber Optic Refractive Index Sensor | Request PDF; Perrotton et al. 2011). D-shaped fiber-based SPR sensors have also been proposed to monitor light from the core to the plasmonic modes efficiently. Traditional D-shaped fiber sensors require the removal of cladding, which results in significant light loss due to the scattering effect. Furthermore, doping the core is necessary to alter the refractive index contrast. However, the resonance wavelength cannot be easily modified to the desired frequency range. Conventional fibers have been replaced with photonic crystal fibers (PCF) for SPR-based sensors. There are several advantages to using this technology for SPR-based sensors, such as adjustable birefringence, configurable geometric parameters, and low-temperature dependence (Chen et al. 2010; Akowuah et al. 2012; Yu et al. 2010; Villatoro et al. 2009; Dash and Jha 2014a, b). Changing the geometrical structure and characteristics can also change the core-guided mode effective index in order to phase match the plasmonic and core-guided modes, which is critical for SPR. In the majority of PCF-based SPR sensors, the analyte is injected into the fiber. To be sure, the components in the PCF's holes can't be easily removed or swapped out. RI sensing has been proposed using a D-shaped PCF-based SPR sensor to solve these challenges (Tian et al. 2012). The normal PCF can be etched or side-polished to produce a D-shaped PCF. Evanescent coupling between core-guided modes and the cladding occurs after polishing. Sensing requires a strong evanescent wave in the D-shaped PCF because of the holes' proximity to the optical core. The flat surface of PCF makes it easier to deposit metal than the inner walls of holes. For sensing purposes, D-PCF-SPR sensors typically have a gold or silver covering (Tian et al. 2012; Tan et al. 2013). A gold-coated hollow-core PCF-based SPR sensor has been disclosed by Tan et al. (2013). Gold exhibits substantial resonance frequency shifts and is chemically stable. As a result, the performance degrades due to the increased absorption coefficient and the resulting wider resonance curve. Silver is susceptible to oxidation, which decreases detection accuracy and results in protein denaturalization and loss of binding activity when a biomolecule is directly adsorbed on a metal surface. D-shaped PCF-SPR sensors coated with graphene are proposed as a solution to the aforementioned challenges. From a manufacturing standpoint, the proposed design is basic and straightforward. When compared to other carbon-based materials, graphene has a lower cost and is easier to make. Graphene may be separated and placed on a silver surface in one or more layers using a variety of methods. Furthermore, the van der Waals force acting between graphene layers allows for precise control over the number of layers that can be formed.

6.4.1 Geometrical Design Variations for Graphene-Based Sensors

6.4.1.1 D-shaped PCF Sensor

The structure of the D-shaped PCF-SPR biosensor is depicted in Fig. 6.11a and b, respectively, in three-dimensional view and cross-section view. Etching or side polishing techniques can be utilized to create the D-shaped flat surface that is desired. h is the length of the PCF that is shaped like a D. The distance between neighboring air holes in the oval configuration of three layers of air holes is shown by the symbol. The diameter of the two center holes on the first layer is denoted by the symbol d_c. In the process of calibrating the coupling between the basic mode and the SPP mode, they play a vital role. To efficiently limit core energy and achieve the birefringence phenomenon, the diameter of the two center holes on the second layer is d_1. The number d represents the diameter of the air hole in the cladding. The number of layers of graphene is L. This is determined by taking into account the enormous surface area ratio, the conjugation structure, and the high level of compatibility. The thickness of the graphene on a single sheet is 0.34 nm. As a result of silver's more distinct resonance and its more sensitive detecting properties in the SPR phenomenon than gold's, the sensor employs silver rather than gold as the SPR excitation material and gives it a thickness of d_g.

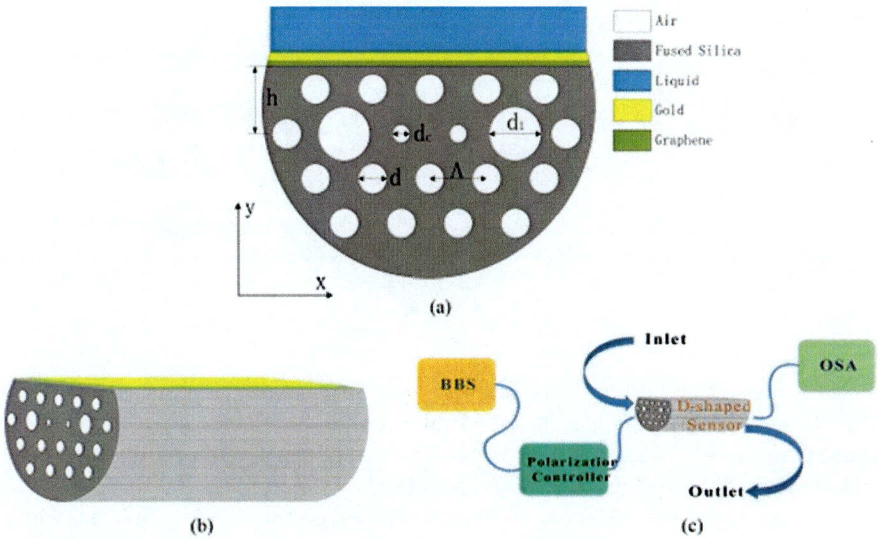

Fig. 6.11 **a** Two-dimensional D-shaped PCF-SPR sensor structure cross-section **b** D-shaped PCF-SPR biosensor 3D model **c** Sensor setup schematic. Reprinted with permission from Optik, Copyright, 2018, Elsevier (Tong et al. 2018)

Figure 6.11c shows a schematic depicting the sensor setup. Adsorption of the determinant onto the horizontal metal surface occurred by simple flow or drop. Light from a wide band light source (BBS) is polarized using a polarization controller before being injected into a D-shaped PCF-SPR biosensor. The final step is for an optical spectrum analyzer (OSA) to pick up the spectrum at the output.

6.4.2 *Multilayer Sensing Mechanism of a PCF-Based SPR Sensor*

An optical fiber of 10 μm in diameter with three rings (air holes) organized in an octagonal lattice was used to build the suggested PCF-based SPR sensor shown in Fig. 6.12. The PCFs' core was 0.5 μm away from the center of the first ring's air openings. There is an octagonal lattice structure in the second and third rings because all of the corner air holes are located at the same distance from each other ($r_2 = 1$ μm, and $r_3 = 1.5$ μm). Figure 6.12 depicts three different diameters for the air holes, which are labeled as d_1–d_3 in the figure and are 0.2 μm, 0.4 μm, and 0.3 μm wide, respectively. To boost the coupling energy between the PCF and SPP modes, which were excited in the graphene layer, tiny air holes with a diameter of d_1 were used. The Sellmeier equation describes the fiber material we employed as fused silica (Fleming 1984).

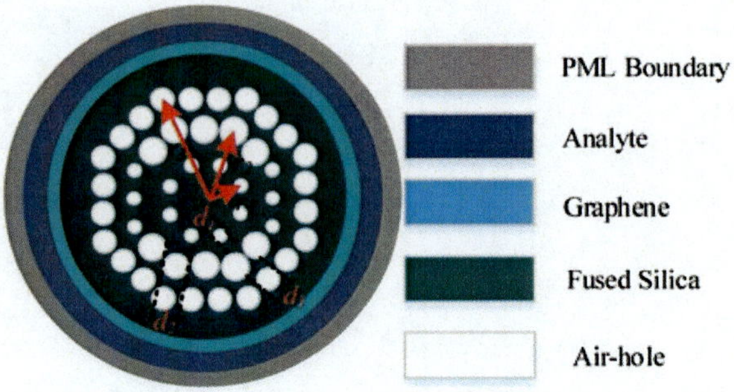

Fig. 6.12 Graphene-coated photonic crystal fiber is used in the multilayer sensing system. Adapted from Paul et al. (2021)

The thickness of plasmonic materials in most SPR-based PCF sensors was stated to be between 30 and 60 nm (Kumar Paul et al. 2018; Rifat et al. 2015; Gomez-Cardona et al. 2020; Saiful Islam et al. 2018). Graphene was used as the single plasmonic material in Fig. 6.12. The formula $t = 0.33$ nm $\times\ t_g$ ($t_g = 1, 2, 3, \ldots$) is used to compute the thickness of structures based on multilayer graphene, where tg is the number of layers (Rifat et al. 2016). We stacked 108, 118, and 128 layers of graphene to get a total thickness of 36.72, 40.12, and 43.52 nm in the range of 30–60 nm when using a multilayer of graphene. In the case of graphene, for example, we first layered 108 layers to a total thickness (t) of 36.72 nm. The graphene layer was covered with a 2 μm thick aqueous analyte layer with a refractive index of n_a. This was used as a medium for sensing information. The numerical simulation was performed using the finite element technique (FEM) and a perfectly matched layer (PML). For improved numerical analysis, a boundary condition known as a PML is used to absorb scattered electromagnetic waves (Saitoh and Koshiba 2005; Reyes-Vera et al. 2018). Because of this, the thickness of the PML is a critical element in the simulation results. Therefore, the simulation with a PML thickness of 0.2 μm and an inner diameter of 4.1 μm got the best results.

6.4.3 Experimental Setup for Liquid Detection Using D-shaped SPR Sensor

As a result of utilizing the full-vectorial FEM, the PCF is drawn in the COMSOL Multiphysics software, where more accurate mode field analysis, electromagnetic wave, and frequency domain physics field findings are generated by finer grid calculation. Figure 6.13 depicts the sensor cross-section in two dimensions, the cross-section of a stack simulation, and the experimental device. All of the structure may be broken down into two sections. The initial section is comprised of blowholes and fused quartz, which serve as a backdrop. Each layer is constructed of six adjacent pores that are each rotated by a factor of 20°, 40°, and 60°, with the coordinate origin in the middle, it has a 10 μm diameter. The first layer's four holes all have the same diameter, $x_1 = 0.5$ μm. As a result of the significant birefringence effect caused by

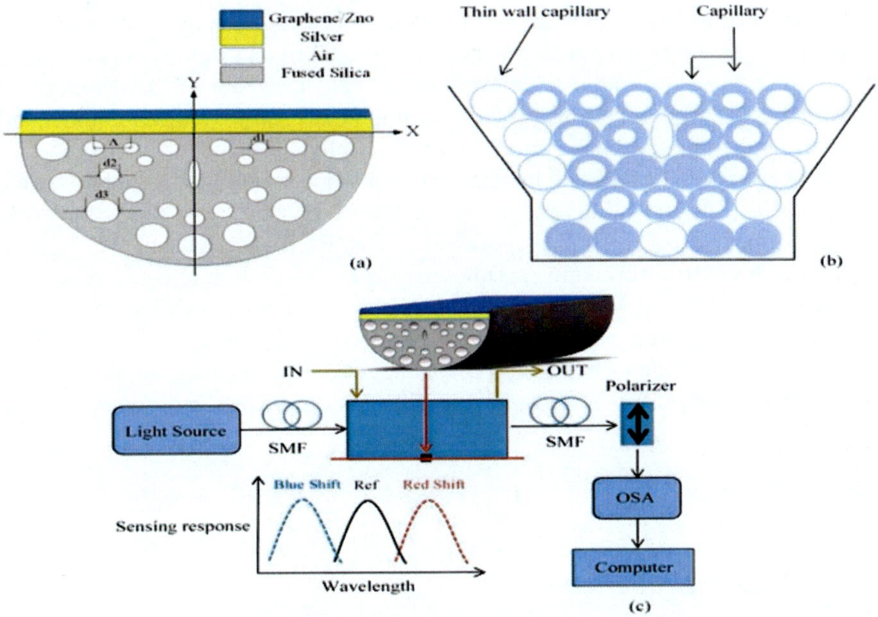

Fig. 6.13 **a** Sensor cross-section, **b** layered rough machining cross-section, **c** the refractive index of liquid analytes is detected using an SPR sensor setup. Adapted from Liang et al. (2021)

the core elliptical pores and the first layer's disappearance on both sides, *y*-polarized light confinement loss and sensitivity are far greater than those of *x*-polarized light. The evanescent wave's interaction with the SPP (surface plasmon polariton) mode is controlled by the two pores above the core and the remaining six pores in the same row. The sizes of $x_2 = 0.7$ μm and $x_3 = 1$ μm, respectively, are the sizes of the blowholes in the second and third layers, which are used for light propagation.

The lattice spacing is given by 2 μm. The SPR excitation layer is the second component. A 30 nm thick layer of silver is employed in SPR as an excitation material. It is covered with graphene or zinc oxide in order to improve the sensing performance and avoid silver oxidation. There are a total of L graphene layers, and the thickness of the graphene is 0.34 nm. The typical stack-stretch method can be used to make the proposed PCF. Quartz sleeves are first pre-treated. Ultra-clean conditions allow for the production of the capillaries, which are subsequently tapered with a hydrogen–oxygen flame in order to seal the holes. Filling the gaps between the capillaries in the quartz sleeve is done with pure quartz rods, according to the design specifications. To create the PCF, a wire drawing process is used on a drawing tower to fuse the quartz and the capillary tube in an oxyacetylene flame. It is then smoothed out, as illustrated in Fig. 6.13b, by etching or side polishing to achieve the desired finish. Wheel polishing and V-groove polishing have gotten more sophisticated in terms of polishing technology. Sputtering, chemical vapor deposition, and high-pressure microfluidic chemical deposition are all options for depositing metal on

PCF's polished surface (An et al. 2014; Sazio et al. 2006; Huang et al. 2020). As described by Kiraly et al. (2013), graphene was produced on single-crystal Ag (111) at ultra-high pressure. Large-scale synthesis of graphene thin films was reported by Kim et al. (2009), who also provided two methods for preparing the films and transferring them to any substrate. As with physically cut graphene, the number of graphene layers can be adjusted by chemical vapor deposition. According to Tiwale et al. (2019), ZnO thin films can be made from new zinc decanoate. Sputtering with a DC magnetron, thermal evaporation, and an RF magnetron can also be used to produce uniform ZnO coatings (Dinovitser et al. 2019). Figure 6.13c shows the schematic design of the RI sensing system. Single-mode fibers (SMFs) and D-type PCFs can be coupled using a wide band light source. It allows analytes to enter and exit through the flow channel. To use an optical spectrum analyzer, connect the D-type PCF to the polarizer via an SMF (OSA). Using an OSA and a computer, we can gather and examine the transmission spectra.

6.4.4 *Analysis of the Silver-Graphene-Coated PCF-SPR Sensor Numerically*

SPP mode may be affected by the thickness of the material layer, resulting in a change in the sensor's performance. With the parameters of $x_1 = 0.5$ μm, $x_2 = 0.7$ μm, and $x_3 = 1$ μm of graphene, the loss spectrum is presented in Fig. 6.14 with the parameters RI = 1.38 and t_s = 30 nm. The resonance peak shifts to a shorter wavelength as the thickness of the graphene layer increases, with a wavelength variation of 72 nm and a loss change of 350.88 dB/cm. More energy is transferred from core mode to SPP mode when graphene is thicker, but this also increases the loss of the resonance curve. After a certain rise in graphene thickness, its properties begin to resemble those of graphite. Coupler efficiency between the core and SPP modes is degraded across a particular wavelength range.

Despite the high sensitivity of the sensor without the graphene layer, the oxidation of silver makes it difficult to utilize efficiently. Sensor sensitivity may be affected by the thickness of the plasmonic substance. The loss spectrum of the silver layer thickness from 30 to 45 nm was examined with all parameters held constant, as illustrated in Fig. 6.15. When the silver layer thickness (t_s) grows from 30 to 45 nm, the resonance wavelength shifts to the short wavelength direction, changing from 1852 to 1600 nm, and the loss peak shifts from 1327.71 to 564.25 dB/cm. There is more damping loss when the silver coating is excessively thick, which diminishes the SPR effect and eventually reduces the loss peak. To investigate the influence of analyte RI on sensor performance, the silver layer thickness is set to 30 nm.

The real fraction of the effective refractive index of the SPP mode changes as the analyte RI changes, resulting in a shift in phase matching wavelength between the core and SPP modes. The sensing capability of the PCF sensor is investigated using a liquid analyte RI range of 1.37–1.41. Figure 6.16 shows how the loss spectra

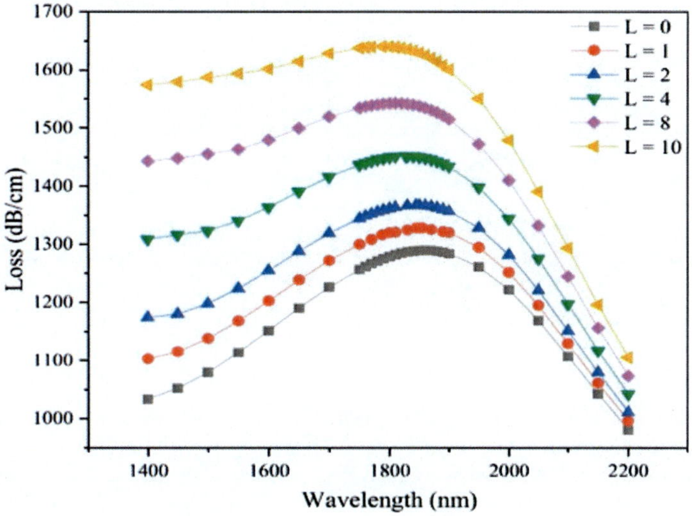

Fig. 6.14 Simulated loss spectra when the number of graphene layers (L) is 0, 1, 2, 4, 8, and 10. ($x_1 = 0.5$ μm, $x_2 = 0.7$ μm, $x_3 = 1$ μm, $\Lambda = 2$ μm, $t_s = 30$ nm, RI = 1.38). Adapted from Liang et al. (2021)

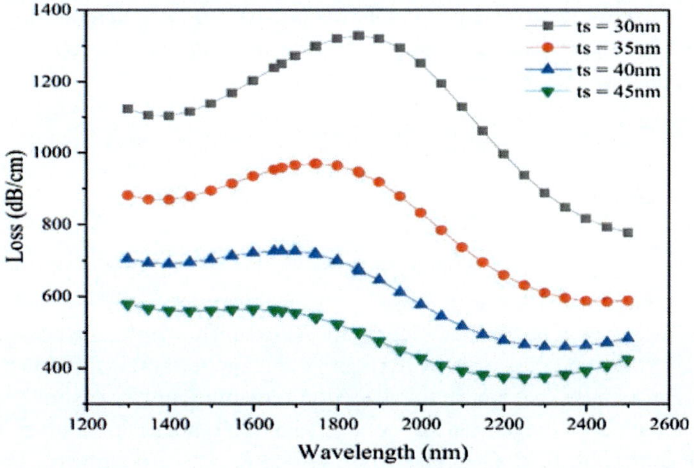

Fig. 6.15 Simulation of the loss spectrum of silver layer thickness as a function of wavelength. ($L = 1$, RI = 1.38, $x_1 = 0.5$ μm, $x_2 = 0.7$ μm, $x_3 = 1$ μm, $\Lambda = 2$ μm, $L = 1$, RI = 1.38). Adapted from Liang et al. (2021)

vary as the analyte RI changes from 1.37 to 1.41, with $x_1 = 0.5$ μm, $x_2 = 0.7$ μm, $x_3 = 1$ μm, $\Lambda = 2$ μm, $t_s = 30$ nm, and $L = 1$. The resonance wavelength shifts from 1812.5 to 1966 nm as the refractive index of the liquid analyte rises, increasing the contact between metal and analyte. The basic mode of the SPP mode exchanges

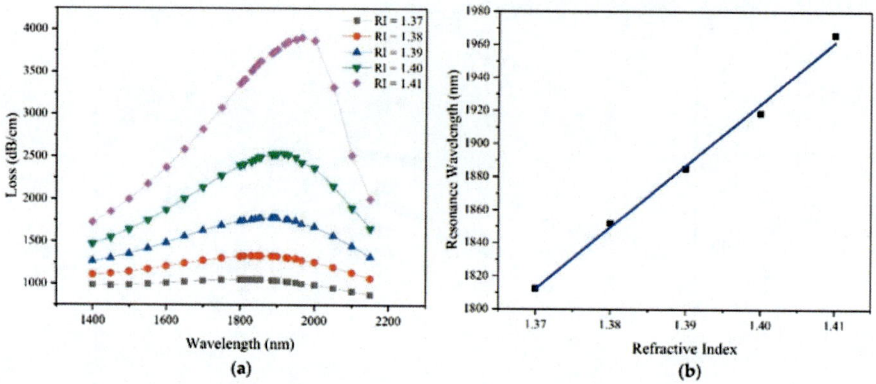

Fig. 6.16 **a** Simulated loss spectra when analyte RI is 1.37–1.41; **b** linear fitting between resonance wavelength and analyte RI. ($x_1 = 0.5$ μm, $x_2 = 0.7$ μm, $x_3 = 1$ μm, $\Lambda = 2$ μm, $t_s = 30$ nm, $L = 1$). Adapted from Liang et al. (2021)

more energy, resulting in increased sensitivity. The loss shift is 2854.87 dB/cm, and the wavelength shift is 153.5 nm. Figure 6.16 displays the resonance wavelength as a function of the liquid analyte's refractive index. When the analyte RI shifts from 1.40 to 1.41, the maximum wavelength sensitivity of sliver-graphene PCF is 4750 nm/RIU and the maximum wavelength resolution is 2.105 105 RIU. The sensor's average sensitivity is 3735 nm/RIU, and the goodness of fit parameter R2 is 0.995, allowing it to detect analytes accurately.

6.5 Summary

Analytical models from the finite element approach are presented in this study on the performance of a D-type photonic crystal fiber used in SPR-based refractive index sensing. Sensitivity enhancements for SPR sensors are made possible by the coupling of SPPs made possible by hyperbolic metamaterial (HMM). HMM and 2D materials can be used to enhance the detection sensitivity of biosensors. A graphene monolayer with a D-shaped plastic optic fiber is a graphene monolayer with a D-shaped plastic optic fiber (a G/HMM/D-POF SPR sensor is shown). Based on a 3D $AuAl_2O_3$ HMM composite structure and a graphene sheet, an optical fiber SPR biosensor was presented. According to a detailed theoretical and experimental investigation, in the visible to near-infrared range, the SPR resonance peak position may be tunable by changing the the number of Au layers in the HMM system. A PCF-based plasmonic sensor with a graphene-coated D shape has been proposed. The sensor's many structural and material parameters have been fine-tuned. To detect biological species such as cells, proteins, and DNA, plasmonic fiber optic biosensors combine the flexibility and compactness of optical fibers with the high sensitivity of nanomaterials. Plasmonic fiber optic biosensors have the potential to revolutionize clinical diagnostics, drug discovery, food process control, illness, and environmental monitoring because of their small size, precision, low cost, and ability to be remotely and distributed sensed. The stacking interaction of graphene not only aids in the adsorption of biomolecules, but also inhibits metals like silver from oxidizing. To model a silver-graphene PCF-SPR refractive index sensor, researchers used the full vector finite element method (FEM). The finite element method is used to do numerical simulations of a surface plasmon resonance (SPR) sensor based on a D-shaped photonic crystal fiber (PCF) to detect changes in the refractive index of liquid analytes (FEM). The performance of the SPR-PCF sensor coated with a graphene layer can be improved by using silver as the plasmonic metal. In order to prevent active plasma material from oxidizing, dielectric materials (such as graphene) are employed (silver). There are several advantages to using the D-shaped PCF-SPR sensor shown here over other types of PCF-SPR sensors, including its ease of manufacture, high sensitivity, low cost, and ability to reuse the sensor.

References

Akowuah EK, Gorman T, Ademgil H, Haxha S, Robinson GK, Oliver JV (2012) Numerical analysis of a photonic crystal fiber for biosensing applications. IEEE J Quantum Electron 48(11):1403–1410. https://doi.org/10.1109/JQE.2012.2213803

An G, Li S, Qin W, Zhang W, Fan Z, Bao Y (2014) High-sensitivity refractive index sensor based on D-shaped photonic crystal fiber with rectangular lattice and nanoscale gold film. Plasmonics 9(6):1355–1360. https://doi.org/10.1007/S11468-014-9749-5

An G, Li S, Hao X, Yan X, Zhang X (2017) D-shaped photonic crystal fiber refractive index sensor based on surface plasmon resonance. Appl Opt 56(24):6988–6992. https://doi.org/10.1364/AO.56.006988

Byun KM, Choi SH, Kim YL (2011) Graphene-on-silver substrates for sensitive surface plasmon resonance imaging biosensors. Opt Express 19(2):458–466. https://doi.org/10.1364/OE.19.000458

Cai W, Shalaev VM (2010) Optical metamaterials. Springer, New York. https://doi.org/10.1007/978-1-4419-1151-3

Cennamo N, Massarotti D, Conte L, Zeni L (2011) Low cost sensors based on SPR in a plastic optical fiber for biosensor implementation. Sensors (Basel) 11(12):11752. https://doi.org/10.3390/S111211752

Cennamo N, D'Agostino G, Pesavento M, Zeni L (2014) High selectivity and sensitivity sensor based on MIP and SPR in tapered plastic optical fibers for the detection of l-nicotine. Sens Actuators B Chem 191:529–536. https://doi.org/10.1016/J.SNB.2013.10.067

Chau YF, Chen MW, Tsai DP (2009) Three-dimensional analysis of surface plasmon resonance modes on a gold nanorod. Appl Opt 48(3):617–622. https://doi.org/10.1364/AO.48.000617

Chau YFC, Chao CTC, Chiang HP, Lim CM, Voo NY, Mahadi AH (2018) Plasmonic effects in composite metal nanostructures for sensing applications. J Nanoparticle Res 20(7):1–13. https://doi.org/10.1007/S11051-018-4293-4

Chau YFC, Chou Chao CT, Lim CM, Huang HJ, Chiang HP (2018) Depolying tunable metal-shell/dielectric core nanorod arrays as the virtually perfect absorber in the near-infrared regime. ACS Omega 3(7):7508–7516. https://doi.org/10.1021/ACSOMEGA.8B00362

Chen S-J, Yeh T-L, Su Y-D (2005) Common-path phase-shift interferometry surface plasmon resonance imaging system. Opt Lett 30(12):1488–1490. https://doi.org/10.1364/OL.30.001488

Chen CH, Tsao TC, Tang JL, Te Wu W (2010) A multi-D-shaped optical fiber for refractive index sensing. Sensors 10(5):4794–4804. https://doi.org/10.3390/S100504794

Chou Chao CT, Chou Chau YF, Chiang HP (2020) Enhancing plasmonic effect in periodic nanometal square prisms with fences and cavities for refractive index and temperature sensing applications. J Nanopart Res 22(9). https://doi.org/10.1007/S11051-020-05023-1

Chou Chau YF et al (2016) Tailoring surface plasmon resonance and dipole cavity plasmon modes of scattering cross section spectra on the single solid-gold/gold-shell nanorod. J Appl Phys 120(9):093110. https://doi.org/10.1063/1.4962175

Chu S, Nakkeeran K, Abobaker AM, Aphale SS, Babu PR, Senthilnathan K (2019) Design and analysis of surface-plasmon-resonance-based photonic quasi-crystal fiber biosensor for high-refractive-index liquid analytes. IEEE J Sel Top Quantum Electron 25(2):1–9. https://doi.org/10.1109/JSTQE.2018.2873481

Cortes CL, Newman W, Molesky S, Jacob Z (2012) Quantum nanophotonics using hyperbolic metamaterials. J Opt 14(6):063001. https://doi.org/10.1088/2040-8978/14/6/063001

Dash JN, Jha R (2014a) Graphene-based birefringent photonic crystal fiber sensor using surface plasmon resonance. IEEE Photonics Technol Lett 26(11):1092–1095. https://doi.org/10.1109/LPT.2014.2315233

Dash JN, Jha R (2014b) SPR biosensor based on polymer PCF coated with conducting metal oxide. IEEE Photonics Technol Lett 26(6):595–598. https://doi.org/10.1109/LPT.2014.2301153

Dinovitser A, Ng BW-H, Abbott D, Sultana J, Islam MS, Islam MR (2019) Exposed-core localized surface plasmon resonance biosensor. JOSA B 36(8):2306–2311. https://doi.org/10.1364/JOSAB.36.002306

Dong J et al (2019) Side-polished few-mode fiber based surface plasmon resonance biosensor. Opt Express 27(8):11348. https://doi.org/10.1364/OE.27.011348

El Barghouti M, Akjouj A, Mir A (2020) Design of silver nanoparticles with graphene coatings layers used for LSPR biosensor applications. Vacuum 180:109497, 11 pp. https://doi.org/10.1016/J.VACUUM.2020.109497

Esfahani Monfared Y (2020a) Refractive index sensor based on surface plasmon resonance excitation in a D-shaped photonic crystal fiber coated by titanium nitride. Plasmonics 15(2):535–542. https://doi.org/10.1007/S11468-019-01072-Y

Esfahani Monfared Y (2020b) Overview of recent advances in the design of plasmonic fiber-optic biosensors

Fleming JW (1984) Dispersion in GeO_2–SiO_2 glasses. Appl Opt 23(24):4486–4493. https://doi.org/10.1364/AO.23.004486

Gao J et al (2021) Enhanced sensitivity of a surface plasmon resonance biosensor using hyperbolic metamaterial and monolayer graphene. Opt Express 29(26):43766–43777. https://doi.org/10.1364/OE.447107

Gomez-Cardona N, Reyes-Vera E, Torres P (2020) High sensitivity refractive index sensor based on the excitation of long-range surface plasmon polaritons in H-shaped optical fiber. Sensors 20(7):2111. https://doi.org/10.3390/S20072111

Heidarzadeh H, Heidarzadeh, Hamid (2020) Analysis and simulation of a plasmonic biosensor for hemoglobin concentration detection using noble metal nano-particles resonances. OptCo 459:124940. https://doi.org/10.1016/J.OPTCOM.2019.124940

Hossen MN, Ferdous M, Abdul Khalek M, Chakma S, Paul BK, Ahmed K (2018) Design and analysis of biosensor based on surface plasmon resonance. Sens Bio-Sens Res 21:1–6. https://doi.org/10.1016/J.SBSR.2018.08.003

Hu S et al (2020) High-performance fiber plasmonic sensor by engineering the dispersion of hyperbolic metamaterials composed of Ag/TiO_2. Opt Express 28(17):25562–25573. https://doi.org/10.1364/OE.397461

Huang H et al (2020) A highly magnetic field sensitive photonic crystal fiber based on surface plasmon resonance. Sensors 20(18):5193. https://doi.org/10.3390/S20185193

Jabin MA et al (2019) Surface plasmon resonance based titanium coated biosensor for cancer cell detection. IEEE Photonics J 11(4). https://doi.org/10.1109/JPHOT.2019.2924825

Jeurgens LPH, Sloof WG, Tichelaar FD, Mittemeijer EJ (2002) Growth kinetics and mechanisms of aluminum-oxide films formed by thermal oxidation of aluminum. J Appl Phys 92(3):1649. https://doi.org/10.1063/1.1491591

Kabashin AV et al (2009) Plasmonic nanorod metamaterials for biosensing. Nat Mater 8(11):867–871. https://doi.org/10.1038/nmat2546

Kim KS et al (2009) Large-scale pattern growth of graphene films for stretchable transparent electrodes. Nature 457(7230):706–710. https://doi.org/10.1038/nature07719

Kiraly B, Iski EV, Mannix AJ, Fisher BL, Hersam MC, Guisinger NP (2013) Solid-source growth and atomic-scale characterization of graphene on Ag(111). Nat Commun 4(1):1–7. https://doi.org/10.1038/ncomms3804

Kumar Paul A, Krishno Sarkar A, Rahman ABS, Khaleque A (2018) Twin core photonic crystal fiber plasmonic refractive index sensor. IEEE Sens J 18(14):5761–5769. https://doi.org/10.1109/JSEN.2018.2841035

Lanza G, Breglio G, Giordano M, Gaddi A, Buontempo S, Cusano A (2011) Effect of the anisotropic magnetostriction on Terfenol-D based fiber Bragg grating magnetic sensors. Sensors Actuators A Phys. 172(2):420–427. https://doi.org/10.1016/j.sna.2011.10.005

Leite IT, Fernandes P, Hierro-Rodriguez A, Teixeira JM, Jorge PAS, Guerreiro A (2014) Analysis of a fibre optic sensor design based on SPR in nanowire metamaterial films. In: 23rd international conference on optical fibre sensors, vol 9157, p 91573C. https://doi.org/10.1117/12.2059766

Li C et al (2020) LSPR optical fiber biosensor based on a 3D composite structure of gold nanoparticles and multilayer graphene films. Opt Express 28(5):6071–6083. https://doi.org/10.1364/OE.385128

Li C et al (2021) Optical fiber SPR biosensor complying with a 3D composite hyperbolic metamaterial and a graphene film. Photonics Res 9(3):379–388. https://doi.org/10.1364/PRJ.416815

Liang H, Shen T, Feng Y, Liu H, Han W (2021) A d-shaped photonic crystal fiber refractive index sensor coated with graphene and zinc oxide. Sensors 21(1):1–16, 71. https://doi.org/10.3390/s21010071

Liu C et al (2017) Mid-infrared surface plasmon resonance sensor based on photonic crystal fibers. Opt Express 25(13):14227–14237. https://doi.org/10.1364/OE.25.014227

Liu C et al (2018) Symmetrical dual D-shape photonic crystal fibers for surface plasmon resonance sensing. Opt Express 26(7):9039–9049. https://doi.org/10.1364/OE.26.009039

Liu A et al (2019) Experimental and theoretical investigation for surface plasmon resonance biosensor based on graphene/Au film/D-POF. Opt Express 27(3):3483–3495. https://doi.org/10.1364/OE.27.003483

Liu R et al (2021) Bulk plasmon polariton in hyperbolic metamaterials excited by multilayer nanoparticles for surface-enhanced Raman scattering (SERS) sensing. Nanophotonics 10(11):2949–2958. https://doi.org/10.1515/NANOPH-2021-0301/ASSET/MEDIA/J_NANOPH-2021-0301_VIDEO_003.WEBM

Lu J, Li Y, Han Y, Liu Y, Gao J (2018) D-shaped photonic crystal fiber plasmonic refractive index sensor based on gold grating. Appl Opt 57(19):5268. https://doi.org/10.1364/AO.57.005268

Luan N, Zhao L, Lian Y, Lou S (2018) A high refractive index plasmonic sensor based on d-shaped photonic crystal fiber with laterally accessible hollow-core. IEEE Photonics J 10(5). https://doi.org/10.1109/JPHOT.2018.2873826

Luo W et al (2019) Simultaneous measurement of refractive index and temperature for prism-based surface plasmon resonance sensors. Opt Express 27(2): 576–589. doi: https://doi.org/10.1364/OE.27.000576

Maier SA, Atwater HA (2005) Plasmonics: localization and guiding of electromagnetic energy in metal/dielectric structures. J Appl Phys 98(1):011101. https://doi.org/10.1063/1.1951057

Melo AA, Santiago MFS, Silva TB, Moreira CS, Cruz RMS (2018) Investigation of a D-shaped optical fiber sensor with graphene overlay. IFAC-PapersOnLine 51(27):309–314. https://doi.org/10.1016/J.IFACOL.2018.11.623

Monfared YE (2018) Transient dynamics of stimulated raman scattering in gas-filled hollow-core photonic crystal fibers. Adv Mater Sci Eng 2018. https://doi.org/10.1155/2018/8951495

Monfared YE, Ponomarenko SA (2019) Extremely nonlinear carbon-disulfide-filled photonic crystal fiber with controllable dispersion. Opt Mater (Amst) 88:406–411. https://doi.org/10.1016/J.OPTMAT.2018.12.010

Monfared YE, Hajati M, Liang C, Yang S, Qasymeh M (2021) Quasi-D-shaped fiber optic plasmonic biosensor for high-index analyte detection. IEEE Sens J 21(1):17–23. https://doi.org/10.1109/JSEN.2019.2945003

Neff H, Zong W, Lima AMN, Borre M, Holzhüter G (2006) Optical properties and instrumental performance of thin gold films near the surface plasmon resonance. Thin Solid Films 496(2):688–697. https://doi.org/10.1016/J.TSF.2005.08.226

Niu L-Y et al (2019) Sensitivity enhanced D-type large-core fiber SPR sensor based on Gold nanoparticle/Au film co-modification. OptCo 450:287–295. https://doi.org/10.1016/J.OPTCOM.2019.06.026

Panda A, Pukhrambam PD, Keiser G (2020) Performance analysis of graphene-based surface plasmon resonance biosensor for blood glucose and gas detection. Appl Phys A 126(3):1–12. https://doi.org/10.1007/S00339-020-3328-8

Paul AK, Mollah MA, Hassan MZ, Gomez-Cardona N, Reyes-Vera E (2021) Graphene-coated highly sensitive photonic crystal fiber surface plasmon resonance sensor for aqueous solution: design and numerical analysis. Photonics 8(5):155. https://doi.org/10.3390/PHOTONICS8050155

Perrotton C et al (2011) Fiber optic surface plasmon resonance sensor based on wavelength modulation for hydrogen sensing. Opt Express 19(106):A1175–A1183. https://doi.org/10.1364/OE.19.0A1175

Poddubny A, Iorsh I, Belov P, Kivshar Y (2013) Hyperbolic metamaterials. Nat Photonics 7(12):958–967. https://doi.org/10.1038/NPHOTON.2013.243

Reyes-Vera E, Usuga-Restrepo J, Jimenez-Durango C, Montoya-Cardona J, Gomez-Cardona N (2018) Design of low-loss and highly birefringent porous-core photonic crystal fiber and its application to terahertz polarization beam splitter. IEEE Photonics J 10(4). https://doi.org/10.1109/JPHOT.2018.2860251

Rifat AA, Amouzad Mahdiraji G, Chow DM, Shee YG, Ahmed R, Adikan FRM (2015) Photonic crystal fiber-based surface plasmon resonance sensor with selective analyte channels and graphene-silver deposited core. Sensors 15(5):11499–11510. https://doi.org/10.3390/S150511499

Rifat AA et al (2016) Copper-graphene-based photonic crystal fiber plasmonic biosensor. IEEE Photonics J 8(1). https://doi.org/10.1109/JPHOT.2015.2510632

Sadeghi Z, Shirkani H, Sadeghi Z, Shirkani H (2020) Highly sensitive mid-infrared SPR biosensor for a wide range of biomolecules and biological cells based on graphene-gold grating. PhyE 119:114005. https://doi.org/10.1016/J.PHYSE.2020.114005

Saiful Islam M, Sultana J, Rifat AA, Dinovitser A, Wai-Him Ng B, Abbott D (2018) Terahertz sensing in a hollow core photonic crystal fiber. IEEE Sens J 18(10):4073. https://doi.org/10.1109/JSEN.2018.2819165

Saifur Rahman M, Rikta KA, Abdulrazak LF, Anower MS (2020) Enhanced performance of SnSe-graphene hybrid photonic surface plasmon refractive sensor for biosensing applications. PhNan 39:100779. https://doi.org/10.1016/J.PHOTONICS.2020.100779

Saitoh K, Koshiba M (2005) Numerical modeling of photonic crystal fibers. J Light Technol 23(11):3580–3590. https://doi.org/10.1109/JLT.2005.855855

Santos DF, Guerreiro A, Baptista JM (2017) SPR optimization using metamaterials in a D-type PCF refractive index sensor. Opt Fiber Technol 33:83–88. https://doi.org/10.1016/J.YOFTE.2016.11.010

Sazio PJA et al (2006) Microstructured optical fibers as high-pressure microfluidic reactors. Science 311(5767):1583–1586. https://doi.org/10.1126/SCIENCE.1124281

Sengupta J, Hussain CM (2021) Graphene-based field-effect transistor biosensors for the rapid detection and analysis of viruses: a perspective in view of COVID-19. Carbon Trends 2:100011. https://doi.org/10.1016/J.CARTRE.2020.100011

Sharma AK, Gupta J (2018) Graphene based chalcogenide fiber-optic evanescent wave sensor for detection of hemoglobin in human blood. Opt Fiber Technol 41:125–130. https://doi.org/10.1016/J.YOFTE.2018.01.012

Song B, Li D, Qi W, Elstner M, Fan C, Fang H (2010) Graphene on Au(111): a highly conductive material with excellent adsorption properties for high-resolution bio/nanodetection and identification. ChemPhysChem 11(3):585–589. https://doi.org/10.1002/CPHC.200900743

SPR based fiber optic refractive index sensor | request PDF

Sreekanth KV et al (2016a) Extreme sensitivity biosensing platform based on hyperbolic metamaterials. Nat Mater 15(6):621–627. https://doi.org/10.1038/NMAT4609

Sreekanth KV et al (2016b) Enhancing the angular sensitivity of plasmonic sensors using hyperbolic metamaterials. Adv Opt Mater 4(11):1767–1772. https://doi.org/10.1002/ADOM.201600448

Stockman MI (2011) Nanoplasmonics: past, present, and glimpse into future. Opt Express 19(22):22029–22106. https://doi.org/10.1364/OE.19.022029

Sun J et al (2019a) Magnetic graphene field-effect transistor biosensor for single-strand DNA detection. Nanoscale Res Lett 14(1):1–8. https://doi.org/10.1186/S11671-019-3048-1/FIGURES/8

Sun J et al (2019b) Sensitive and selective surface plasmon resonance sensor employing a gold-supported graphene composite film/D-shaped fiber for dopamine detection. J Phys D Appl Phys 52(19). https://doi.org/10.1088/1361-6463/AB08C1

Tabassum R, Gupta BD (2016) SPR based fiber-optic sensor with enhanced electric field intensity and figure of merit using different single and bimetallic configurations. Opt Commun 367:23–34. https://doi.org/10.1016/J.OPTCOM.2016.01.014

Tadepalli S et al (2015) Peptide functionalized gold nanorods for the sensitive detection of a cardiac biomarker using plasmonic paper devices. Sci Rep 5(1):1–11. https://doi.org/10.1038/srep16206

Tan Z, Li X, Chen Y, Fan P (2013) Improving the sensitivity of fiber surface plasmon resonance sensor by filling liquid in a hollow core photonic crystal fiber. Plasmon 9(1):167–173. https://doi.org/10.1007/S11468-013-9609-8

Tian M, Lu P, Chen L, Lv C, Liu D (2012) All-solid D-shaped photonic fiber sensor based on surface plasmon resonance. Opt Commun 285(6):1550–1554. https://doi.org/10.1016/J.OPTCOM.2011.11.104

Tiwale N, Senanayak SP, Rubio-Lara J, Alaverdyan Y, Welland ME (2019) Optimization of transistor characteristics and charge transport in solution processed ZnO thin films grown from zinc neodecanoate. Electron Mater Lett 15(6):702–711. https://doi.org/10.1007/S13391-019-00173-4/FIGURES/5

Tong K, Wang F, Wang M, Dang P, Wang Y, Sun J (2018) D-Shaped photonic crystal fiber biosensor based on silver-graphene. Optik (Stuttg) 168:467–474. https://doi.org/10.1016/j.ijleo.2018.04.119

Ubeid MF, Shabat MM (2015) Numerical investigation of a D-shape optical fiber sensor containing graphene. Appl Phys A Mater Sci Process 118(3):1113–1118. https://doi.org/10.1007/s00339-014-8925-y

Unterhuber A, Andreana M, Le T, Drexler W (2019) Ultrashort pulse Kagome hollow-core photonic crystal fiber delivery for nonlinear optical imaging. Opt Lett 44(7):1588–1591. https://doi.org/10.1364/OL.44.001588

Vermisoglou E et al (2020) Human virus detection with graphene-based materials. Biosens Bioelectron 166. https://doi.org/10.1016/J.BIOS.2020.112436

Villatoro J et al (2009) Photonic crystal fiber interferometer for chemical vapor detection with high sensitivity. Opt Express 17(3):1447–1453. https://doi.org/10.1364/OE.17.001447

Wang SF, Chiu MH, Chang RS (2006) Numerical simulation of a D-type optical fiber sensor based on the Kretchmann's configuration and heterodyne interferometry. Sensors Actuators B Chem. 114(1):120–126. https://doi.org/10.1016/j.snb.2005.04.012

Wang L et al (2017) Characteristics of D-shaped photonic crystal fiber surface plasmon resonance sensors with different side-polished lengths. Appl Opt 56(5):1550–1555. https://doi.org/10.1364/AO.56.001550

Wang Q, Jing JY, Wang XZ, Niu LY, Zhao WM (2020a) A D-shaped fiber long-range surface plasmon resonance sensor with high Q-factor and temperature self-compensation. IEEE Trans Instrum Meas 69(5):2218–2224. https://doi.org/10.1109/TIM.2019.2920187

Wang Q, Niu LY, Jing JY, Zhao WM (2020b) Barium titanate film based fiber optic surface plasmon sensor with high sensitivity. Opt Laser Technol 124. https://doi.org/10.1016/J.OPTLASTEC.2019.105899

Wu L, Chu HS, Koh WS, Li EP (2010) Highly sensitive graphene biosensors based on surface plasmon resonance. Opt Express 18(14):14395. https://doi.org/10.1364/OE.18.014395

Wu Z et al (2013) Room temperature methane sensor based on graphene nanosheets/polyaniline nanocomposite thin film. IEEE Sens J 13(2):777–782. https://doi.org/10.1109/JSEN.2012.2227597

Wu T et al (2017) Surface plasmon resonance biosensor based on gold-coated side-polished hexagonal structure photonic crystal fiber. Opt Express 25(17):20313–20322. https://doi.org/10.1364/OE.25.020313

Xi X et al (2020) An Au nanofilm-graphene/D-type fiber surface plasmon resonance sensor for highly sensitive specificity bioanalysis. Sensors 20(4):991. https://doi.org/10.3390/S20040991

Xu S et al (2017) Real-time reliable determination of binding kinetics of DNA hybridization using a multi-channel graphene biosensor. Nat Commun 8(1):1–10. https://doi.org/10.1038/ncomms14902

Yang W et al (2021) High performance D-type plastic fiber SPR sensor based on a hyperbolic metamaterial composed of Ag/MgF2. J Mater Chem C 9(39):13647–13658. https://doi.org/10.1039/D1TC02217B

Ying Y, Si G, Luan F, Xu K, Qi Y, Li H (2017) Recent research progress of optical fiber sensors based on D-shaped structure. Opt Laser Technol 90(September 2016):149–157. https://doi.org/10.1016/j.optlastec.2016.11.021

Yu X et al (2010) A selectively coated photonic crystal fiber based surface plasmon resonance sensor. J Opt A Pure Appl Opt 12(1). https://doi.org/10.1088/2040-8978/12/1/015005

Zainuddin NAM, Ariannejad MM, Arasu PT, Harun SW, Zakaria R (2019) Investigation of cladding thicknesses on silver SPR based side-polished optical fiber refractive-index sensor. Results Phys 13:102255. https://doi.org/10.1016/J.RINP.2019.102255

Zeni L et al (2018) [INVITED] Slab plasmonic platforms combined with plastic optical fibers and molecularly imprinted polymers for chemical sensing. OptLT 107:484–490. https://doi.org/10.1016/J.OPTLASTEC.2018.06.028

Zhang C et al (2017a) U-bent fiber optic SPR sensor based on graphene/AgNPs. Sens Actuators B Chem 251:127–133. https://doi.org/10.1016/J.SNB.2017.05.045

Zhang NMY et al (2017b) Hybrid graphene/gold plasmonic fiber-optic biosensor. Adv Mater Technol 2(2):1600185. https://doi.org/10.1002/ADMT.201600185

Chapter 7
Interferometric-Based SPR Sensors

7.1 Introduction

The general needs of our daily lives are persistently in touch with science and technology and are greatly influenced by the technological advancement of sensing applications. The inherent capabilities of optical fiber have led to exponential growth in the commercialization of fiber-based technology. In the last few decades, a large variety of sensing applications based on optical fiber interferometric sensors have emerged. The array of applications includes wireless communication systems, satellite communication, modern radar systems, healthcare instruments, and electronic warfare. Beside low loss and high bandwidth, fiber-based solutions avoid electromagnetic interference, corrosion, and information leakage through the information-carrying channel to the outside as there is no short circuit possibility, unlike conducting copper wire. This enhances safety and establishes a secured transmission to prevent eavesdropping on the information between the transmitter and receiver. There are some other requirements of the military, like information on surrounding temperature, radiation, severe vibration, shock, and mechanical stress, which is very useful for viability. This information can be extracted with the help of a fiber optic sensor.

The sensing concept can be broadly classified as interferometry, scattering, and fiber Bragg grating-based detections. These techniques can be found suitable based on a specific application. For example, scattering and interferometry techniques are popularly used in security systems, defense, and remote sensing applications. Fiber Bragg grating-based sensing is widely applicable in temperature sensing, strain sensing, and physical intrusion detection systems. Interferometric-based sensing systems have good accuracy with very high resolution and exhibit multipurpose uses (Giallorenzi et al. 1982). However, such systems are limited by the cost of signal processing electronics and the relative nature of measurement (Claus et al. 1992; Wang et al. 2001). Nowadays, optical fiber-based interferometric SPR sensors are very popular for sensing different physical parameters such as strain, temperature, pressure, and refractive index. It takes the advantage of fiber optics along with SPR

S. K. Raghuwanshi et al., *Geometric Feature-Based Fiber Optic Surface Plasmon Resonance Sensors*, Springer Tracts in Electrical and Electronics Engineering,
https://doi.org/10.1007/978-981-99-7297-5_7

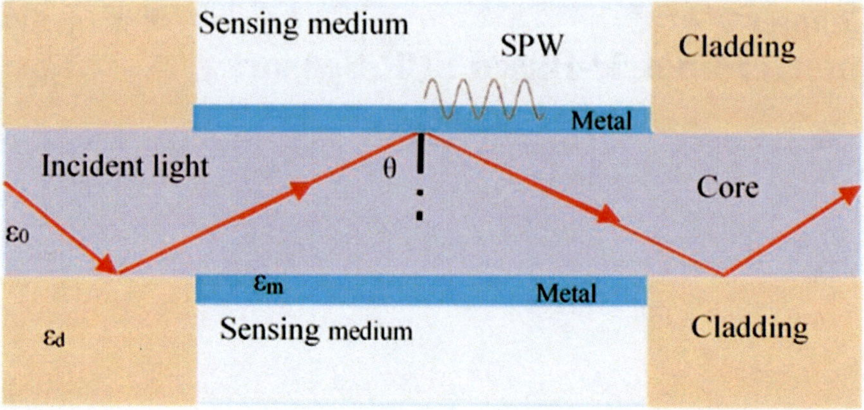

Fig. 7.1 Schematic representation of the SPR sensing mechanism. Adapted from Zhao et al. (2023)

sensing with a special type of interferometric structure. Such sensor structures have easy manufacturing along with easy handling of measurement errors caused by environmental effects (Chen et al. 2021). There are four different types of interferometers: Fabry–Perot, Michelson, Sagnac, and Mach–Zehnder interferometers. Before going into these interferometric structures' details, the basic SPR sensing phenomena are given in Fig. 7.1 (Zhao et al. 2023). In Fig. 7.1, ϵ_d, ϵ_m, and ϵ_o represent dielectric constants for the cladding region, metal, and core, respectively. When an incident optical signal falls at a specific angle at the core–metal interface, total internal reflection takes place, except for the evanescent modes. These evanescent modes in the metal region excite plasmon waves at the surface of the metal. Resonance occurs in the transmission characteristics of SPR at the phase matching condition, i.e., equal value of phase constant and evanescent mode.

In this chapter, the working principles, fabrication protocols, and application of various interferometric SPR sensors are discussed. An interferometer works on the different optical paths of a fiber to create interference between two beams. It means there should be beam splitting and beam combining components in the interferometric configuration. In that case, one of the optical paths should be arranged in such a way that it can vary with external variations. Several types of temporal and spectral information can be obtained from an interferometric signal, in which the target can be detected through the changes in wavelength, intensity, phase, frequency, and bandwidth of the signal. These variations can improve the performance of the device in terms of high sensitivity and accuracy as well as dynamic range. Nowadays, microscale applications are trending and can only be possible with the miniaturization of fiber optic interferometers. For this, new small-sized fiber devices are developed to operate on fiber scales to replace the conventional bulky fiber optic components such as combiners, beam splitters, and objective lenses. For the fiber optic interferometer miniaturization process, inline structures are very popular and investigated, in which one physical line contains two optical paths. The inline fiber structures have several merits, such as high stability, high coupling efficiency, and easy alignment.

7.2 Interferometers with Various Fabrication Techniques and Their Detection Capability

7.2.1 Mach–Zehnder Interferometer

Monitoring of various parameters like liquid level detection, refractive index measurement, temperature, and strain measurement have drawn great attention to the Mach–Zehnder structure of optical sensing applications in the areas of biological, environmental, and chemical industries. It takes advantage of the cross-sensitivity of multi-sensing environments based on long-period fiber gratings and fiber Bragg gratings (Lu and Chen 2008; Swart 2004; Zhang et al. 2005). The advantages of MZI structures are easy fabrication, simplicity, and robustness (Jiang et al. 2011; Choi et al. 2007). In the past, various fabrication techniques like taper fiber, special fiber, and core-offset structures based on MZI structures have been reported. In Choi et al. (2007), core-cladding mode of interference occurs, where the fundamental mode inside the core excites cladding modes as depicted in Fig. 7.2. Two fabrication methods are adopted for the MZI structure: one is set through splicing, and the other is based on the collapsing of photonic crystal fiber (PCF) regions. In the upper part of Fig. 7.2, three pieces of solid-core PCF are considered and spliced together. The core mode of the first PCF is excited into multiple modes at the spliced region with the middle PCF. The reverse of this occurs at the second splicing region, where interference takes place similar to MZI interference. Special care has been taken for the various parameters like fusion time and arc power, along with the length of the PCF, during the splicing. The proposed structure claims potential applications in strain and very high-temperature measurement.

In 2006, Villatoro et al. fabricated a microstructured MZI-based modal interferometer with tapered optical fiber for various sensing applications where the guided mode inside the core was excited. However, through this process, low-order modes of the cladding are excited, which can be observed from the frequency spectrum of the interference pattern. Li et al. in (2012), developed an innovative strategy based on the MZI structure employing a thinned fiber with a tiny core and cladding diameter with a single-mode-multimode-thinned-single-mode (SMTS) fiber for the sensor application of liquid level, refractive index, temperature, and axial strain. Because the multimode fiber's mode field width is substantially larger than that of the thinned fiber, the core mode and cladding modes of the thinned fiber might be activated simultaneously.

The schematic of the fabricated structure is represented in Fig. 7.3, where a thinned fiber (TF) is spliced with multimode fiber (MMF), and the resulting unit is further spliced between two pieces of single-mode fiber (SMF). When light enters the MMF

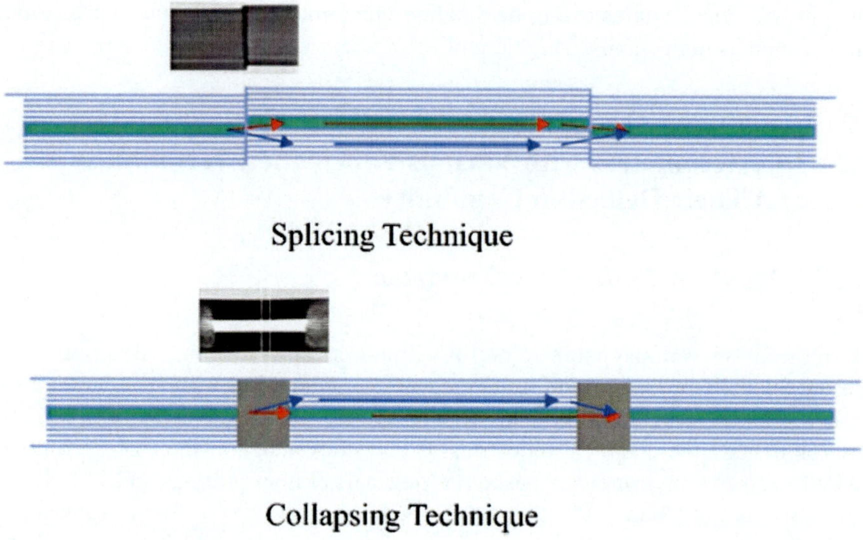

Fig. 7.2 Schematic diagram for core-cladding modes of excitation and interference through splicing and collapsing techniques. Reprinted with permission from Optics Express, Copyright, 2007, Optica (Choi et al. 2007)

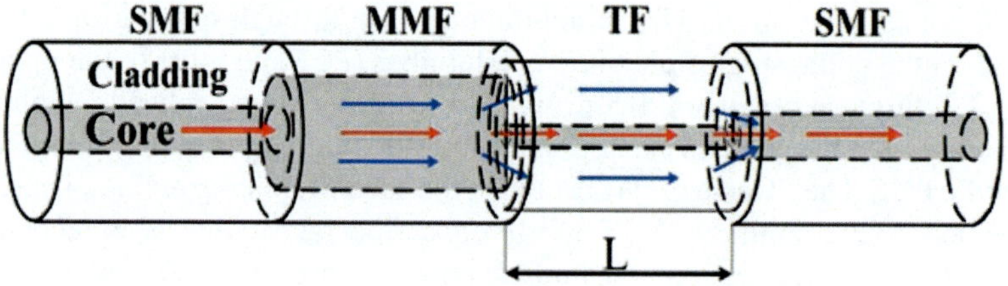

Fig. 7.3 Schematic of the fabricated MZI structure using SMF, MMF, and TF; L is the length of thinned fiber. Reprinted with permission from Optics Express, Copyright, 2012, Optica (Li et al. 2012)

through the lead-in SMF, a portion of the power is linked to the TF's cladding modes due to a mode field mismatch at the MMF-TF spliced point. It will stimulate various cladding modes that propagate throughout the TF's cladding. Similarly, at the TF-SMF splice point, some of the TF cladding modes that are connected back into the core of the SMF interfere with the TF core mode. As a result, they are linked into the basic mode of the lead-out fiber. Fringe visibility and transmission loss of the spectrum are affected by coupling loss, which depends on the cleaving of fiber and the splicing process associated with a fusion splicer. The reduced diameter of TF causes enhanced interference of light between the core mode and the cladding mode. The evanescent wave of the basic guided mode will interact with the surrounding environment, causing a change in the transmission spectrum.

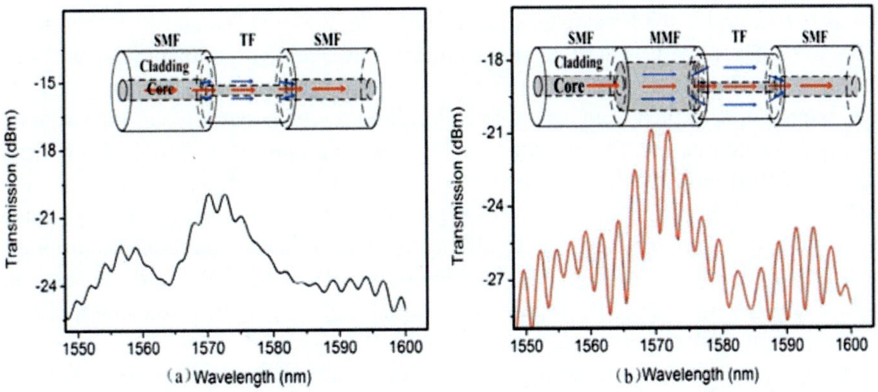

Fig. 7.4 Transmission spectrum of fabricated MZI structures with **a** SMF-TF-SMF and **b** SMF-MMF-TF-SMF. Reprinted with permission from Optics Express, Copyright, 2012, Optica (Li et al. 2012)

The improvement in the interference pattern with the use of MMF is depicted in Fig. 7.4. Figure 7.4 shows the transmission spectrum of two structures, i.e., the SMF-TF-SMF structure and the SMF-MMF-TF-SMF structure. In the latter case, a sharp peak and deep fringe with highly improved fringe visibility are found with small power loss due to an increase in the number of splices. In both cases, a thinned fiber of 48.38 mm is used, and interference patterns are investigated for various lengths of MMF.

Here, the length of the thinned fiber acts as the length of the MZI structure, and the change in effective refractive index of the core and cladding modes is Δn_{eff}. The mth deep of the transmission spectrum λ_m occurs at:

$$\lambda_m = \frac{2\Delta n_{eff} L}{2m+1}. \tag{7.1}$$

The change in environmental conditions changes the length of the MZI structure or the effective refractive index, which shifts the wavelength position of mth deep (Lu et al. 2009). For example, when the fabricated probe is placed in such a way that the thinned fiber is surrounded by a liquid, there is a decrease in the effective refractive index due to the length L_n of the thinned fiber inside the liquid and the remaining part outside liquid. The wavelength position of mth order deep shifts is given in Eq. (7.2)

$$\lambda_m = \frac{2\Delta n_{eff}(L - L_n)}{2m+1} + \frac{2\Delta n_{eff} L_n}{2m+1}. \tag{7.2}$$

The first term on the right side of Eq. (7.2) is the contribution of TF outside the liquid, and the second term is the contribution due to the length surrounded by the liquid. With the vertical placement of the fabricated probe inside the liquid, it can be used to identify the level of liquid. When the liquid level increases, the change in effective refractive index will decrease, and the wavelength position of deep will shift to the left of the spectrum. With the prior knowledge of liquid level sensitivity, the refractive index of the liquid can also be identified. In the simultaneous measurement of multiple parameters, like refractive index and temperature variation, a fiber Bragg grating (FBG)-based MZI structure is given (Cao et al. 2015) and different parameters are discussed. In this work, the MZI structure is developed using the core-offset technique, and the structure is cascaded with FBG. The schematic of the proposed architecture is shown in Fig. 7.5. The deep position of the MZI and the resonant wavelength of the FBG are separately used for refractive index and temperature measurement. Through coupled mode theory, the Bragg wavelength (λ_B) of FBG is a function of the grating period Λ and effective refractive index n_{eff} of the core given by Othonos (1997)

$$\lambda_B = 2n_{eff}\Lambda. \tag{7.3}$$

The Bragg wavelength of FBG is a function of temperature variation and is insensitive to changes in refractive index. This variation due to the change in temperature ΔT can be observed by taking the differentiation of Eq. (7.2)

$$\frac{\Delta\lambda_B}{\Delta T} = \left(\frac{1}{\Lambda}\frac{\partial\Lambda}{\partial T} + \frac{1}{n_{eff}}\frac{\partial n_{eff}}{\partial T}\right)\lambda_B. \tag{7.4}$$

In Eq. (7.4), $\frac{1}{\Lambda}\frac{\partial\Lambda}{\partial T}$ and $\frac{1}{n_{eff}}\frac{\partial n_{eff}}{\partial T}$ are known as coefficients of thermal expansion and thermo-optics, respectively. The advantage of the MZI structure is that it can be used to determine various physical parameters with the deep position mentioned in Eq. (7.1). Here, dip position due to MZI structure is a function of temperature as well as change in refractive index; however, the center wavelength of FBG is only

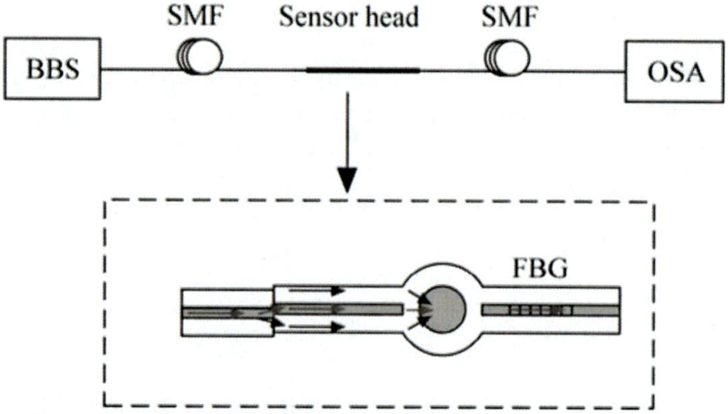

Fig. 7.5 Schematic diagram of the sensor for multiparameter sensing. Reprinted with permission from Optoelectronics Letters, Copyright, 2015, Springer Nature (Cao et al. 2015)

sensitive to the variation in temperature. With certain changes in temperature (ΔT) and effective refractive index (Δn_{eff}), the wavelength shifts in the deep of MZI ($\Delta\lambda_1$) and FBG ($\Delta\lambda_B$) can be demultiplexed using the following relation:

$$\begin{bmatrix} \Delta\lambda_1 \\ \Delta\lambda_B \end{bmatrix} = \begin{bmatrix} S_{T1} & S_{n1} \\ S_{TB} & S_{nB} \end{bmatrix} \begin{bmatrix} \Delta T \\ \Delta n_{eff} \end{bmatrix}, \tag{7.5}$$

where S_{T1} and S_{TB} represent the thermal sensitivity of MZI and FBG, respectively, and S_{n1} and S_{nB} are the refractive index sensitivity of MZI and FBG, respectively.

Apart from the above detecting parameters, the MZI structure is also used in food decomposition detection (Razak et al. 2018) to monitor the liquid concentration, 3-aminopropyl-triethoxysilane detection (Deng et al. 2021), and humidity sensing (Lou et al. 2016). In some other types of fabrication approaches, the MZI structure is preserved inside a film of Polydimethylsiloxane (PDMS) (Gong et al. 2019) for temperature monitoring, MZI structure is sealed in a magnetic fluid-filled capillary (Luo et al. 2018) to use as a magnetic sensor, MZI with few mode fibers (Liu et al. 2021) is used for high-temperature sensing, and peanut shape structure (Yu et al. 2019) of MZI is used for RI sensing.

7.2.2 *Michelson Interferometer*

The easy fabrication and simple working principle of the Michelson interferometer (MI) structure have attracted great attention from researchers in the design of optical fiber sensors for various applications. These sensors can be used to detect changes in temperature, strain, liquid level, refractive index, vibration, stress, etc. As discussed earlier, MI structures measure phase induced in two arms terminated by reflecting surfaces. One arm is used as a measuring arm, and the other arm acts as a reference arm, which is kept constant from environmental or external disturbance. In Cubik et al. (2014), A MI sensor with polarization-maintaining fiber is reported for vibration detection. The arms are of 2 m long polarization-maintaining fiber and terminated by a reflecting surface. A DFB laser source was used to provide an optical signal of 0.03 nm linewidth at 1550 nm. The reflected signal is detected by an InGaAs photodetector, and the detected signal is processed through a signal processing unit. Li et al. (2018) reported core-offset-based MI structure optic sensor for liquid level detection. The core-offset structure is developed by splicing a piece of SMF between two SMFs, and the end face is coated with reflecting material. Assuming L_1 and L_2 as the lengths of SMF at the middle and reflection-ended SMF, and λ as the wavelength of the optical signal, the free spectral range is given by

$$\text{FSR} = \frac{\lambda^2}{2(L_1 + L_2)\Delta n_{eff}}. \tag{7.6}$$

The phase difference $\Delta\varphi$ induced between core and cladding modes after traveling through middle SMF is given by

$$\Delta\varphi = \frac{4\pi(L_1 + L_2)\Delta n_{eff}}{\lambda}. \tag{7.7}$$

As the length of SMF inside (L_{liq}) liquid increases, the phase change is given by Li et al. (2018)

$$\Delta\varphi = \frac{4\pi\left(L_1 + L_2 - L_{liq}\right)\left(n_{eff}^{co} - n_{eff}^{cl1}\right)}{\lambda} + \frac{4\pi\left(L_2 + L_{liq}\right)\left(n_{eff}^{co} - n_{eff}^{cl2}\right)}{\lambda}. \tag{7.8}$$

In Eq. (7.8), n_{eff}^{co} represents the effective core index of SMF, and n_{eff}^{cl1} and n_{eff}^{cl2} are cladding effective refractive indices in air and liquid surrounding medium, respectively. There are two offset points located at the tip of the reflecting surface, which is dipped inside the liquid, and another one is placed at the top of the water surface. The phase-induced wavelength shift in the interferogram Another work is reported in Zhou et al. (2015) based on the core-offset technique for refractive index sensing. In this technique, an intensity-modulated refractive index sensor is proposed based on MI structure, where a thin core fiber (TCF) is spliced with some misalignment to a conventional single-mode fiber. The TCF section acts as a cavity, and light through SMF splits into two parts at the misaligned joint between SMF and TCF. The intensity of light is distributed between the core and cladding of TCF. The light is reflected through the cleaved end of TCF and interferes with the core of SMF at the spliced point. Interference pattern contrast depends on the ratio of light power coupled in the core and cladding regions and on transmission loss, which is affected by the length of TCF and the offset at the spliced region. The variation in the free spectral range is shown in Fig. 7.6 for various lengths of TCF. The FSR of the reflection characteristics decreases with an increase in the length of TCF, and a greater number of peaks appear in the interferogram. Further, fringe visibility of the interference pattern is investigated with the core offset between the core and cladding by adjusting the two-directional movement of the fibers through the splicer motor. There are various modes of travel in the cladding region (let I_{cl}^{m} represents intensity of mth order mode traveling in cladding), and single mode travels in core with intensity of I_{co}. The resultant intensity of the interfered light through superimposition of core and cladding modes is given by (Zhou et al. 2015)

$$I = \left[I_{co} + \sum I_{cl}^{m} + 2\sum I_{co} I_{cl}^{m} \cos(\varnothing_m)\right] \frac{(n_{co} - n)^2}{(n_{co} + n)^2}, \tag{7.9}$$

where n_{co} and n are refractive indices of the core and external region, respectively. The phase delay between the core mode and mth order cladding is represented by $\varnothing_m$.

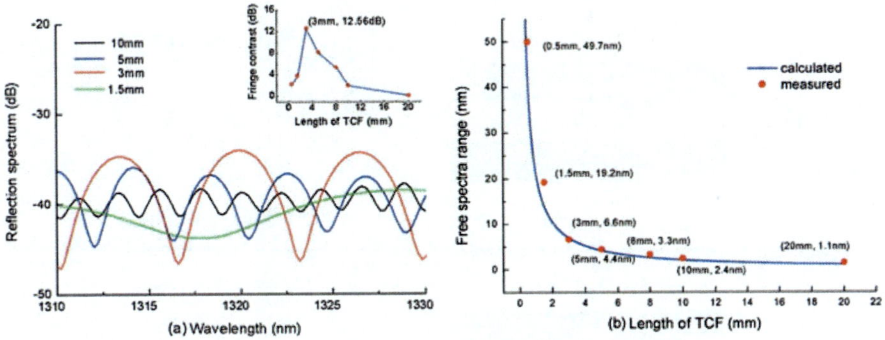

Fig. 7.6 **a** Reflection spectrum of offset core Michelson interference for various lengths of thin core fiber and **b** variation in the free spectral range of reflection interference with change in length of TCF. Reprinted with permission from Sensors and Actuators B: Chemical. Copyright, 2015, Elsevier (Zhou et al. 2015)

In Guo et al. (2021), Taper fiber is used with a reflection-ended layer to form MI. The high-order modes were triggered at the taper, causing distinct modes to be propagated forward in the fiber, reflected by the fiber's end face, and then recoupled back to the fiber core to generate MI. Figure 7.7 shows a schematic of the tapering process used in the fabrication of the MI-structured sensor. The discharge time and intensity of the splicing machine are 1500 and 200 ms, respectively, during the tapering process. After cleaning, one portion of the SMF was inserted in the fiber splicing machine; the end face of the fiber surpassed the discharge electrode, and discharge was done to build the taper structure. It should be noted that only the left-over fiber is tapered during the splicing process. The distance of fiber after the tapering section is L with the other end terminated by a reflecting film as indicated in Fig. 7.7.

The broadband optical signal from a super luminescent laser diode is passed to the fabricated sensor, and the interference pattern of the reflected signal is observed on the optical spectrum analyzer for various lengths of the fiber after the tapering section, as indicated in Fig. 7.8. It appears from the interference pattern shown in Fig. 7.8 that the free spectral range decreases with an increase in the length of the fiber section, due to which more peaks appear in the interferogram. Apart from temperature and refractive index sensing, interferometric fiber optic sensors are frequently used in experimental studies of ultrasonic transmission detection in structural materials because of their high sensitivity, vast dynamic area, and anti-electromagnetic interference (Yuan et al. 2004; Liang 2007; Jang et al. 2004). In Liang et al. (2013), the paper proposes a Michelson interferometer with two optical fiber loop reflectors and an acoustic emission sensor to detect vibrations caused by ultrasonic waves traveling through a solid body. Instead of the usual Michelson interferometer end reflector, two optical fiber loop reflectors serve as the sensing and reference arms. The optimal working point of an optical fiber sensor is determined by theoretical derivation and computer simulation, and the signal frequency recorded by the sensor is the frequency of the

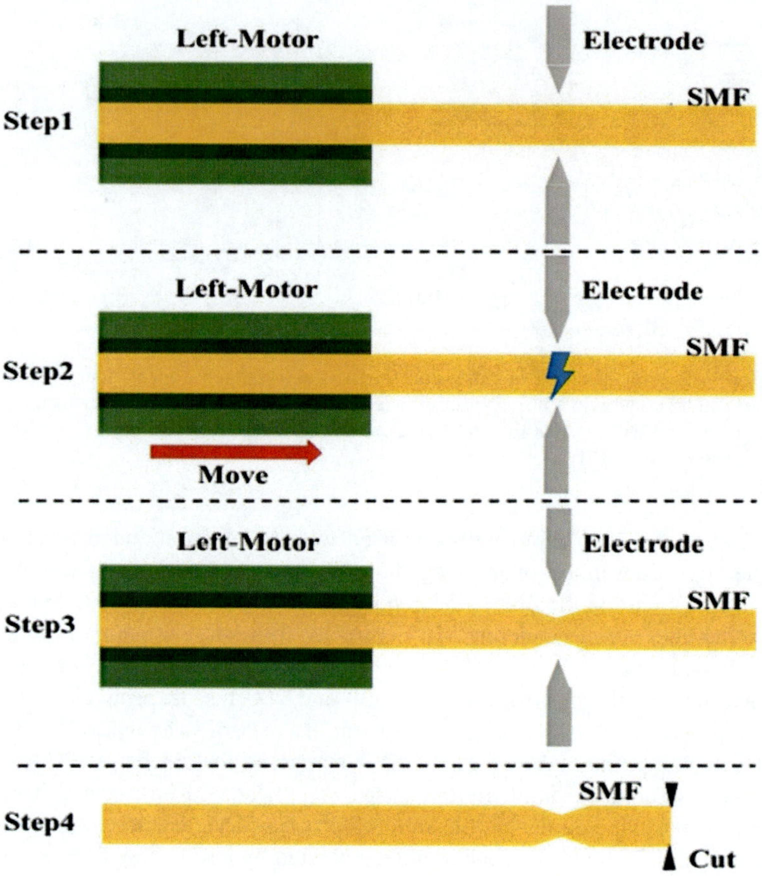

Fig. 7.7 Schematic for the various steps during the tapering process. Adapted from Guo et al. (2021)

input signal. The PZT (Piezoelectric Ceramic) is powered by a signal generator as a known ultrasonic source, and the polarization controller is used to generate reflected light interference. The fiber length is changed by adjusting the DC voltage on the PZT with the fiber loop to produce a sensor system response that changes phase to 90°.

The deep position λ_{dip} of reflected interference for the *m*th order cladding mode is given by

$$\lambda_{dip} = \frac{4\left(n_{eff}^{co} - n_{eff}^{cl}\right)L}{(2m+1)}. \tag{7.10}$$

To improve the reflectivity of the MI, gold is used as a reflecting material. The fabricated probe is used to detect temperature variation with the wavelength shift

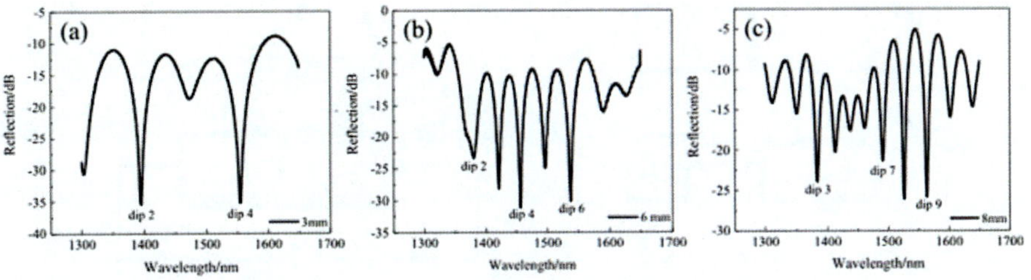

Fig. 7.8 Reflection spectrum of tapered MI for various lengths L of reflection-ended fiber **a** 3 mm **b** 6 mm **c** 8 mm. Adapted from Guo et al. (2021)

of the deep position in the spectral interferogram. Another type of fiber tip with a MI structure is reported in Wang et al. (2021a) to detect temperature variation. A combination of single-mode fiber and solid-core fiber (SCF) is used to form the MI structure. In this work, polymer-UV glue is partially filled in the fiber from the tip side (covering a small portion of the solid-core fiber). This forms an MI cavity inside the fiber, which is sensitive to the ambient temperature variation. Temperature variations may cause this coated area to change simultaneously, but because of the barrier of the silica coating, the internal change of the SCF is still the deciding element that dictates the interferometer's output. Consider the length and refractive index of L_g and n_g, respectively, for glue material filled inside SCF. Further, n_{SCF} and L_{SCF} denotes the refractive index and length of the core of SCF, and the temperature sensitivity of the fabricated material is given by Wang et al. (2021a)

$$\frac{d\lambda}{dT} = \lambda \frac{L_{\text{SCF}} \frac{\partial n_{\text{SCF}}}{\partial T} + n_{\text{SCF}} \frac{\partial L_{\text{SCF}}}{\partial T} - L_g \frac{\partial n_g}{\partial T} - n_g \frac{\partial L_g}{\partial T}}{n_{\text{SCF}} L_{\text{SCF}} - n_g L_g}. \tag{7.11}$$

Partial differentiation in Eq. (7.11) represents thermo-optic variation. Assuming that α represents the linear coefficient of temperature variation, the relation between the change in cavity length and the change in temperature is $\Delta L = \alpha \Delta T$.

Chen et al. in (2015), gave a voltage sensor based on MI structure using a Fabry–Perot-based tunable filter and demodulation technique for phase-modulated optical signal. In this work, a piezoelectric transducer is used to convert applied voltage into mechanical strain, which induces a phase difference in the Michelson interferometer, as shown in Fig. 7.9. Further, the phase difference is demodulated to obtain the applied voltage. The intensity fluctuation of the optical sensor is insensitive to the voltage since phase modulation is involved. Further, the fluctuations due to temperature and pressure fluctuations at low frequencies can be eliminated. This avoids polarization fading, and there is no need for a polarization controller. Phase induced is a function of the center wavelength (λ_c) of the interference signal, the differential length (d) of two arms of MI, and refractive index n of the arms, and the offset phase ($\varnothing_0$) induced due to the environmental effect. The phase induced is given by Chen et al. (2015)

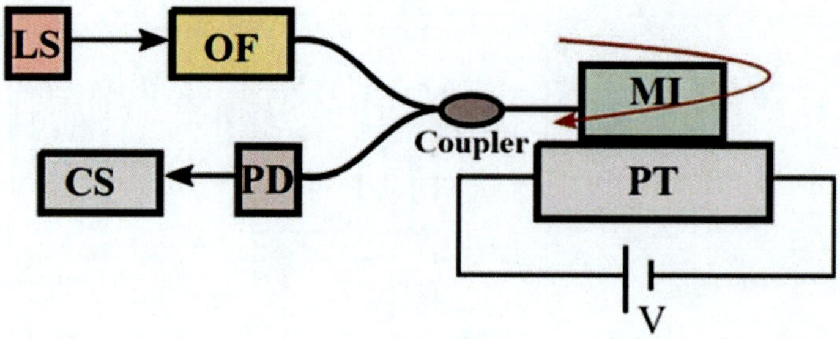

Fig. 7.9 Schematic diagram of a piezo-transducer-based MI; LS: laser source; OF: optical filter; MI: Michelson interferometer; PT: piezo-transducer; PD: photodetector; CS: computer system

$$\varnothing(\lambda_c) = \frac{4\pi nd}{\lambda_c} + \varnothing_0. \tag{7.12}$$

An alternating applied voltage of V to the PT, as indicated in Fig. 7.9 will cause the change in differential length Δd of MI and phase fluctuation $\Delta\varnothing_0$, which gives a change in phase induced as

$$\Delta\varnothing(\lambda_c) = \frac{4\pi n\Delta d}{\lambda_c} + \Delta\varnothing_0. \tag{7.13}$$

Let, V_0 and f_0 are the respective applied magnitude and frequency of applied voltage, $V = V_0\cos(2\pi f_0 t)$. The variation in differential length is in linear proportion to V with strain constant K and can be written as

$$\Delta d = KV_0\cos(2\pi f_0 t). \tag{7.14}$$

The randomly generated phase noise $\Delta\varnothing_0$ is suppressed through demodulation techniques. The phase noise is removed by applying carrier signal $c(t) = A\cos(2\pi ft)$ of amplitude A (controlled by applied voltage V) and frequency f to an optical filter with a tunable center frequency. The center wavelength of the output light is $\Delta\lambda = A\cos(2\pi ft)$. The phase variation of MI structure is given by

$$\Delta\varnothing(\lambda_c) = \frac{4\pi nd}{\lambda_c}\left(\frac{A\cos(2\pi ft)}{\lambda_c} + \frac{KV_0\cos(2\pi f_0 t)}{d} + \Delta\varnothing_0\right). \tag{7.15}$$

In Eq. (7.15), the coefficient of $\cos(2\pi ft)$ represents modulation depth, and the coefficient of $\cos(2\pi f_0 t)$ is the coefficient of fluctuation of the sensor. The resultant interference of the MI structure is given by

$$I = I_1R_1 + I_2R_2 + 2\sqrt{I_1R_1I_2R_2}\cos(\Delta\varnothing(\lambda_c)). \tag{7.16}$$

Jacobi anger expansion of $\cos(\Delta\varnothing(\lambda_c))$ in Eq. (7.16) gives fundamental and higher-order harmonics. The higher-order harmonics can be suppressed with the help of a low-pass filter, which contributes to the phase fluctuation. In Zhou et al. (2014), A Michelson interferometer-based optical fiber sensor is reported for refractive index measurement. It is temperature-independent, and cross-sensitivity between temperature and refractive index is eliminated. Similarly, Sahu et al. in (2016) has given us a biosensor that is compact in size and easily integrated with other optical devices.

7.2.3 Sagnac Loop Interferometer

Sagnac loop interferometer (SLI) structured optical sensors have easy fabrication and integration on a chip with low cost and high sensitivity. Various sensing parameters like temperature, strain, refractive index, and liquid level can be achieved with a suitable choice of optical components and architectures. In Ge et al. (2022), the SLI structure is developed using polarization-maintaining fiber (PMF) for temperature sensing. In this structure, a piece of PMF is spliced between two SMFs. The sensitivity is improved through the excitation of the cladding modes of PMF. Consider L as the length of PMF and I_0 as the injected intensity in SLI. The reflected intensity out of SLI due to mth order excited mode of cladding and core is (Ge et al. 2022)

$$I = \frac{I_0}{2}\left\{1 - \cos\left[\frac{2\pi L}{\lambda}\left(B + \sum_m \Delta n_m^{eff}\right)\right]\right\}. \tag{7.17}$$

In Eq. (7.17), B is the birefringence, and Δn_m^{eff} is the difference in effective refractive index between the mth order excited mode of cladding and core. Here, fiber loss in the Sagnac loop is neglected. Resonance wavelength position of deep (λ_n) can be found, when the phase induced is an integral multiple of 2π, and can be given by

$$\lambda_n = \frac{(B + \sum_m \Delta n_m^{eff})L}{n}. \tag{7.18}$$

The wavelength shift in λ_n with environmental temperature change (ΔT) is

$$\Delta\lambda = \lambda_n\left(\frac{1}{(B + \sum_m \Delta n_m^{eff})}\frac{\partial B + \partial \sum_m \Delta n_m^{eff}}{\partial T} + \frac{1}{L}\frac{\partial L}{\partial T}\right)\Delta T, \tag{7.19}$$

where $\Delta\lambda/\Delta T$ indicates the temperature sensitivity of the proposed sensor, which can be improved with a greater number of cladding modes. A similar configuration is reported in Omar et al. (2018) for temperature sensing. In this work, a tunable laser source in combination with a pump laser is utilized, and the phase difference

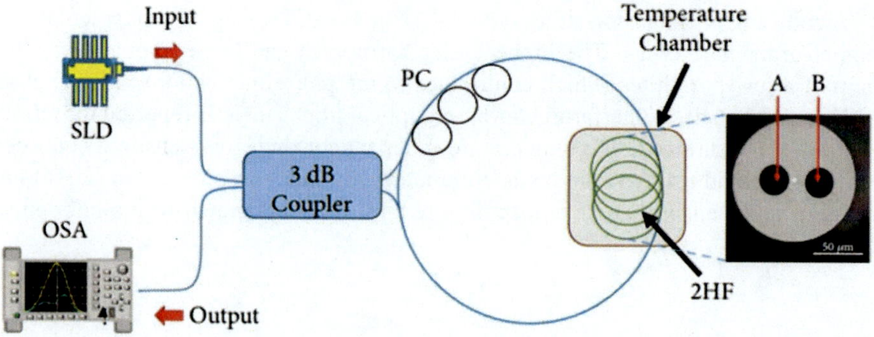

Fig. 7.10 Schematic diagram of a two-hole fiber-based Sagnac loop for temperature sensing: SLD: super luminescent laser diode; PC: polarization controller; 2HF: two-hole fiber; OSA: optical spectrum analyzer. Adapted from Domínguez-Cruz et al. (2018)

between optical signals from these two laser sources is utilized for the change in temperature. In Domínguez-Cruz et al. (2018), A two-hole asymmetric fiber is used in a Sagnac loop for temperature sensing, as shown in Fig. 7.10. The two-hole fiber is placed inside a temperature chamber, which provides a temperature difference between the holes and results in a change in birefringence. Because of improved features such as strong birefringence, broad dispersion, and increased reactivity to external thermal impacts, the existence of air holes in fiber and the ability to be filled with other materials provide some significant engineering potential.

The broadband optical signal from the super luminescent laser diode is split equally between the two propagating paths of the Sagnac loop in clockwise and anticlockwise directions. Two-hole fiber is spliced between two fibers and acts as a sensing element. The change in temperature difference gives rise to the change in the birefringence pattern observed on OSA. The linear response, simple fiber transversal construction, and usage of larger pores, which allow for the insertion of additional materials, make this a perfect platform for the development of very sensitive sensors. In order to achieve high birefringence, different types of fibers are investigated in the Sagnac loop as a sensing part. For example, the performance of photonic crystal fiber (PCF) and PANADA fibers is investigated in the Sagnac loop for improved birefringence, and FSR is analyzed for different operating conditions (Seraji and Pazooki 2017). The optical signal is passed through a 3 dB coupler to the Sagnac loop, which incorporates PCF or PANADA fibers. The presence of birefringence in the fiber results in a velocity differential and optical path variation related to the polarization state. As a result, when these fields enter the directional coupler again, they will have a relative phase difference. As a result, the reflection becomes zero, and the fields in the coupler recombine, resulting in light at the interferometer output. Depending on the birefringence, the interference of the propagating fields in the loop might be constructive or destructive. It is found that the increased loop length leads to a decrease in the free spectral range with a steeper slope of the decreasing trend. The loop length can be optimized to have high birefringence and integration over a chip due to its compact size. Further, the metal-filled side hole of PCF is used in the fabrication of SLI for temperature sensing (Reyes-Vera et al. 2017). Due to the high-temperature sensitivity of the fabricated sensor, the wavelength shift of the interferogram is used for temperature sensing through the metal-filled side-hole PCF. Metal-filled side-hole PCF is spliced between single-mode fibers, forming SLI. The optical signal from the supercontinuum laser source is fed to the SLI through a 3 dB coupler. Longitudinal heat transfer takes place in fiber due to the thermal conductivity of metal. A polarization controller is inserted in SLI to improve the transmission of signals. The propagating wave and counterpropagating wave inside the loop are combined after traveling the loop, and the resulting birefringence is observed on the optical spectrum analyzer. Transmission of the sensor is given by

$$T = \frac{(1 - \cos\varnothing_t)}{2}, \tag{7.20}$$

where $\varnothing_t = 2\pi \Delta n_g L_T/\lambda$ is the phase difference induced between the polarization modes of PCF due to the propagating wavelength λ, group refractive index Δn_g and length of the side hole L_T of side-hole PCF. As discussed previously, the wavelength position of the deeps (λ_d) are resonant wavelengths and exist when $\varnothing_t$ is an integral multiple of 2π. Let Δn_{gt} and ΔL_T are change in group refractive index and length of the fiber when exposed to temperature, the wavelength shift of the dip position with respect to the change in temperature is given by Reyes-Vera et al. (2017)

$$\frac{d\lambda_d}{dT} = \frac{\Delta\lambda}{\lambda}\left(\frac{d\Delta n_{gt}}{dT} L_T + \Delta n_{gt}\frac{dL_T}{dT}\right), \tag{7.21}$$

where L_T is the length of fiber exposed to the thermal environment. The contribution from the modulation of the fiber birefringence caused by metal expansion would be much greater than that from the variation of the fiber length caused by thermal expansion (Reyes-Vera et al. 2014), so Eq. (7.21) can be simplified further by ignoring the second term. In Shao et al. (2016), the Lyot SLI temperature and torsion sensor are reported in which two sections of high birefringence fiber are inserted into a Sagnac loop, as shown in Fig. 7.11.

In Fig. 7.11, a broad-wavelength optical signal is passed to the Sagnac loop through a 3 dB coupler. There are two polarization controllers placed before high birefringence fibers. One HiBi fiber of length L_1 is used for temperature sensing, and another one of length L_2 is used as a reference to measure the phase induced. There is a rotating region on a single-mode fiber between two fibers, which investigates the torsion effect on the Sagnac loop. Assume that θ_1 and θ_3 are the angles formed between the polarization directions of opposing beams and the fast axes of the two sections of HiBi fiber. θ_2 is the angle formed by the fast axes of the two HiBi fiber sections. The transmission spectrum of the SLI is given by

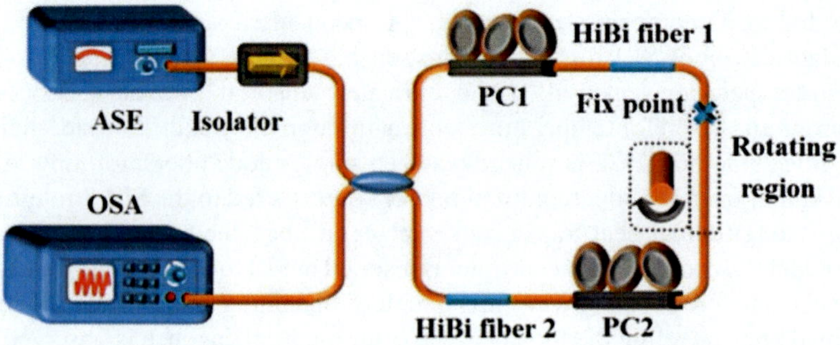

Fig. 7.11 Schematic for the complete experimental setup of temperature and torsion sensor. Adapted from Shao et al. (2016)

$$T = \left[\cos\left(\frac{\varnothing_1 + \varnothing_2}{2}\right)\sin(\theta_1 + \theta_3)\cos\theta_2 + \cos\left(\frac{\varnothing_1 - \varnothing_2}{2}\right)\cos(\theta_1 + \theta_3)\sin\theta_2\right], \tag{7.22}$$

where $\varnothing_1$ and $\varnothing_2$ are phase differences caused by orthogonal modes of HiBi fibers. The proposed Lyot–Sagnac interferometer has an advantage over conventional SLI in terms of a closely spaced interference pattern and sharp peaks in the interferogram at a fixed temperature, as shown in Fig. 7.12. When the temperature increases, the dip position shifts linearly with temperature variation.

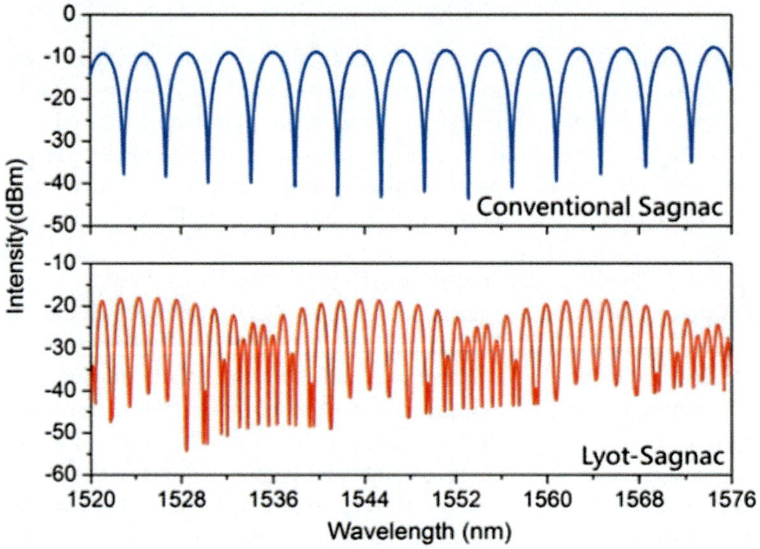

Fig. 7.12 Resultant interference pattern of conventional SLI and Lyot–Sagnac interferometer. Adapted from Shao et al. (2016)

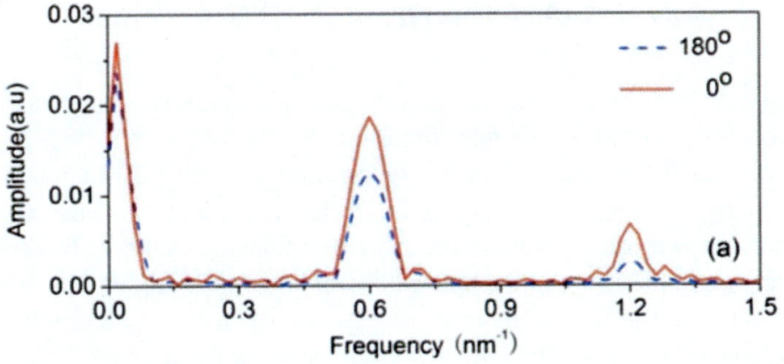

Fig. 7.13 Spatial spectral representation of the interference pattern for the twist angle of 0° and 180°. Adapted from Shao et al. (2016)

The author examined the torsion responses of the Lyot–Sagnac interferometer at room temperature to verify its torsion-sensing capacity and torsion characteristics. The SMF was fastened securely between the two lengths of HiBi fiber, and the other end was twisted by the torsion mechanism. When torsion is applied to the SMF, the angle between the fast axes of the two sections of HiBi fiber changes, resulting in changes in the spectrum's fringe visibility in the interferometer. As a result, the visibility of the fringe varies with the torsion applied, and the applied torsion may be observed and analyzed by observing the evolution of fringe visibility. To clearly notice the shift in fringe visibility, we performed a quick Fourier transform on the interferometer spectra to produce the spatial frequency spectra. Figure 7.13 shows the spatial spectra of the interferometer by taking a fast Fourier transform of the interference pattern for the twist angles 0° and 180°.

Practically, multiple physical parameters act simultaneously on many sensing systems. Therefore, simultaneous measurements of multiple sensing parameters can be made by analyzing the cross-influence of these parameters. Like torsion, the influence of torsion is investigated under various operating temperatures. Then, the influence of temperature on the torsion for different twist angles can be investigated, and the cross-influence can be demultiplexed for the simultaneous measurement of temperature and torsion. Other sensing applications, like strain sensing, are also reported in Liu et al. (2023) based on SLI. In this work, sensitivity and range of measurement are enhanced by using PMF and fiber Bragg grating. PMF has high strain sensitivity with a small range, and FBG has low strain sensitivity with a large measuring range. Together, they produce high sensitivity and range for strain measurement. Sagnac loop interferometers are also used in the refractive index measurement (Li et al. 2022) and intrusion localization (Esmail et al. 2022). Apart from the single Sagnac loop-based sensing system, multiple Sagnac loop-based systems can be used in distributed sensing applications. Zhirnov et al. in (2022) has proposed distributed acoustic sensing based on demultiplexing of double Sagnac loop interferometers.

7.2.4 *Fabry–Perot Interferometer*

Fabry–Perot interferometers play an important role in the implications and design of laser sources, optical filters, and fiber optic sensors. In the last few decades, tremendous growth has been observed in the fabrication of Fabry–Perot interferometer-based fiber optic sensors (Weimin et al. 2018; Islam et al. 2014). The measurement of displacement is a basic requirement in various sensing systems. In Zhou and Yu (2010), the FPI sensor is fabricated with a fiber end face and a highly reflective surface with air in between for displacement measurement. In this work, demodulation techniques are given for displacement measurement and improved resolution for demodulation. FPI-based displacement sensors with ultra-high sensitivity are reported in Yang et al. (2021) and Tian et al. (2020), where the working principle is based on the Vernier effect. It contains a movable microsphere with controlled forward and backward movement inside the capillary. Recently, FPI-based displacement sensors with precise measurement have been applied for health monitoring systems, geotechnology, oil well monitoring, and industry construction. For example, Zhu et al. in (2017) reported three-dimensional displacement sensing for structural health monitoring systems. In this work, the fabricated sensor is an extrinsic FPI sensor placed in a spatial arrangement with dedicated inclined mirrors. The schematic of the reported technique is shown in Fig. 7.14.

In Fig. 7.14, there is a roof-like structure of reflecting surfaces with the inclination θ_1,θ_2, and θ_3 in a rectangular coordinate system. OF_1, OF_2, and OF_3 are respective

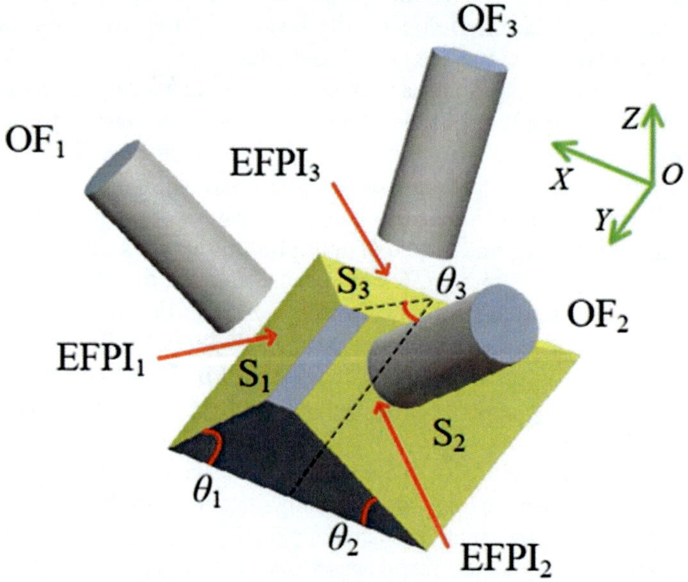

Fig. 7.14 Schematic diagram of a 3D displacement sensor with the spatial arrangement of three FPIs. Adapted from Zhu et al. (2017)

end surfaces of S_1, S_2, and S_3. If L_1, L_2, and L_3 are the length of cavities formed by end face OF_1, OF_2, and OF_3, respectively, and ΔL_1, ΔL_2, ΔL_3 are respective changes in cavity change, the respective displacement changes Δx, Δy, and Δz are given by

$$\begin{bmatrix} \Delta x \\ \Delta y \\ \Delta z \end{bmatrix} = \begin{bmatrix} \sin\theta_1 & 0 & \cos\theta_1 \\ -\sin\theta_2 & 0 & \cos\theta_2 \\ 0 & -\sin\theta_3 & \cos\theta_3 \end{bmatrix}^{-1} \begin{bmatrix} \Delta L_1 \\ \Delta L_2 \\ \Delta L_3 \end{bmatrix}. \tag{7.23}$$

The measurement of change in displacement is a function of the inclination angle of the roof-like surface, and these angles will change with a change in the 3D orientation of the roof-like structure. In Eq. (7.23), the interference pattern depends on the cavity length of the respective FPI, and the intensity I_i of the resultant signal is given by

$$I_i = I_{i1} + I_{i2} + 2\sqrt{I_{i1} I_{i2}} \cos\left(\frac{2\pi L_i}{\lambda} + \varnothing_i\right), \tag{7.24}$$

where I_{i1} and I_{i2} are reflected intensities from end faces and reflecting mirrors, respectively. $\varnothing_i$ represents the offset phase induced in the ith FPI. Free spectral range FSR (difference between two adjacent peaks or valley λ_a and λ_b) of the resultant interference in Eq. (7.24) is governed by the following expression (Zhu et al. 2017):

$$\text{FSR}_i = \lambda_{ib} - \lambda_{ia} = \frac{\lambda_{ia}\lambda_{ib}}{2L_i}. \tag{7.25}$$

Using the demodulation process given in Eq. (7.25), the length of ith cavity L_i can be determined. A schematic diagram of the complete setup is shown in Fig. 7.15, where the system works in reflection mode. The optical signal from a laser source is coupled to three input–output channels controlled through an optical switch.

The output of individual channels is observed on a computer-controlled four-channel optical interrogator, with the interference pattern of each FPI shown in Fig. 7.16. Apart from the air gap between the fiber end facet and the reflecting surface, other fabrication techniques are also reported. Liu et al. in (2022), fabricated FPI based on photopolymer material for temperature sensing applications. In this work, cylindrical photopolymer material with length and diameter of 150 and 20 μm is grown after the end face of single-mode fiber, the fabrication process is illustrated in Fig. 7.17. As depicted in Fig. 7.17, once the polymer material comes to a solid state, the fiber at the other end of the material is removed. The reflecting interfaces are SMF-photopolymer material and photopolymer–air interfaces with a reflectivity of R_1 and R_2, respectively. Here, the thermo-optic effect of photopolymer material is utilized for temperature sensing. Consider n_1, n_2, and n_3 as the refractive indices of single-mode fiber, photopolymer material, and air; the reflectivity at both interfaces is given by (Liu et al. 2022)

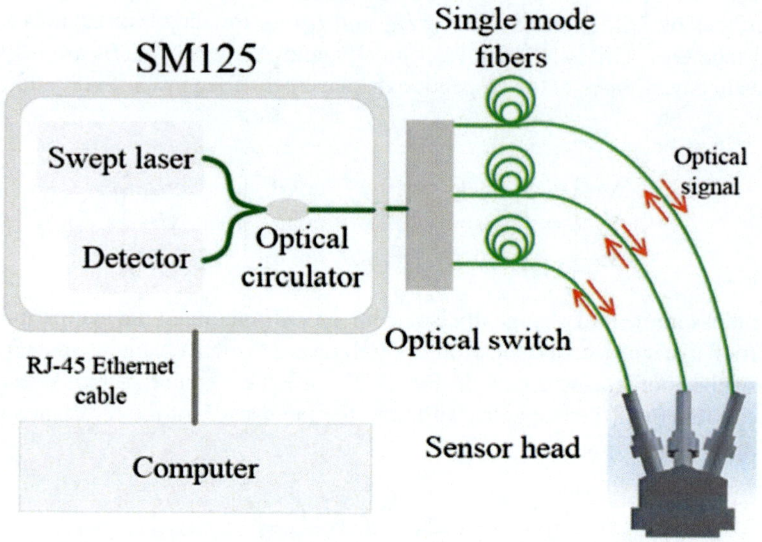

Fig. 7.15 Schematic of the complete setup, including transmission, modulation, and detection of spatially arranged FPIs. Adapted from Zhu et al. (2017)

$$R_1 = \left(\frac{n_1 - n_2}{n_1 + n_2}\right)^2$$
$$R_2 = \left(\frac{n_2 - n_3}{n_2 + n_0}\right)^2. \tag{7.26}$$

In the proposed FPI structure, multiple reflection with insignificant power strength is neglected, and the reflected optical field E_R is given by

$$E_R = E_0\left[\sqrt{R_1} + (1 - R_1)(1 - \alpha_1)(1 - \alpha_2)\sqrt{R_2}e^{j\varnothing}\right], \tag{7.27}$$

where, α_1 and α_2 are coefficients of transmission loss at the reflecting interface, which depend on the surface roughness of the reflecting interface. In Eq. (7.27), $\varnothing$ represents phase induced due to FPI cavity of length L, and it depends on the wavelength of the optical signal as well as on the refractive index of the photopolymer material, given by

$$\varnothing = \frac{4\pi n_2 L}{\lambda}. \tag{7.28}$$

The resulting normalized intensity $I_R = \frac{|E_R|^2}{E_0^2}$ of the superimposed signal due to FPI structure is given by

$$I_R = R_1 + [(1 - R_1)(1 - \alpha_1)(1 - \alpha_2)]^2 + 2R_1(1 - R_1)(1 - \alpha_1)(1 - \alpha_2)\cos\varnothing. \tag{7.29}$$

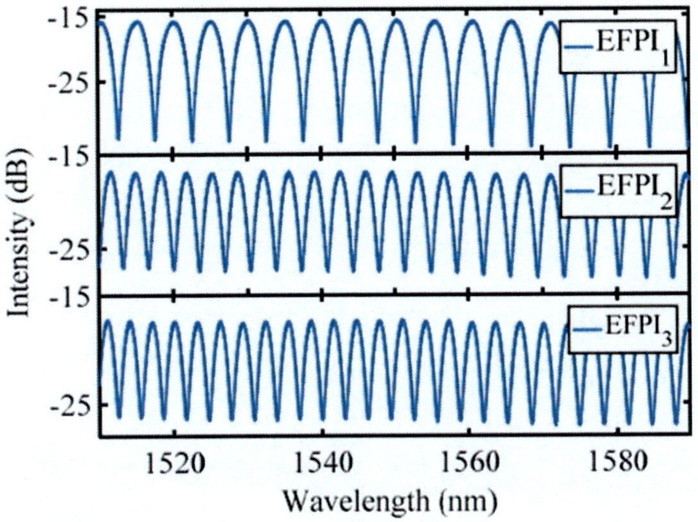

Fig. 7.16 Interference pattern of individual extrinsic FPI; EFPI1, EFP2, and EFP3 correspond to the first, second, and third extrinsic FPI, respectively. Adapted from Zhu et al. (2017)

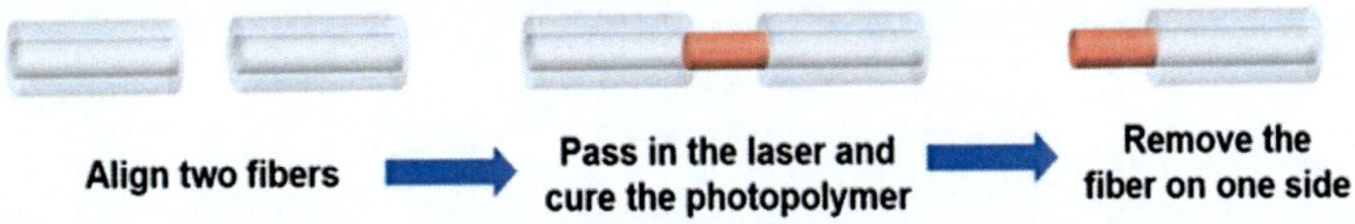

Fig. 7.17 Schematic diagram showing the fabrication process of photopolymer material-based FPI. Reprinted with permission from Sensors and Actuators B: Physical. Copyright, 2022, Elsevier (Liu et al. 2022)

The thermal sensitivity of the photopolymer material is enhanced by inscribing it inside a PDMS. The constructive interference takes place with $\varnothing = (2p + 1)\pi$, where p is an integer. The temperature sensitivity of the material is given by

$$\frac{d\lambda}{dT} = \frac{4}{2p+1}(LC + nE), \tag{7.30}$$

where C represents the coefficient of thermal expansion; E is the coefficient of transformation from thermal to optical. A liquid-level detector sensor based on the FPI structure is fabricated in Roldán-Varona et al. (2021) using hollow-core fiber and a silica capillary tube. The designed sensor claims its potential application in the level detection of immiscible liquids. In this work, dual cavities are demonstrated with more than two reflective interfaces. There are three sections, namely air, liquid, and hollow-core fiber, and their intensities of reflected signals are I_1, I_2, and I_3, respectively, as shown in Fig. 7.18.

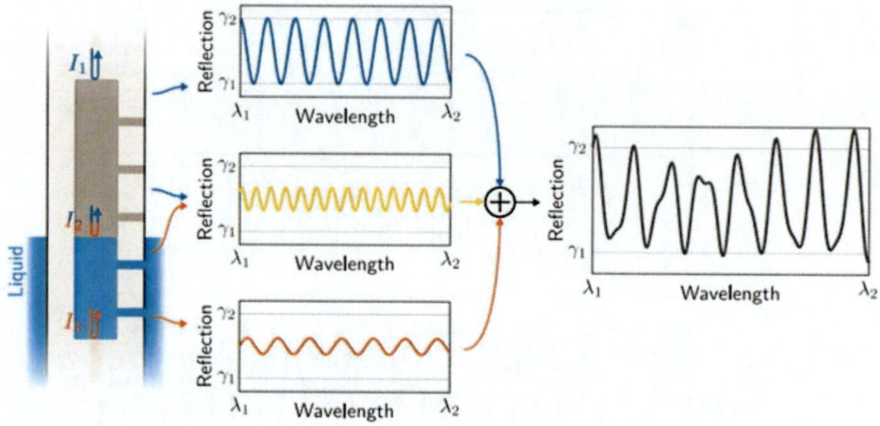

Fig. 7.18 Schematic diagram showing the interference pattern of a multi-cavity FPI-based liquid level detector. Adapted from Roldán-Varona et al. (2021)

Suppose that the cavity lengths of air, liquid, and HCF are L_a, L_l, and L_{HCF}, respectively, and their respective refractive indices are n_a, n_l, and n_{HCF}. The resultant interference of the superimposed signal with wavelength λ from all interfaces is given by Roldán-Varona et al. (2021)

$$I_R = I_1 + I_2 + I_3 + 2\sqrt{I_1 I_2}\cos\left(\frac{2\pi n_a \cdot 2L_a}{\lambda} + \varnothing_a\right) + 2\sqrt{I_2 I_3}\cos\left(\frac{2\pi n_l \cdot 2L_l}{\lambda} + \varnothing_l\right) + 2\sqrt{I_3 I_1}\cos\left(\frac{2\pi n_{\text{HCF}} \cdot 2L_{\text{HCF}}}{\lambda} + \varnothing_{\text{HCF}}\right), \tag{7.31}$$

where effective refractive indices inside a medium depend on the time for which the light remains in that medium. In Eq. (7.31), the fourth, fifth, and sixth terms on the right side represent interference contributions due to HCF in air, liquid, and air–liquid. In this manner, the fabricated probe can be used to detect the level of liquid along with its refractive index. There have been some other techniques used to enhance the sensitivity of the probe reported in the past. In Salunkhe et al. (2020), the temperature sensitivity of an FPI-based probe is enhanced and optimized through a coating of polystyrene. In this technique, various thicknesses of air-free bubbles are coated on the edge of single-mode fiber. The thickness of the coating material depends on the type of solution used for temperature detection. Apart from temperature, liquid level, and refractive index detection, it has been seen that hydrogen is used as fuel and has potential applications in the fields of healthcare and industry as well. Hydrogen detection and measurement are very important from a safety point of view. In Luo et al. (2021), FPI is proposed based on nanofilm and ultrashort fiber Bragg grating for the detection of hydrogen. Firstly, an HCF is spliced with single-mode fiber, and in the second step, a multilayer graphene film is transferred to the end position of the HCF. The reflection from the graphene layer and SMF creates the FPI structure. Further, other metallic layers like Au and Pd are deposited on the tip of graphene to ensure higher reflectivity. Liu et al. in (2020), fabricated an FPI sensor based on a femtosecond laser inside water and investigated different cavity lengths through pulse energy. Similarly, fiber optic photonic crystal hollow-core fiber is reported in the FPI structure. In other types of sensing, like strain measurement, it is required to have a large detection range of strain with high resolution. In Huang et al. (2010), high resolution for large-scale detection of strain is reported through extrinsic FPI structure in reflection mode. In this work, frequency, phase, and period tracking for high-resolution measurement of strain are reported through data processing algorithms. In the frequency tracking method, the Fourier transform of the spectral interferogram (a function of wave numbers) gives harmonics of interference frequency, whose dominant frequency corresponds to a cavity length. The cavity length l is given by

$$l = \frac{n\pi}{\vartheta_E - \vartheta_s}, \tag{7.32}$$

where ϑ_s and ϑ_s represent starting and end wavenumbers in frequency domain representation of the spectral interferogram, and n is the integer that corresponds to different cavity lengths. The least cavity length is found for $n = 1$. The phase tracking is based on the intensity of interfered beams; the minimum intensity of the resulting signal occurs with the phase difference condition given by

$$\frac{4\pi l}{\lambda_v} = (2m + 1)\pi, \tag{7.33}$$

where m represents an integer and the change in cavity length can be estimated by taking the derivation of the above equation with the center wavelength location (λ_v) of the wavenumber

$$dl = \frac{d\lambda_v}{\lambda_v} l. \tag{7.34}$$

Further, cavity length change can be addressed through period tracking between two adjacent deeps in the spectral interferogram, given by

$$l = \frac{\lambda_{d1}\lambda_{d2}}{2(\lambda_{d2} - \lambda_{d1})}, \tag{7.35}$$

where λ_{d1} and λ_{d2} are wavelength locations of two consecutive deeps in the spectral interferogram.

7.3 Interferometric SPR Sensors

Surface plasmons are free-charge oscillations that occur at the interface of a thin metal layer and a dielectric overcoating with an accompanying surface plasmon wave propagating along the interface. Based on the attenuated total reflection coupler approach, light incident at a certain angle larger than the critical angle can generate surface plasmon excitation. Surface plasmon resonance occurs when the wave vector of SPW matches that of the incident transversely magnetically polarized light, resulting in a drop in the angular or wavelength spectrum involved. If the propagation constant of surface plasmons (SPs) equals that of incident photons, a portion of the signal energy is converted into electrons, resulting in SP resonance. SPR sensing gives a precise and accurate measurement of the refractive index of the material adjacent to the metal. SPR sensor technology has tremendous promise in light of the growing demand for the detection and analysis of chemical and biological compounds in a variety of critical fields such as medicine, environmental monitoring, biotechnology, drug monitoring, and food monitoring (Homola 2008). In this section, various interferometric SPR sensors are discussed in terms of their architecture, content, performance, and application.

7.4 Mach–Zehnder Interferometer SPR Sensors

7.4.1 Various Fibers in MZI Configuration

MZI's concept is straightforward, light is directed to a two-way power divider, where it is evenly divided into two portions, propagates in two distinct arms, and then merges at the end point. Deposition of a multilayer thin film behaves like SPR in one arm of MZI. The MZI configuration is the most suitable and simple technique due to the transformation of phase to amplitude variation. The added advantage of this configuration is the presence of a reference arm, which is easily accessible to compensate for variations in any other physical parameters. Some common techniques for MZI structure-based SPR are shown in Fig. 7.19.

In Akbarpour et al. (2022), a simple multimode-single-mode-multimode (MSM) sensor with an MZI structure is proposed for refractive index sensing, which is based on SPR. In this work, the sensing area is effectively exposed to external media to enhance sensitivity. This has been done by reducing the radius of the core of single-mode fiber so that a greater number of excited modes of the clad will be in contact with the environmental perturbation, as shown in Fig. 7.19a. First, an SMF is spliced between two multimode fibers (MMFs), and thereafter, the sensing area is etched and coated with gold material. MZI configuration is developed due to the excitement of multimodes in cladding mode. The partial removal of the cladding medium and replacement with the coating of gold gives the sensor high sensitivity. The various thicknesses of the coated layer and the length of the SMF are investigated

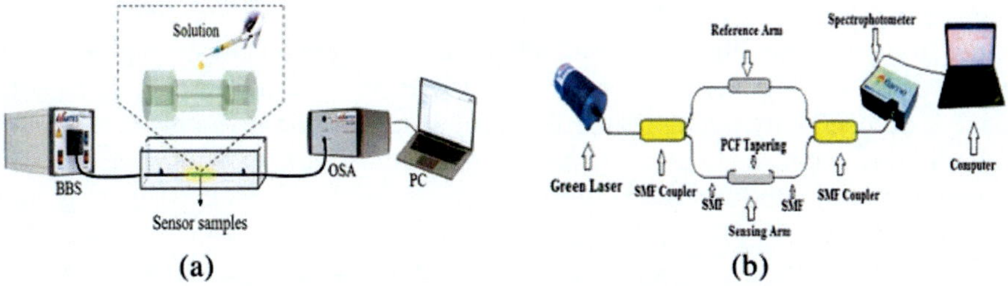

Fig. 7.19 Mach–Zehnder structured SPR sensor based on **a** MMF-SMF-MMF. Reprinted with permission from Optical Fiber Technology. Copyright, 2022, Elsevier (Akbarpour et al. 2022) and **b** SMF-PCF-SMF with tapered PCF. Reprinted with permission from Journal of Optics, Copyright, 2023, Springer Nature (Jassam and Ahmed 2023)

in relation to the sensitivity of the sensor. The proposed MSM sensor works on the modal interference of core mode and multiple excited modes. The core diameters of single-mode fiber and MMF are 8.2 and 50 μm, respectively, and both have the same diameter of 125 μm. Due to the collision of excited modes with the outer surface of the fiber, there is leakage of energy out of the fiber in the form of evanescent waves (EWs). Finally, these EWs reach the surrounding medium of the coated material, which improves the sensing capability. So, the free spectral range is given by

$$\Delta\lambda_{\text{FSR}} = |\lambda_{m+1} - \lambda_m| \approx \frac{\lambda_m^2}{l\Delta n_{eff}^m}, \tag{7.36}$$

where λ_m represents the dip position of the interference signal, l is the length of SMF, and Δn_{eff}^m is the difference in effective refractive index between fundamental and mth order mode of the excited signal. As Eq. (7.36) indicates that the free spectral range decreases with an increase in the length of SMF, instead of MSM configuration, photonic crystal fiber is used as a sensing element between two SMFs in MZI-structured SPR (Jassam and Ahmed 2023). In this structure, one arm containing an SMF-PCF-SMF section is used as the reference arm, and another arm contains a tapered PCF between two SMFs acting as the sensing arm, as shown in Fig. 7.19b. The tapered section of PCF is coated with gold material. After coating, it is surrounded by toxic lead ions to change the propagating properties of the excited mode. The broad-wavelength optical signal is split equally between the two arms of the MZI structure. Light from the core of SMF excites multiple modes in the collapsed region between SMF and PCF. The propagation property of the modes traveling in the PCF section is influenced by the refracting index of the surrounding tapered region.

Fiber Bragg grating is also used in the MZI structure, where one arm of the structure contains optical integration of the solution to avail SPR sensing (Nemova et al. 2008). This solution measures the phase variation in the waveguide. Each MZI arm has five layers: substrate, waveguide, buffer, metal layer, and sensing medium. The medium detected is placed on top of the metal layer. There are two operating modes in this five-layer structure: one is guided mode existing inside the waveguide, and another is pure surface plasmon polariton (SPP) existing at the dielectric-metal layer. In guided mode, there is an oscillation in the waveguide, contributing weak SPP, whereas pure surface plasmon gives a strong existence of electromagnetic waves and decays exponentially through other dielectric layers. A Bragg grating is imprinted into the waveguide, and the buffer layer acts to excite pure plasmon polariton. The refractive index of the waveguide (w) and buffer layer (b) is modulated by the grating of period Λ given by the expression:

$$\Delta n_{w,b} = n_{w,b}\left[1 + \cos\left(\frac{2\pi}{\Lambda}z\right)\right]. \tag{7.37}$$

7.4.2 *Tunable Coupling in MZI*

Apart from the above two techniques, the MZI structure is developed with a tunable coupling coefficient for SPR (Levy et al. 2006). Using the power splitting ratio, the proposed SPR sensor can be used to sense the full range of parametric behavior of the sensed material. The proposed methodology claimed a 20% increase in sensitivity over the conventional approach for the MZI structure. As depicted in Fig. 7.20a, SPR sensing results in a change in attenuation ($\Delta\alpha$) and phase ($\Delta\varnothing$) of the optical signal, where α is the attenuation ratio of the two arms of MZI. The output intensity of SPR is given by

$$I_{\text{out}} = \alpha^2 I_{\text{in}}. \tag{7.38}$$

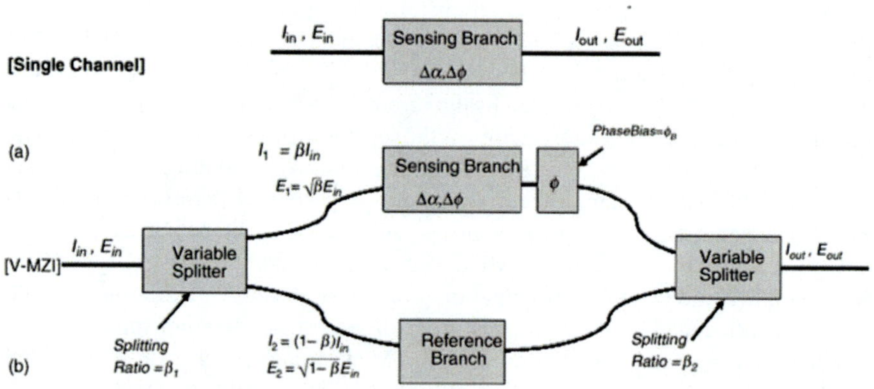

Fig. 7.20 Schematic of MZI structured SPR sensor based on tunable coupling ratio of power splitter. Reprinted with permission from Sensors and Actuators B: Chemical, Copyright, 2006, Elsevier (Levy et al. 2006)

Suppose that K_1 is the power coupling coefficient for the sensing arm and $(1 - K_1)$ is for the reference arm, the incoming intensity I_{in} will be distributed as $K_1 I_{in}$ and $(1 - K_1)I_{in}$ in sensing and reference arm, respectively, shown in Fig. 7.20b. It should be noted that $K_1 \neq 1/2$, otherwise the MZI structure will be symmetric. Consider K_2 as the power coupling coefficient at the combining end of MZM. The output of a variable MZI structure can be written as

$$I_{\text{out, MZI}} = I_{\text{in}}\left[\alpha^2 K_1 K_2 + (1 - K_1)(1 - K_2) + 2\alpha\sqrt{K_1 K_2 (1 - K_1)(1 - K_2)}\cos(\varnothing)\right]. \tag{7.39}$$

The sensitivity of the variable MZI structure with respective to refractive index n_{eff} is

$$\frac{\partial I_{\text{out,MZI}}}{\partial n_{eff}} = \frac{2\left[\alpha K_1 K_2 \Delta\alpha + \sqrt{K_1 K_2 (1 - K_1)(1 - K_2)}(\cos(\varnothing)\cdot\Delta\alpha - \alpha\sin(\varnothing)\cdot\Delta\varnothing)\right]}{\Delta n_{eff}}. \tag{7.40}$$

As appears from Eq. 7.40, a change in sensitivity can be obtained with a variable power coupling ratio, and the range of sensitivity can be further enhanced with a change in phase and attenuation ratio.

7.4.3 *Vertical Plasmonic MZI*

The above-mentioned sensors are based on the detection of the refractive index of chemical solutions. Because of its vast range of uses, the mid-infrared region has recently gained a lot of interest. Thermal imaging, infrared spectroscopy, and chemical and biological sensing are examples of these uses. This spectral band is significant because many chemical and biological substances have their unique absorption inside it. In Shamy et al. (2020), An MZI-structured mid-infrared plasmon waveguide-based metal–insulator layer is proposed for the on-chip design of a gas sensor. The proposed design has a metal–insulator layer acting as the reference arm and a metal–insulator layer acting as the sensing arm of the MZI structure, as shown in Fig. 7.21. The metal layer is filled with silver, and the insulator layer is sapphire (Al_2O_3) forming a vertical MZI structure. There are two slots of width W_1 and W_2 acting like power splitters and combiners, respectively. The optical signal from W_1 splits into two parts and travels through the reference and sensing arms, finally combining at W_2. The separation between two slots is L and the propagation constant is β. The intensity of the combined signal is given by Shamy et al. (2020)

$$I_{\text{out}} = I_{\text{in}}\left[\left(a_1 a_2 e^{-\alpha_1 L} + b_1 b_2 e^{-\alpha_2 L}\right)^2 - 4 a_1 a_2 b_1 b_2 e^{-(\alpha_1+\alpha_2)L}\sin^2\left(\frac{2\pi\left(n_{eff,\text{MI}} - n_{eff,\text{MIM}}\right)L}{2\lambda}\right)\right], \tag{7.41}$$

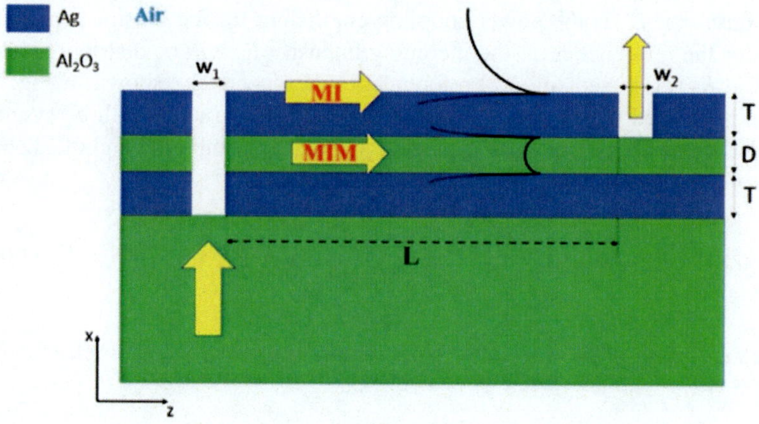

Fig. 7.21 Schematic diagram for the on-chip design of an optical gas sensor based on the MZI structure. Adapted from Shamy et al. (2020)

where a_1, a_2, and b_1, b_2, and α_1, α_2 are coupling coefficients of input and output, and losses of metal–insulator-metal and metal–insulator layers, respectively. The resonance wavelength is $\lambda_{res} = \Delta n_{eff} L/m$; and free spectral range is $\lambda_{\mathrm{FSR}} = \lambda^2/(\Delta n_{eff} L)$, where m is an integer and $\Delta n_{eff} = n_{eff,\mathrm{MI}} - \Delta n_{eff,\mathrm{MIM}}$.

In theory, MIM waveguides support two TM modes: one with symmetric and the other with antisymmetric transversal electric field components (Ex). However, the symmetric mode is more interesting since the antisymmetric mode has poor confinement and is cut off at tiny slot widths. The refractive index of the reference arm is greater than that of the sensing arm due to the high confinement factor. The reference arm refractive index further increases with a decrease in the wavelength and slot width. Two major obstacles exist in the construction of a high-performance plasmonic MZI gas sensor in the MIR region employing a MI waveguide as the sensing arm: (1) The sensitivity of the MI waveguide is low for two reasons: first, when the insulator index drops, so does the sensitivity. As a result, the sensitivity of the MI waveguide for gas sensing is lower than that of higher-index biomolecules and liquid sensing, as shown in Fig. 7.22. Another reason is that when the operating wavelength moves away from the metal plasma resonance wavelength, the sensitivity of the waveguide diminishes. (2) MI waveguides with low-index gas as the insulator material have substantially lower MI waveguide losses than MIM waveguides.

In order to enhance the transmission spectra of the proposed sensor, a grating structure incorporated at the substrate-metal interface (shown in Fig. 7.23a) will improve the power coupling to the layer, which in turn enhances the power of the transmission spectra as shown in Fig. 7.23b. It should be noted that the working of MIM in single mode can be designed with a value lower than the threshold value of slot thickness and operating wavelength.

Further, a metal–insulator-metal nanostructure is reported in Shen et al. (2017) based on the plasmon MZI structure. This simple interferometric sensor is made up

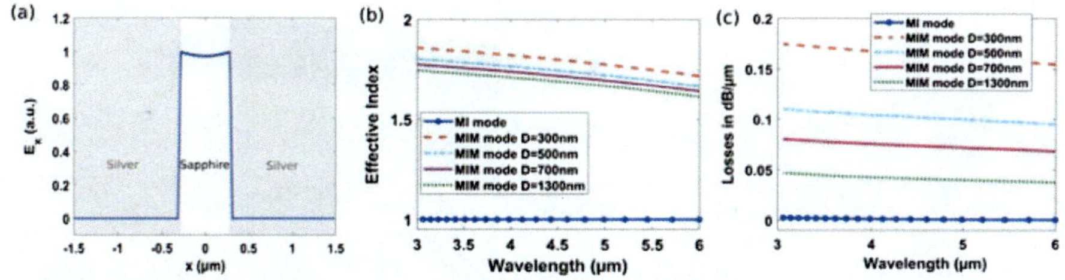

Fig. 7.22 **a** Metal–insulator-metal with $D = 600$ nm **b** effective refractive index and **c** variation in losses of layers as a function of wavelength for different values of D; D: slot width, MI: metal–insulator; MIM: metal–insulator-metal. Adapted from Shamy et al. (2020)

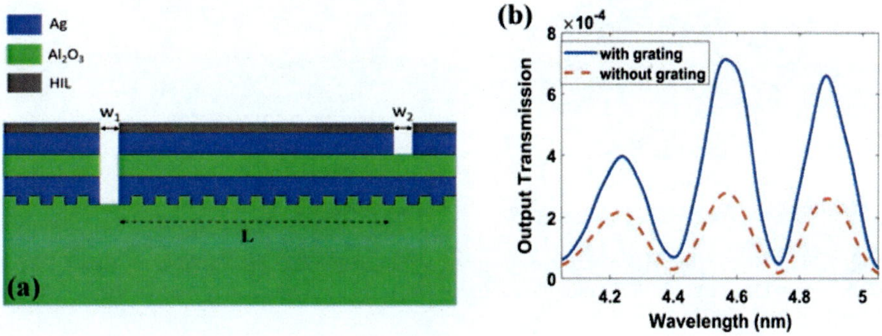

Fig. 7.23 **a** Vertical MZI structure incorporated with a grating substrate-metal layer and **b** improvement in the power spectra of transmission. Adapted from Shamy et al. (2020)

of a semi-circular aperture-slit nanostructure patterned on a multi-layered film. This nanostructure scatters a normally incident light beam into multi-frequency surface plasmon polariton propagating at the metal/dielectric interface and in the metal–insulator-metal (MIM) waveguide, causing interference at the aperture in the far-field scattering. In this structure, a glass substrate is taken, and three layers of Au-Sio_2-Au are deposited on it. A schematic diagram of the proposed architecture is depicted in Fig. 7.24. The Au layer has a thickness of 300 nm, while the core SiO_2 layer has a thickness of 100 nm. Through all three layers, a single deep semi-circular nano slit with a width of $w = 300$ nm is printed. Only the first metal layer is used to design an aperture with a radius of $r = 150$ nm. When irradiated by a collimated light beam from the glass substrate side, the semi-circular nanoslit scatters a portion of the incident radiation into SI-SPPs (sensing arm) and MIM-SPPs (reference arm). These two SPP modes travel through the light channels and are dispersed by the central aperture, resulting in far-field interference. This plasmonic MZI architecture keeps the sensor arm from being lit by incoming light, removing the need for two SPP modes of interference.

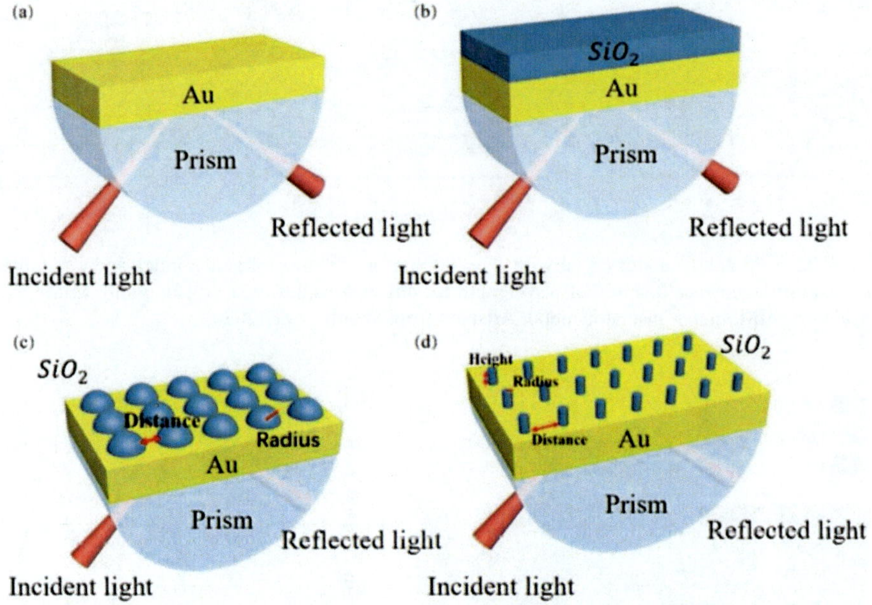

Fig. 7.24 Various structures of surface plasmon resonance are based on **a** gold films, **b** gold silicon dioxide thin films, **c** gold silicon dioxide nanospheres, and **d** gold silicon dioxide nanorods. Adapted from Sun et al. (2023)

7.4.4 Polarization Controlled MZI

To analyze the phase change generated by the SPR effect, typical prim-coupled systems employ a Mach–Zehnder interferometer. Mechanical drift and vibrations produce a significant degree of unwanted noise in a typical situation because the probe beam and the reference beam must be divided into two independent optical channels. In the new scheme, the differential phase detection technique is based on an equal optical path for the reference beam as well as the sensing beam (Ho et al. 2004). In a surface plasmon resonance-based phase-sensitive interference imaging platform, a split Mach–Zehnder interferometer is presented for demonstrating phase modulation of p- and s-polarized beams (Banerjee et al. 2017). For a p-polarized beam, a significant phase change has been observed in an Al-coated prism-based SPR configuration, whereas practically there is no phase change for an s-polarized wave. In this work, a prism-plasmon metal film-dielectric analyte (three layers each with a respective dielectric constant) is used. ϵ_1, ϵ_2, ϵ_3. Kretschmann configuration is carried out for phase sensing in ambient air environments. The resonance condition of the Kretschmann–Raether configuration is

$$\sqrt{\epsilon_1}\frac{\omega}{c}\sin\theta_1 = \frac{\omega}{c}\sqrt{\frac{\epsilon_2\epsilon_3}{\epsilon_2 + \epsilon_3}}. \tag{7.42}$$

In Eq. (7.42), θ_1 represents the angle of incidence of light on the first layer of the proposed configuration. Let, d be the thickness of the plasmon layer, the overall reflectivity of a three-layer configuration is (Banerjee et al. 2017)

$$R = |r_{123}|^2 = \left| \frac{r_{12} + r_{23}e^{jk_2 d}}{1 + r_{12}r_{23}e^{jk_2 d}} \right|^2, \tag{7.43}$$

where k_2 is the propagation constant of the plasmon medium, and r_{12} and r_{23} are coefficients of reflection at prism-plasmon and plasmon-dielectric layers, respectively. A beam expander and collimator are used to spatially extend and collimate the light beam from a He–Ne laser source. After passing through a polarizer with a transmission axis of 45° and a quarter-wave plate with a fast axis of vertical, the collimated beam is turned into a circularly polarized beam. This circularly polarized beam strikes the birefringence lens, which has an optic axis perpendicular to the primary axis. For an s-polarized beam, there is no phase change. The spatial transformation of transmission (t) and reflection (r) of the polarization beam splitter is done with two polarization states: horizontal (with subscript H) and vertical (with subscript V). The general matrix model for polarization beam splitters is:

$$M_{\text{PBS}} = \begin{bmatrix} t_H & jr_H & 0 & 0 \\ jr_H & t_H & 0 & 0 \\ 0 & 0 & t_V & jr_V \\ 0 & 0 & jr_V & t_V \end{bmatrix}. \tag{7.44}$$

Kashif et al. in (2016) gave SPR phase sensor based on MZI structure to investigate the performance of drift, sensitivity, induction of the phase solution, and repeatability of the technique through polarization detection of s and p-polarized signals, which depends on the phase shift. The schematic diagram of the technique is shown in Fig. 7.25. The optical signal from a laser source is passed through a half-wave plate and split into two parts using a beam splitter. One of the split parts is passed through the sensing arm, which contains a three-layer prism of glass substrate, gold film, and a flow solution of glycerine. Another part of the split signal is passed through the reference arm and combined with the sensed signal at the second beam splitter. Thereafter, a polarized beam splitter is used to convert the resultant signal into an s- and p-polarized signal, which is finally detected by a high-speed photodetector. The thickness of the gold layer is 40 nm, and the flow rate of the solution along with the flow of air is optimized for high sensitivity. High sensitivity is observed for smaller glycerine samples. The performance of the SPR sensor also depends on the flow of solution in the flow cell, which causes changes in the differential phase of MZI.

Apart from the above-reported techniques, Ravikumar et al. (2018) proposed a CS–Ni film-based fiber optic interferometer sensor for polyhistidine detection. They have used wavelength interrogation for effective detection of polyhistidine with a sensitivity of 0.0308 nm/(ng/ml). Its selectivity analysis has also been performed in the presence of other biomolecules such as full proteins (trypsin, HSA, and BSA).

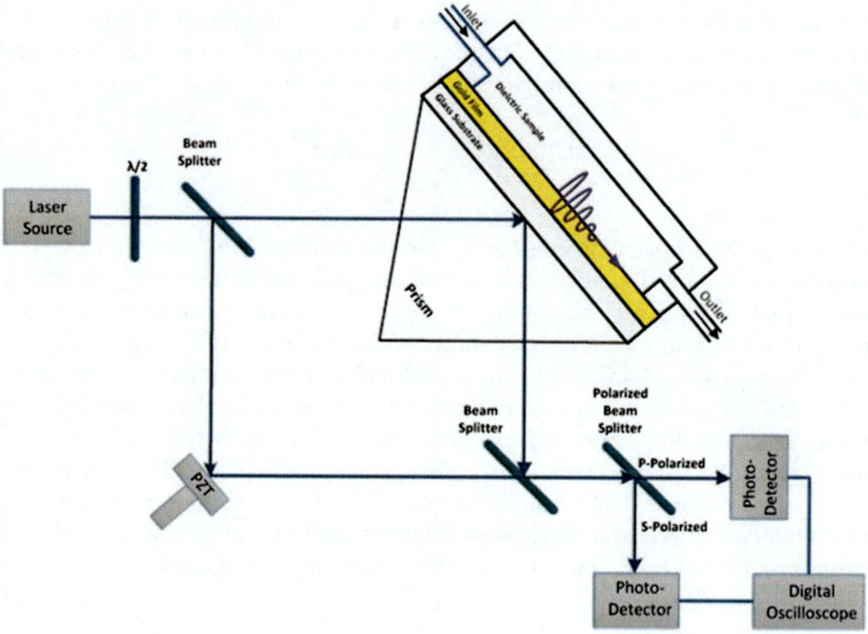

Fig. 7.25 Schematic diagram of the complete SPR sensor based on MZI structure through s-polarized and p-polarized detection of the sensed signal. Reprinted with permission from Optics Communications, Copyright, 2016, Elsevier (Kashif et al. 2016)

It has also been found that its selectivity is high due to the affinity of His-Tag to CS–Ni film. This sensor is very effective in comparison to conventional techniques such as SDS-PAGE and western blot; its size is small, and its response is too fast. The proposed fiber optic sensor was fabricated by splicing 20 mm of no-core fiber (NCF) with two single-mode fibers of (8.2/125 μm, SMF-28) using a fusion splicer as shown in Fig. 7.26b. For further processing, 2 g of chitosan was taken and mixed in a 4% acetic acid solution to get 2% chitosan solutions whose pH is around 6. Further, to get 20 ml of chitosan solution, 0.1 g of $NiCl_2$ was mixed. The mixed solution was stirred for 24 h at room temperature and then filtered to remove the excess $NiCl_2$. For the functionalization of $NiCl_2$ over the NCF structure, it needs to be cleaned first using piranha solution (it is a combination of hydrogen peroxide and concentrated sulphuric acid in a ratio of 1:3), then dipped into DI water and then dried using the nitrogen gas. In this process, organic residues were removed from the fiber surface, and then it was hydroxylated due to piranha treatment. Thereafter, the fiber surface was immersed into the polydimethylsiloxane (PDMS) microchannel, and then a chitosan–nickel blend was added to the channel and dried in an oven at 60 °C overnight. This CS-Ni film-functionalized NCF surface works as the sensing element for a histidine-tagged protein sensor. The experimental system consists of the Amplified Spontaneous Emission (ASE) broadband light source (center wavelength: 1570 nm), an optical spectrum analyzer (OSA), and a NCF sensor, as shown in

Fig. 7.26a. The proposed sensor is a type of inline fiber optic sensor based on the transmission mode. During measurement, an optical signal was launched through the ASE light source into the inline optical fiber NCF sensor, which collected the transmitted spectrum into the OSA for further analysis. Both the sensing arm and reference arm are inline within the NCF and have the same physical length but different optical path lengths due to modal dispersion.

Figure 7.26b shows the image of the spliced points of SMF and NCF. In this NCF region. The excitation of the fundamental mode of incoming light leads to exciting the higher-order modes and produces the interference between the fundamental and higher-order modes leading to the formation of self-images of the input field along the length of NCF. This process mainly excites the LP_{0m} modes into the NCF region as light propagates from the SMF to the NCF region due to the circular symmetry of the input field.

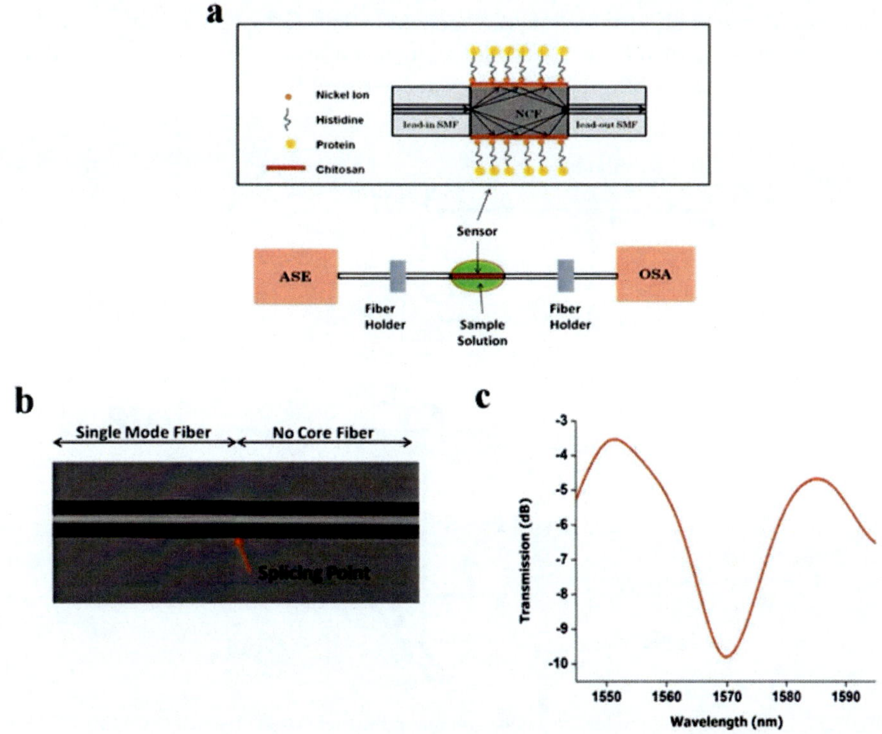

Fig. 7.26 **a** Schematic of sensing system and (inset) sensor configuration; **b** after splicing image of lead-in SMF with NCF; and **c** spectral response of the bare sensor. Reprinted with permission from Biosensors and Bioelectronics, Copyright 2018, Elsevier (Ravikumar et al. 2018)

7.5 Michelson Interferometer SPR Sensors

7.5.1 Prism-Based Pasmon Michelson Interferometer

A simple MI SPR sensor structure is depicted in Fig. 7.27, Wu et al. added a complete reflection device to their SPR sensor in 2003, which helps to improve the phase shift between p- and s-components while also making optical beam alignment easier (Wu et al. 2003). In Fig. 7.27, a He–Ne laser source is used to provide an optical signal for the interferometry. The optical signal is split into two polarization states, namely s-polarized and p-polarized, using a polarization beam splitter (PBS1). The split is driven by an acoustic optic modulator and further split by a beam splitter. One part of the split signal is detected by PD1, and another part is sensed through a plasmonic prism. A commercial phase meter was then used to determine the phase difference between the signal and reference arms. Experiments on the phase response with varied incidence angles for methanol, water, and ethanol have predicted a refractive index resolution of up to 2×10^{-7} RIU for their technology. Instead of prism coupling, the sensing arm can be single-mode or specialty-based fibers.

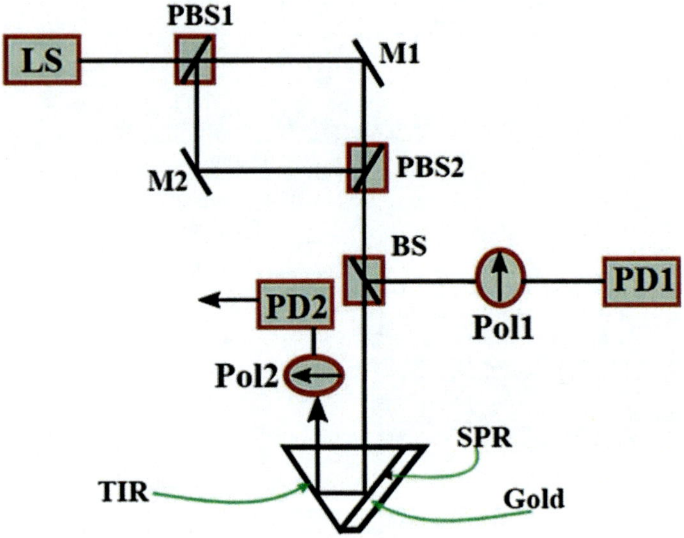

Fig. 7.27 Schematic diagram for the Michelson interferometer-based prism SPR sensor

7.5.2 Specialty Fiber-Based Plasmon Michelson Interferometer

Different interaction mechanisms can be used in optical biosensing to transform variations in the concentration of the biological substance into an optical signal. For example, waveguide or PCF-based evanescent field sensors have been proposed to detect hydrocarbons or methadone in contaminated water. The detection of volatile organic compounds (VOCs) is crucial in industry to monitor and maintain environmental air quality. Pawar et al. proposed a sensor for volatile gas sensing that is based on a dual-cavity Fabry–Perot interferometric (DFPI)-based sensor developed using graphene, PMMA, and graphene/PMMA-immobilized polarization-maintaining fiber (PMF) (Pawar et al. 2020). The optical properties of these materials change due to gas exposure. In this process, mainly wavelength and power shifts occur in the interference spectrum. It occurs due to an effective change in the refractive index of graphene and PMMA layers. In this study, authors have used different versatile sensing materials on special fibers to observe the physical and chemical interaction and explored it further for remote detection and air quality monitoring.

The DFPI principle helps in the sensing of different gases in this case. DFPI has been fabricated using a 5-mm length of PMF that was spliced with a fusion splicer to the end of single-mode fiber (SMF). The end of PMF was immobilized with the sensing material through the dip coating method. In this case, three types of sensing probes were developed: (i) graphene-immobilized DFPI; (ii) PMMA-immobilized DFPI; and (iii) graphene/PMMA-immobilized DFPI. The probes were dried at room temperature after sensing material immobilization. Figure 7.28a shows the polymer/nanomaterial functionalized DFPI. This consists of the three reflecting surfaces that help in the generation of an interference pattern. These produce three reflection coefficients R_1, R_2, and R_3 at SMF-PMF, PMF-sensing film surface, and sensing film-air, respectively. In this case, most of the light is coupled into the solid core of PMF, as shown in Fig. 7.28b, and further coupling of light takes place in the cladding part of PMF, as shown in Fig. 7.28c. Similarly, at the R_2 surface, the majority of light travels into an FP cavity, and a small amount of light reflects. In this setup, a C-band laser source (wavelength range: 1525–1565 nm) is connected at one end of a 3 dB coupler, and the other end is connected to the optical spectrum analyzer (OSA), whose wavelength range is 600–1700 nm. In this case, various cavities have been formed, and their responses have been tested with different volatile compounds such as methanol, acetone, toluene, isopropanol, and chloroform at concentrations ranging from 5 to 200 μL/L. For sensing purposes, the sensor probe was inserted into the glass gas chamber (volume 100 ml) with a known quantity of VOCs (in the range of 0.5–20 μL) added into the glass chamber. Thereafter, different concentrations of liquid, such as 5, 20, 50, 100, 150, and 200 μL/L were added and recorded on the spectrum. During measurement, when the sensor response achieves saturation during one measurement, the gas outlet of the gas chamber is opened.

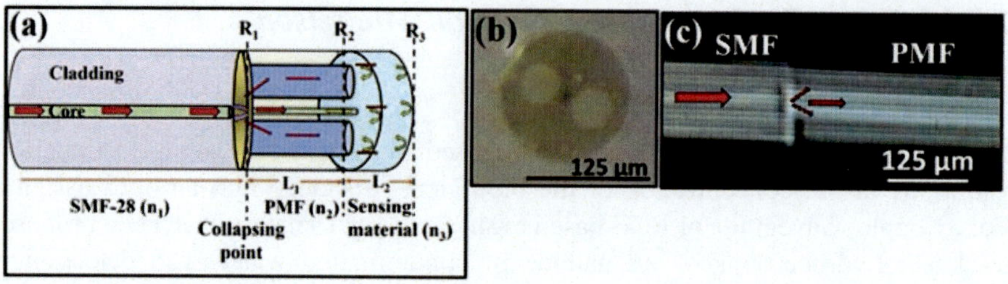

Fig. 7.28 **a** Schematic of a DFPI sensor, **b** microscopic image of PMF, and **c** image shows splicing of SMF with PMF. Reprinted with permission from Sensors and Actuators B: Chemical, Copyright 2020, Elsevier (Pawar et al. 2020)

7.5.3 Coherence Emission-Based Plasmon Michelson Interferometer

In Calo et al. (2012), polymeric waveguide-based integrated optical biosensors are constructed into MI topologies. Interferometric biosensors use the biological agent's refractive index shift and the resulting change in the interference state to produce high sensitivity values with simple design guidelines and inexpensive production costs. Creating intense, coherent, and ultra-fast light sources with nanoscale dimensions is a critical challenge for many nanophonics applications. To date, plasmonic nanolasers are one of the most promising nanophotonic devices with this extraordinary capability. In Piccotti et al. (2023), The emission characteristics of two-dimensional Au hexagonal nanodot arrays, manufactured using nanosphere lithography and linked with a dye liquid solution employed as the gain medium, are reported. Spectral and angle-resolved photoluminescence experiments as a function of pump fluence show low-threshold stimulated emission at room temperature. The polarization properties of stimulated emission are examined, exposing a strong linear polarization character managed by the polarization orientation of the pumping beam, while first-order temporal coherence characteristics are determined using tilted-mirror Michelson interferometer measurements performed as a function of pump fluence. Nanosphere lithography (NSL) was used to create hexagonal nanodome arrays (HNDA) using the multi-step procedure seen in Fig. 7.29a, with measurements made as a function of pump fluence. In the first step, a monolayer of 522 nm diameter polystyrene (PS) nanoparticles (by MicroParticles GmbH) is self-assembled on a silica glass substrate to produce a tightly packed hexagonal array. Following that (step 2), a reactive ion etching (RIE) treatment is done on the PS nanosphere monolayer in an Ar + O_2 environment to lower the nanospheres' diameter (down to around 330 nm) while maintaining their ordered organization. Magnetron sputtering (step 3) is used to deposit a silica layer (thickness 200 nm) to create a continuous, nanostructured coating and a more mechanically stable device. Finally, a 70 nm thick Au film is sprayed onto the silica layer to create the plasmonic nanodome array (step 4). A second set of samples was created for comparison by depositing 70 nm of silica instead of Au to generate a completely dielectric nanodome array. COMSOL Multiphysics was used to perform finite element method (FEM) electrodynamic simulations of the near-field and far-field optical characteristics of the nanodome arrays. Figure 7.29b shows a close-up plane view of the nanoarray (the gray color highlights the rhombic unit cell). A cross-sectional view of the gold nanodome array is shown in Fig. 7.29c, and a sketch of the hexagonal array in the reciprocal space is shown in Fig. 7.29d. The red region highlights the first Brillouin zone. A map of the Au nanodome array in ethanol taken in TM mode is depicted in Fig. 7.29d.

A modified Michelson interferometer with two angled mirrors was set up to characterize the temporal coherence of our devices' laser output, as shown in Fig. 7.30. The sample is pumped by the Nd:YAG laser's second harmonic and the emission passes through an iris in front of a beam splitter. After that, the beam is divided and directed toward two inclined mirrors, one stationary and the other adjustable along the beam direction. The interference pattern obtained by stimulating the sample over the threshold is shown in Fig. 7.30b. The picture has a significant number of vertical fringes, suggesting that the emitted radiation is coherent. The intensity profile as

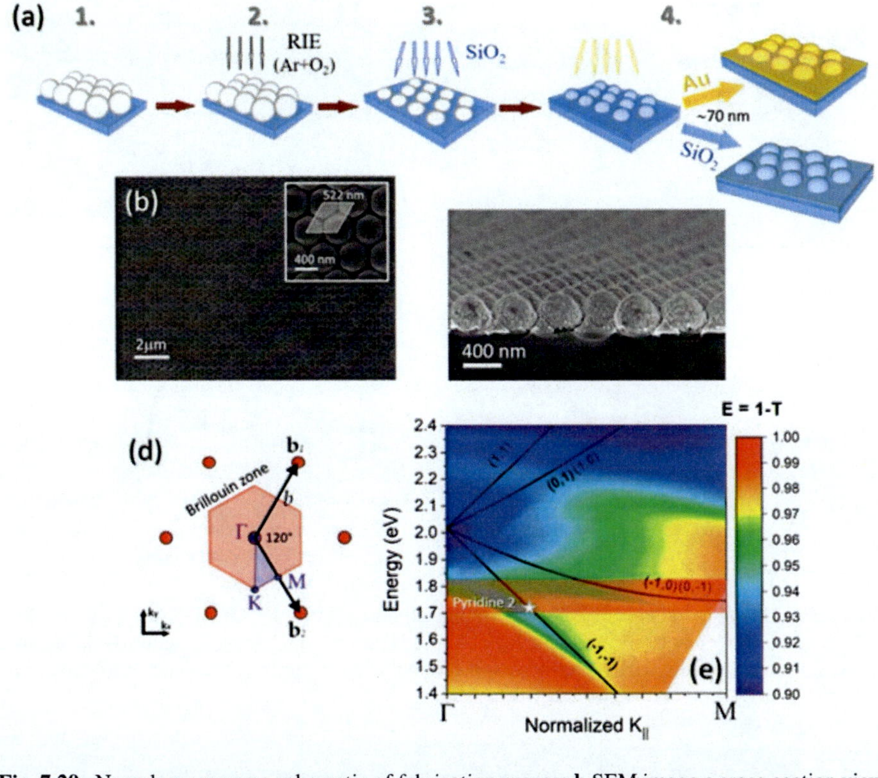

Fig. 7.29 Nanodome arrays **a** schematic of fabrication process **b** SEM image **c** cross-section view of Au nanodome **d** hexagonal array in reciprocal space **e** experimental extinction map of the Au nanodome array in ethanol taken in TM mode. Adapted from Piccotti et al. (2023)

a function of the pixel coordinate (x) normal to the fringes, as recovered from the interference pattern, is shown in Fig. 7.30c. To eliminate any non-constant background contributions to the recorded intensity, a fast Fourier transform (FFT) filter was used. Then, from the intensity profile, the visibility V as a function of OPD was derived and displayed in Fig. 7.30d. The Gaussian fit of the visibility data is shown by the orange curve in Fig. 7.30d. The first-order correlation of light beams traveling through the two arms of a Michelson interferometer is determined as a function of the optical delay τ. The first-order correlation function is given by

$$g(\tau) = \frac{< E^*[t]E[t+\tau] >}{< |E[t]|^2 >}, \tag{7.45}$$

where E is the electric field and <.> indicates temporal average. The coherence length is estimated by $L_c = c\tau_c$; c is the speed of light and τ_c is the coherence time.

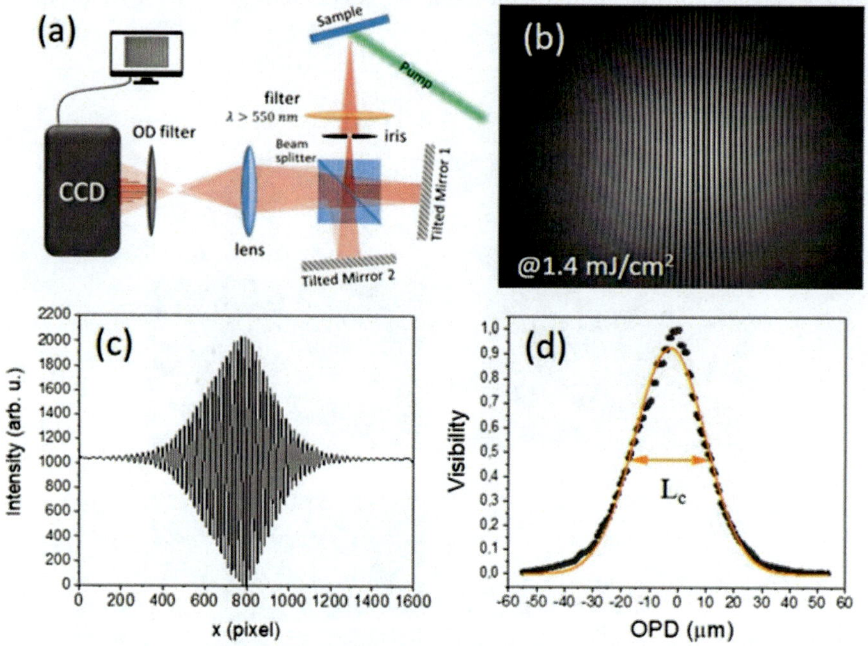

Fig. 7.30 The temporal coherence of the stimulated emission of the Au-HNDA + Py2 sample is measured using a Michelson interferometer with angled mirrors: **a** schematic diagram; **b** interference pattern; **c** intensity profile of interference pattern; and **d** fringe visibility. Adapted from Piccotti et al. (2023)

7.6 Sagnac Interferometer SPR Sensors

Recently, authors have given extensive attention to optical interferometers like the Sagnac loop interferometer based on SPR. To enhance the sensitivity of chemical sensing, various approaches are reported, like cascading of multiple interferometers, the vernier effect, microstructured fibers, and couplers. However, the use of the vernier effect in multiple sensor systems requires identical free spectral range and extinction ratios for the individual interferometers, which is a little challenging in its fabrication. The type of fibers used in the fabrication also determines their ease of handling in the fabrication process and their identical free spectral range. As mentioned before, in plasmon sensing, a part of the light in the Sagnac loop should be leaked outside to interact with the external coating. May this provision be availed of through the tapering process or some other means? However, these processes are complex in their implications.

NCF is a kind of silica fiber having the same refractive index (RI) in the core and cladding. Because NCF has no RI difference, the external substance may act as the "cladding" of NCF and seldom requires tapering and chemical corrosion, which is advantageous for the optic fiber biochemical sensor. In Zhao et al. (2021), For sucrose concentration measurement, a Sagnac interferometer with no-core fiber (SI-NCF) based on the Vernier effect is proposed. The best length of NCF is decided by sensor calibration. The SI-NCF is then cascaded with a standard Sagnac interferometer (SI) to induce the Vernier effect and improve sensor performance. The schematic diagram of the cascaded Sagnac loop system is shown in Fig. 7.31.

As shown in Fig. 7.31, a broad-wavelength optical signal is passed through a 3 dB coupler to two cascaded Sagnac loop interferometers. SLIs consist of polarization

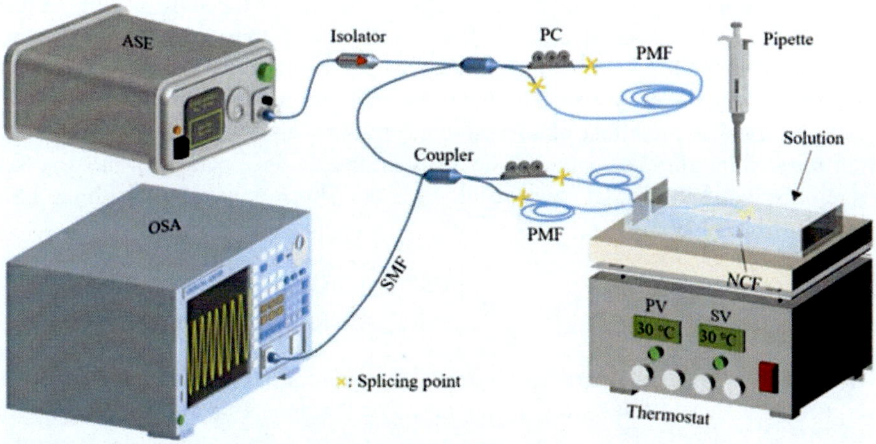

Fig. 7.31 Schematic diagram of two cascaded Sagnac interferometers based on vernier effect. Reprinted with permission from Biomedical Optics Express, Copyright 2020, Optica (Zhao et al. 2021)

controllers and polarization-maintaining fibers. The first SLI in the path of the optical signal is treated as a reference arm, and the second SLI incorporates no-core fiber and acts as a sensor arm in the detection of sucrose concentration. The sensing part of the system is kept inside a thermostat to isolate it from environmental temperature variation. The transmission intensity of simple SLI with PMF of length L, birefringence Δn, and wavelength λ is given by:

$$I = \frac{1}{2}\left[1 - \cos\left(\frac{2\pi \Delta n L}{\lambda}\right)\right] \tag{7.46}$$

The free spectral range of the Sagnac loop interferometer can be derived as follows:

$$FSR_{SI} = \frac{\lambda^2}{\Delta n \cdot L} \tag{7.47}$$

The no-core fiber structure of SI consists of three parts, i.e., PMF-NCF-PMF, as depicted in Fig. 7.32a. The optical signal coming from the first SLI is passed through the PMF-NCF-PMF section. The cladding of NCF is the external medium (sucrose concentration). To get a proper understanding of external refractive index and NCF interaction, the beam propagating method is used for the simulation. The lengths of the NCF and each PMF section are 64 and 10 mm, respectively. The diameter, numerical aperture, and bi-refractive index of PMF are 10 μm, 0.125, and 4.5×10^{-4}, respectively. Diameter and core refractive index of the NCF are 125 μm and 1.4440, respectively. The simulating wavelength is 1550 nm, and the simulated results are depicted in Fig. 7.32b, c for the grid points of 200×200 on 150×150 μm^2. Light traveling in the core of PMF excites multiple modes at the junction of PMF and NCF due to a core mismatch. There is the formation of a four-cycle self-imaging pattern of periodic optical power distribution. The size of the self-imaging pattern varies with the change in refractive index of NCF from 1.45 to 1.33. It indicates that the proposed structure can be used to detect various refractive index analytes in a biochemical detection system. The optical signal propagating through PMF-NCF-PMF experiences the effect of sucrose concentration and NCF, which results in a transmission interference pattern. The resultant interference pattern of the first SLI (reference) and second SLI (sensing) gives another type of envelope with the FSR given by Shao et al. (2015),

$$\mathrm{FSR}_{env} = \frac{\mathrm{FSR}_R \cdot \mathrm{FSR}_S}{|\mathrm{FSR}_R - \mathrm{FSR}_S|} \tag{7.48}$$

where, FSR_R and FSR_S are free spectral range of reference and sensing SLI, respectively. The shift in FSR_{env} is magnified by the impact factor of,

$$M = \frac{\mathrm{FSR}_R}{|\mathrm{FSR}_R - \mathrm{FSR}_S|} \tag{7.49}$$

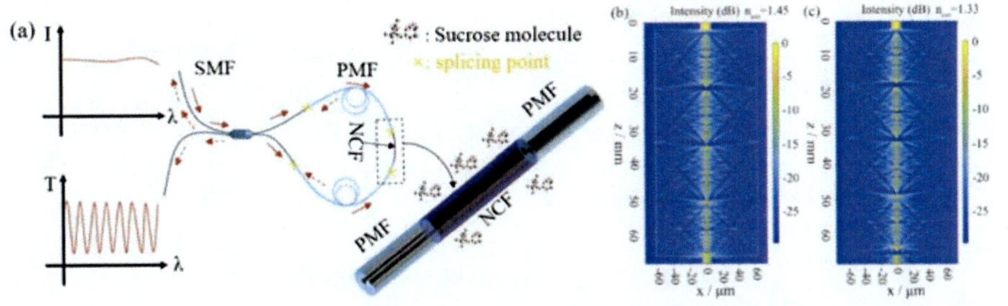

Fig. 7.32 **a** Schematic diagram of the no-core fiber SLI structure and simulated result of beam propagation method for **b** refractive index of 1.45 and **c** refractive index of 1.33. Reprinted with permission from Biomedical Optics Express, Copyright 2020, Optica (Zhao et al. 2021)

In Eq. (7.49), the sign of the denominator determines the direction of wavelength shift; the positive sign indicates no change in the slope of the envelope, and the negative sign indicates the change in the slope of the envelope.

Integration of integrated optics and microfluidics has emerged as optofluidics, which gives flexibility in the design of waveguides and sensors. PCF-based optofluidic sensors based on SLI can be implemented for the detection of the refractive index of the chemicals. The micrometer-sized air holes in PCF act as channels for the flow of fluids, whose refractive index can be measured. The various properties of PCF, like modal interference, transmission loss, and dispersion, are greatly influenced by the refractive index of the fluid passing through microchannel-like structures. In Wu et al. (2014), A C-shaped microchannel refractometer is fabricated and characterized using PCF. In this structure, C-shaped thinned fiber is spliced between PCF. Single-mode fiber is used on either side of the structure to allow the flow of analytes through the proposed structure. The proposed structure works as a Sagnac loop interferometer, as shown in Fig. 7.33. The working principle is similar to the method discussed above: the birefringence of the fiber loop gives an interference pattern to the resulting signal at the second arm of the 3 dB coupler, which induces phase change. The phase change is a function of the length of the PMF-PCF section. As shown in Fig. 7.33b, an ultra-thinned piece of C-shaped fiber is incorporated between PM-PCF and SMF. The optical signal coming from the 3 dB coupler transmits through the central hollow core region of the C-shaped fiber, and fluid in and out takes place through the opening of the C-shaped side. The effective refractive index of the guided modes changes with the change in the refractive index of the hollow channel due to the presence of fluid. The transmission spectrum due to the polarimetric interferometry of the Sagnac interferometer is already incorporated in the previous discussion. The phase birefringence $B(\lambda)$ and group birefringence $G(\lambda)$ of PM-PCF is given by,

$$\begin{aligned} B(\lambda) &= n_{eff}^{\text{slow}}(\lambda) - n_{eff}^{\text{fast}}(\lambda) \\ G(\lambda) &= B(\lambda) - \lambda \frac{dB(\lambda)}{d\lambda} \end{aligned} \quad (7.50)$$

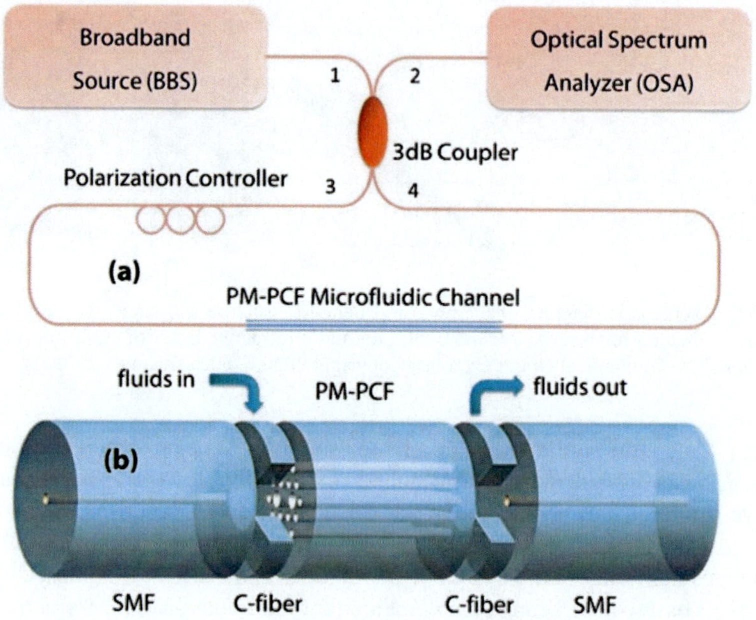

Fig. 7.33 Schematic diagram of **a** complete setup based on PMF-PCF microfluidic channel based on Sagnac loop interferometer and **b** C-shaped sensor for the detection of various analytes. Adapted from Wu et al. (2014)

where, $n_{eff}^{slow}(\lambda)$ and $n_{eff}^{fast}(\lambda)$ represents effective refractive indices of the two polarized states of PM-PCF along the slow and fast axes, respectively. Transmission dips of the interference pattern occur, when the phase difference due to phase birefringence of the PM-PCF is an integral multiple of 2π.

In Li et al. (2011), The surface plasmon on the Sagnac loop interferometer is explored to investigate the interfacial reaction of optical activity between two different refractive indices. The phase difference due to left and right circularly polarized light leads to the optical rotation of linearly polarized light inside optical activity. This relationship exists between the geometrical phase (the Berry phase due to the helical path) and the optical cavity. To measure the accumulated geometrical phase, the complete interaction zone of the optical activity medium is passed through linear polarization. In such a technique, SPR sensors exhibit excellent performance in terms of sensitivity.

The optical path length in the proposed technique is kept constant, and to enhance the intensity of the interference signal, the BS arrangement is based on Fig. 7.34a, in which a traveling light signal repeatedly travels through the path as indicated in Fig. 7.34c. However, the arrangement shown in Fig. 7.34b allows the light signal to travel only once, as indicated in Fig. 7.34d.

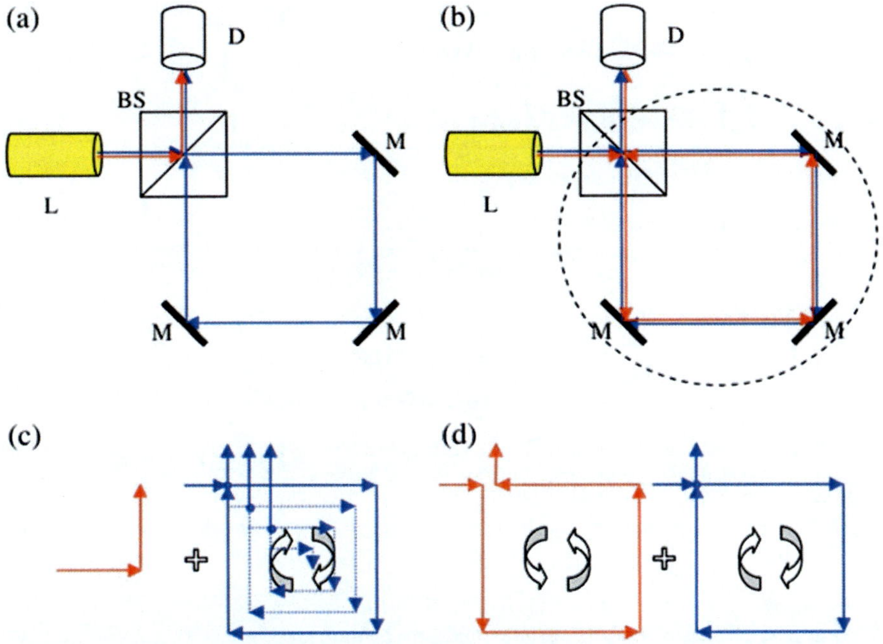

Fig. 7.34 Schematic diagram showing the effect of beam splitter orientation: **a** repeated closed loop optical path; **b** overlapping of single-time close loop optical path and single-time open loop optical path; **c** single-time open loop path and repeated close loop path; and **d** separated single-time close loop optical path and single-time open loop optical path. L: source; D: detector; BS: beam splitter; M: reflecting surface. Reprinted with permission from Optics Communications, Copyright 2011, Elsevier (Li et al. 2011)

7.7 Fabry–Perot Interferometer SPR Sensors

Many recent research studies have concentrated on creating new configurations and processes while preserving the simplicity of typical SPR sensors in order to increase the sensitivity and practicality of SPR sensors and expand their applications. Among these configurations, Fabry–Perot interferometer-based SPR sensors are generally used to enhance the sensitivity of the detection of biochemical. In Xiao et al. (2012), FPI structure with the interaction of SPR is reported to observe the phase response and optical intensity of SPR. The proposed sensor is a prism-based structure with a curved surface and a film coating on it, where a metal layer of thickness d and dielectric constant $\in_m$ is coated on it. Refractive indices of the prism and sensing material are n_p and n_s, respectively. The schematic diagram of the proposed structure is shown in Fig. 7.35. As indicated in Fig. 7.35, a p-polarized optical signal is normally incident on the curved surface of the prism and reflected at the prism-metal interface. A part of the light on the interface is transmitted to the dielectric layer, which experiences SPR sensing; the remaining part of the light is reflected and emerges from the curved surface of the prism.

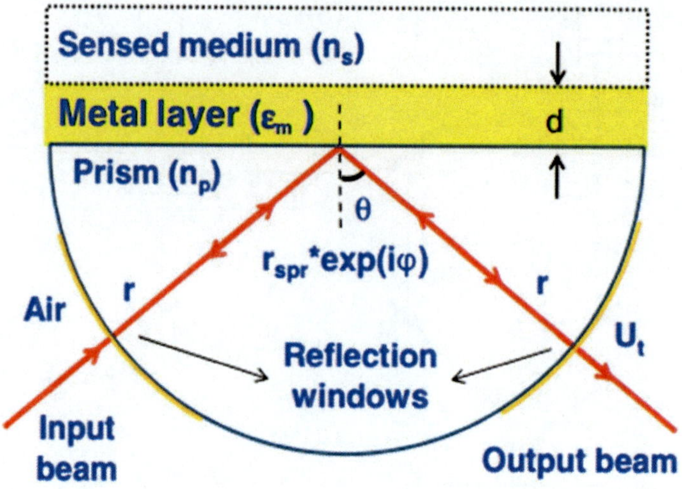

Fig. 7.35 Prism-based Fabry–Perot interferometry through SPR sensing. Reprinted with permission from Biomedical Optics Letters, Copyright 2020, Optica (Xiao et al. 2012)

Due to the reflection at the prism-metal layer interface, phase change φ induced in the reflected signal. The reflectance at the reflecting interface is given by,

$$R = r_{\mathrm{SPR}} E^{j\varphi} \tag{7.51}$$

where, r_{SPR} is the magnification in the amplitude of the reflected signal due to SPR. The reflectance R is influenced by various parameters like the dielectric constant and thickness of metal, the sensing medium and its refractive index, and the resonance frequency. The incident optical signal has multiple reflections in the Fabry–Perot cavity, and the emergent signals with various phase changes have different path lengths and intensities. Assume that E_o is the amplitude of the incident wave, r is the reflectance at the film prism, and δ is the round-trip phase change caused by the cavity length. The intensity of the transmitted signal is (Xiao et al. 2012):

$$I = \frac{A_o^2 r_{\mathrm{SPR}}^2 (1 - r^2)^2}{(1 - r^2 r_{\mathrm{SPR}}^2)^2 + 4r^2 r_{\mathrm{SPR}}^2 \sin^2(\delta + \varphi)} \tag{7.52}$$

The phase shift in SPR has a greater influence than its amplitude counterpart and has demonstrated improved sensing sensitivity and easier SPR imaging fulfillment. However, most modern SPR sensors are based on angular or wavelength interrogation, owing to the complexity of optical phase shift detection. A similar structure is reported in Zain et al. (2022) for food safety purposes through food preservative detection in the solution. A He–Ne laser was employed as a light source in this experiment. The light reflected from the prism's surface was monitored using a light detector. To guarantee appropriate light polarization was incident on the prism surface, a polarizer was used. The prism was covered with a 50-nm gold layer using a pre-set thickness sputter coating method. The prism was coated with sputter coating equipment, which regulates the thickness of the gold layer. The thickness of the gold coating was estimated via a simulation using COMSOL. The resonance dip in air was studied using angle interrogation. The rotatable stage was used to change the angle of incidence, while the optical power meter measured the light output. When light energy was linked into the plasmons of the metal dielectric contact, a resonance dip was seen. This resonance happens when the incident light's wave vector matches the plasmons' wave vector.

In Uddin et al. (2017), Sugar concentration measurement is reported based on Fabry–Perot et al. In this structure, two glass layers are coated with silver and silicon dioxide with a separation distance of d. This structure with coated material enhances the sensitivity of the interference pattern with the additional advantages of low absorption loss and adjustable transmittivity with a variable thickness of the coated layer. The transmission spectra with air and a sugar solution in the gap were measured and used to calculate the gap thickness and the liquid's refractive index. First, the gap thickness was calculated using two successive interference peaks with an air gap (refractive index 1.000). The refractive index of a liquid was estimated using the gap thickness and two matching peaks in the transmission spectrum of a sugar solution. The refractive index was estimated at a wavelength of roughly 590 nm. The transmittance spectra of the sensed signal for various concentrations of sugar are shown in Fig. 7.36. The sensor construction for the Fabry–Perot etalon described in this study is seen in the bottom rightmost position of Fig. 7.36. The top and bottom reflection layers were coated with Ag thin films on glass slides. The Ag film thickness was around 10–15 nm, which resulted in very minimal absorption and enabled enough light to flow through the etalon while creating substantial interference. On the Ag films, a thin SiO_2 layer (5 nm) was deposited. It considerably improved the surface's hydrophilicity, allowing the liquid to easily flow into the hollow. The reported work can be used to detect glucose levels in the blood sample and has potential application in the detection of various chemicals in fluids for disease diagnosis.

Lu et al. in (2006) proposed FPI-based biomolecular detection structure, which concerned the sensitivity of the system. In this structure, silver, $BaTiO_3$, and island-like gold layers are developed on quartz material, which forms the FPI structure. Thereafter, gold particles are immobilized through 1,6-hexanedithiol. $BaTiO_3$ is placed between two thin layers of reflective surfaces of Ag and Au with the help of pulsed laser deposition techniques. Ag, $BaTiO_3$, and Au were chosen due to their good reflection properties, stable aqueous solutions, and better sensitivity, respectively.

In this case, the nanostructured Au thin film "recognizes" organic molecules through Au–S covalent attachment while also acting as a key transducer via SPR. The enhanced interference sensor gadget is used as an optical sensor to detect the adsorption of molecules onto a gold surface. A flow cell was used to house the sensor. The cell was then placed on the absorbance instrument's sample stage. The sensor was rinsed multiple times with ethanol after being treated with the analyte solution and then dried under nitrogen flow. Before molecule adsorption, we test the sensor's

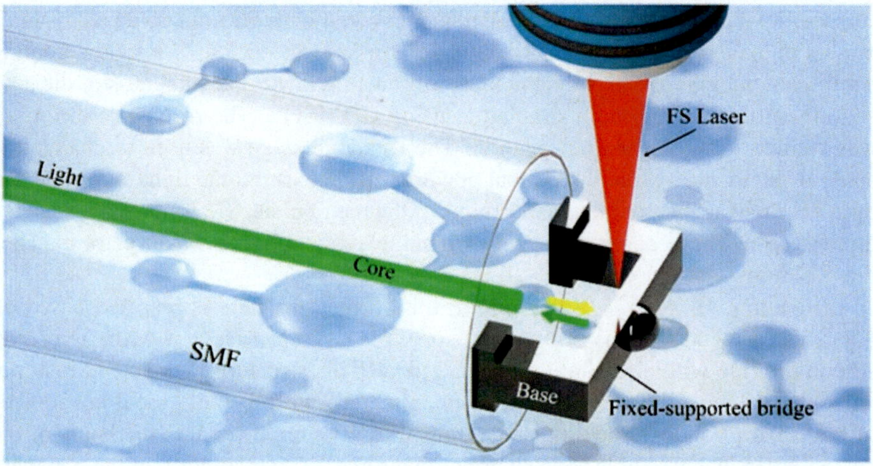

Fig. 7.36 Schematic diagram of a two-photon-based Fabry–Perot glucose detector. Adapted from Tang et al. (2022)

dependability by measuring the extinction spectrum numerous times under identical settings. after 4 h of immersion in the solution. The spectrum suffers a minor blueshift and a slight shape change after immersion, and the difference spectrum between the treated and untreated sensors now shows a noticeable interference fringe pattern. When the immersion time increases, the blueshift, shape variation, and spectrum difference of the spectrum increase. In addition, Au nanoparticles were used to boost the signal.

Further in Tang et al. (2022), two-photon printing technique is used to fabricate Fabry–Perot-based high-speed glucose concentration detection. The schematic diagram of the reported FP sensor is shown in Fig. 7.36. The polymer fixed-support bridge was manufactured on the single-mode fiber's end face. A basic FP cavity was created, and it was reactive to changes in the refractive index of the analytes. The concentration of glucose in the solution was then determined using the FP interferometer with the fiber-tip fixed-supported bridge.

A variety of fiber-tip fixed-supported micro-bridges with varying thicknesses and heights were manufactured and evaluated to get the greatest glucose sensing capability. The authors measured the FSR of the fiber-tip FP interferometer because the FSR of interference reflection may be used to compute the effective cavity length of the FP cavity sensor.

The reflection spectrum of various cavity lengths and microscopic images of the fabricated cavity are shown in Fig. 7.37. As shown in Fig. 7.37, the free spectral range is 50.18, 25.38, and 15.04 nm with respective Fabry–Perot cavity lengths of 20, 40 and 60 μm at dip location of 1383, 1375, and 1324.5 nm, respectively. The change in reflection spectrum with refractive index is depicted in Fig. 7.38. The reflection spectrum was observed by placing a fabricated FP cavity on a glucose solution on a three-directional platform.

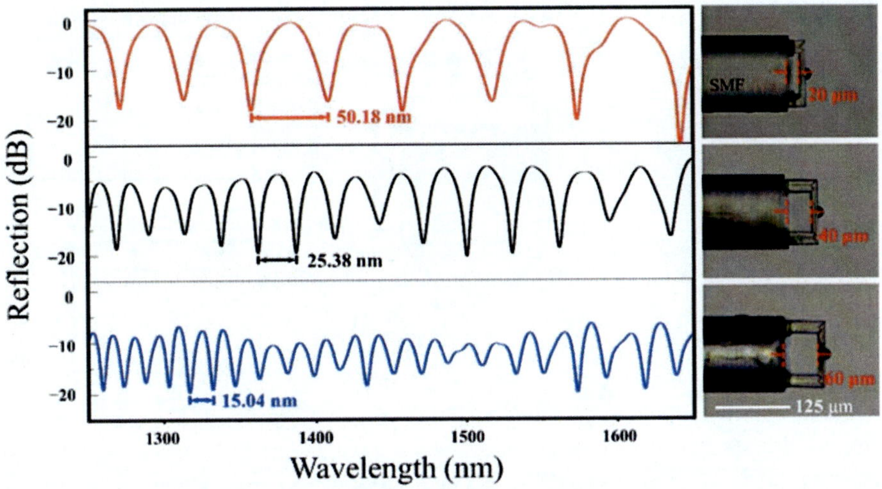

Fig. 7.37 Reflection spectrum and microscopic images of the FP glucose detector for various cavity lengths. Adapted from Tang et al. (2022)

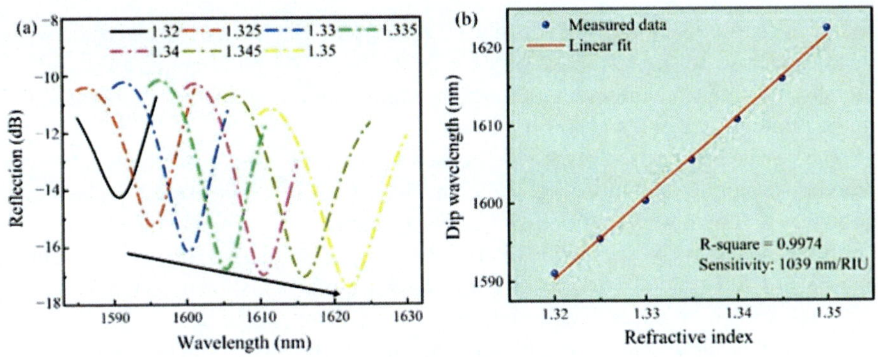

Fig. 7.38 **a** Change in the dip position with the change in refractive index and **b** dip wavelength position as a function of refractive index. Adapted from Tang et al. (2022)

The transmitter was a broadband light source, and the reflection spectrum was recorded using an optical spectrum analyzer. The reflection spectrum drift was instantly recorded, and the glucose concentration was estimated once the FP cavity was filled with the glucose solution. The wavelength position of the dip in the reflection spectrum increases with the increase in refractive index of the sensing material, and this variation is almost linear, as appears from Fig. 7.38.

The need for far-field angular resolved measurement of specular reflection, which increases the size and requires accurate calibration of the optical equipment, is a typical downside of surface plasmon resonance detection. In Allison et al. (2021), an alternate optoelectronic technique in which the plasmonic sensor is embedded into a

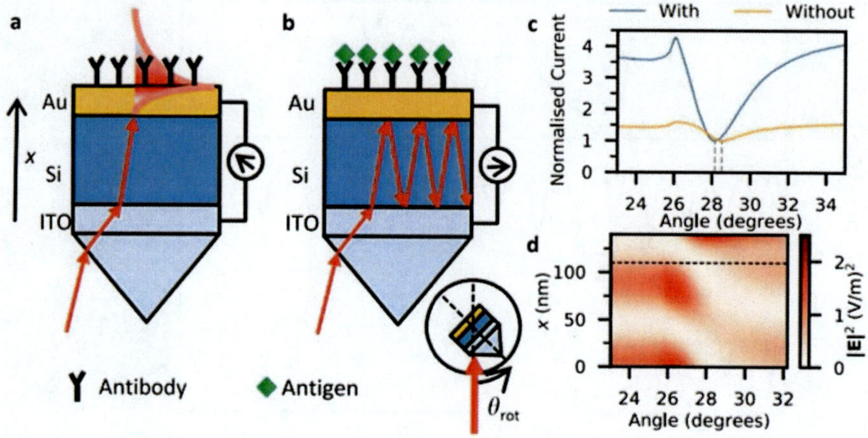

Fig. 7.39 Schematic diagram of **a** FPI surface plasmon sensor with immobilized antibody on gold surface **b** FPI surface plasmon sensor when antigen is bound to immobilized antibody **c** photocurrent as a function of rotation angle and **d** electric variation as a function of angle and thickness. Adapted from Allison et al. (2021)

solar cell. Through contact with the plasmon, incident light creates an electrical signal that is sensitive to the refractive index of a solution. The photogenerated current is increased owing to the interaction of the plasmon mode with the Fabry–Perot modes in the photovoltaic cell's absorbing layer.

The device is proposed for the SPR based on protein–protein interaction, with the working principle shown in Fig. 7.39. In Fig. 7.39a, b, prism-based Kretschmann geometry is shown with immobilized antibody and antigen bound to the immobilized antibody, respectively, for the SPR Fabry–Perot interferometer. The proposed device can be utilized for its working in angle sweep mode, as shown in Fig. 7.39c. In this case, the photocurrent has a clear angle dependency, comparable to the normal reflection measurements of traditional optical SPR biosensors. The rotation angle is used to predict the normalized current. This angular can be further extended to predict the electric field intensity, as mentioned in Fig. 7.39d.

7.8 Future Prospects

The chapter on interferometric-based surface plasmon resonance (SPR) sensors provides an overview of different types of interferometric optical sensors and their applications (Agrawal et al. 2020a, b, c; Kumar et al. 2022a, b). The future prospects of this chapter revolve around further advancements and developments in the field of interferometric SPR sensors, offering potential opportunities for various applications and improvements. (1) Enhanced Sensing Capabilities: Future research can focus on improving the sensitivity, dynamic range, and accuracy of interferometric

SPR sensors (Kumar et al. 2022b, 2023; Kumar and Singh 2021; Kumari et al. 2022; Li et al. 2021; Liu et al. 2023a; Raghuwanshi et al. 2021). By optimizing the design of the interferometric setup, including beam splitting and combining elements, and exploring new signal processing techniques, the sensors can achieve higher precision and detection limits. (2) Miniaturization and Integration: The trend toward miniaturization of optical sensors is expected to continue. Future efforts can focus on developing compact and integrated interferometric SPR sensor systems that can be easily incorporated into micro-scale applications. This includes the exploration of new fiber devices and components that operate on smaller scales, replacing conventional large components. (3) Multimodal Sensing: Interferometric SPR sensors have the potential to provide information beyond refractive index measurements. Future research can explore the integration of additional sensing modalities, such as temperature, pressure, strain, or stress, into interferometric SPR sensors (Raghuwanshi et al. 2021; Singh et al. 2019, 2023; Singh and Kumar 2020; Wang et al. 2021b). This integration can enable the simultaneous detection of multiple parameters, expanding the range of applications. (4) Biomedical and Environmental Applications: Interferometric SPR sensors hold promise in biomedical and environmental fields. Future prospects include the development of interferometric SPR sensors for real-time monitoring of biomolecular interactions, medical diagnostics, environmental monitoring, and food safety assessment. These sensors can contribute to advancements in medicine, biotechnology, pharmaceuticals, and environmental management. (5) Advanced Signal Processing and Analysis: Further advancements can be made in signal processing and analysis techniques for interferometric SPR sensors. This includes exploring advanced algorithms, machine learning approaches, and data fusion methods to extract valuable information from complex interferometric signals. These techniques can enhance the accuracy and reliability of the sensor measurements. (6) Development of New Interferometric Structures: Research efforts can focus on the design and development of novel interferometric structures, including modified Mach–Zehnder interferometers, Michelson interferometers, Sagnac loop interferometers, and Fabry–Perot interferometers, tailored for specific applications. These new structures can offer improved performance, sensitivity, and compatibility with different sensing scenarios (Wang et al. 2021b, c, 2022; Zhang et al. 2023b; Zhu et al. 2022).

In conclusion, the future prospects of the chapter on interferometric-based SPR sensors lie in enhancing the sensing capabilities, miniaturization, and integration of the sensors, exploring multimodal sensing approaches, advancing biomedical and environmental applications, improving signal processing and analysis techniques, and developing new interferometric structures. These prospects will contribute to the continued progress and utilization of interferometric SPR sensors in various fields, enabling accurate and precise measurement of physical and chemical parameters.

7.9 Summary

In this chapter, the working principles and applications of various types of interferometric optical sensors are discussed. The main categories of interferometric sensors are Mach–Zehnder type interferometers, Michelson interferometers, Sagnac loop interferometers, and Fabry–Perot interferometers. These sensors are used to sense various physical parameters like temperature, pressure, stress, strain, etc. An interferometer works on the different optical paths of a fiber to create interference between two beams. It means there should be beam splitting and beam combining components in the interferometric configuration. In that case, one of the optical paths should be arranged in such a way that it can vary with external variations. Several types of temporal and spectral information can be obtained from an interferometric signal, in which the target can be detected through the changes in wavelength, intensity, phase, frequency, and bandwidth of the signal. These variations can improve the performance of the device in terms of high sensitivity and accuracy as well as dynamic range. Nowadays, micro-scale applications are trending and can only be possible with the miniaturization of fiber optic interferometers. For this, new small-sized fiber devices are developed to operate on fiber scales to replace the conventional bulky fiber optic components such as combiners, beam splitters, and objective lenses. Apart from conventional optical interferometers, SPR sensing gives a precise and accurate measurement of the refractive index of the material adjacent to the metal. SPR sensor technology has tremendous promise in light of the growing demand for detection and analysis of chemical and biological compounds in a variety of critical fields such as medicine, environmental monitoring, biotechnology, drug monitoring, and food monitoring. In the last section of this chapter, various interferometric SPR sensors are discussed in terms of their architecture, content, performance, and application.

References

Agrawal N, Saha C, Kumar C, Singh R, Zhang B, Jha R, Kumar S (2020a) Detection of L-cysteine using silver nanoparticles and graphene oxide immobilized tapered SMS optical fiber structure. IEEE Sens J 20:11372–11379

Agrawal N, Saha C, Kumar C, Singh R, Zhang B, Kumar S (2020b) Development of uric acid sensor using copper oxide and silver nanoparticles immobilized SMSMS fiber structure-based probe. IEEE Trans Instrum Meas 69:9097–9104

Agrawal N, Zhang B, Saha C, Kumar C, Pu X, Kumar S (2020c) Ultra-sensitive cholesterol sensor using gold and zinc-oxide nanoparticles immobilized core mismatch MPM/SPS probe. J Lightwave Technol 38:2523–2529

Akbarpour Z, Ahmadi V, Roghabadi FA (2022) Enhanced Mach-Zehnder interferometer multimode–single-mode–multimode fiber optic refractive index sensor based on surface plasmon resonance. Opt Fiber Technol 73:103035

Allison G, Sana AK, Ogawa Y, Kato H, Ueno K, Misawa H et al (2021) A Fabry-Pérot cavity coupled surface plasmon photodiode for electrical biomolecular sensing. Nat Commun 12(1):6483

Banerjee J, Bera M, Ray M (2017) Split Mach-Zehnder interferometer for surface plasmon resonance based phase modulation. Opt Commun 403:55–61

Calo G, Farinola A, Petruzzelli V (2012) Design and optimization of high sensitivity photonic interferometric biosensors on polymeric waveguides. Prog Electromagn Res Lett 33:151–166
Cao Y, Zhao C, Tong ZR (2015) All fiber sensor based on Mach-Zehnder interferometer for simultaneous measurement of temperature and refractive index. Optoelectron Lett 11(6):438–443
Chen X, He S, Li D, Wang K, Fan Y, Wu S (2015) Optical fiber voltage sensor based on Michelson interferometer using phase generated carrier demodulation algorithm. IEEE Sens J 16(2):349–354
Chen X, Li X, Yi D, Hong X, Chen Y (2021) Plasmonic tapered-fiber interference sensor for simultaneously detecting refractive index and temperature. Opt Lett 46(24):6071–6074
Choi HY, Kim MJ, Lee BH (2007) All-fiber Mach-Zehnder type interferometers formed in photonic crystal fiber. Opt Express 15(9):5711–5720
Claus RO, Gunther MF, Wang A, Murphy KA (1992) Extrinsic Fabry-Perot sensor for strain and crack opening displacement measurements from-200 to 900 degrees C. Smart Mater Struct 1(3):237
Cubik J, Kepak S, Liner A, Papes M, Kajnar T, Zboril O et al (2014) Optical vibration sensor based on Michelson interferometer arrangement with polarization-maintaining fibers. In: Fiber optic sensors and applications XI. vol 9098. SPIE, pp 100–107
Deng H, Chen X, Huang Z, Kang S, Zhang W, Li H et al (2021) Optical fiber based Mach-Zehnder interferometer for APES detection. Sensors 21(17):5870
Domínguez-Cruz R, May-Arrioja DA, Martínez-Manuel R, Lopez-Cortes D (2018) Temperature sensor based on an asymmetric two-hole fiber using a sagnac interferometer. J Sens 2018:1–6
El Shamy RS, Khalil D, Swillam MA (2020) Mid infrared optical gas sensor using plasmonic Mach-Zehnder interferometer. Sci Rep 10(1):1293
Esmail MA, Ali J, Almohimmah E, Almaiman A, Ragheb AM, Alshebeili S (2022) Sagnac loop based sensing system for intrusion localization using machine learning. In: Photonics, vol 9, no 5. MDPI, p 275
Ge Q, Zhu J, Cui Y, Zhang G, Wu X, Li S et al (2022) Fiber optic temperature sensor utilizing thin PMF based Sagnac loop. Opt Commun 502:127417
Giallorenzi TG, Bucaro JA, Dandridge A, Sigel GH, Cole JH, Rashleigh SC, Priest RG (1982) Optical fiber sensor technology. IEEE Trans Microw Theory Tech 30(4):472–511
Gong J, Shen C, Xiao Y, Liu S, Zhang C, Ding Z et al (2019) Optical fiber temperature sensor based on Mach-Zehnder interferometer coated with a film of PDMS. In: 2019 18th international conference on optical communications and networks (ICOCN). IEEE, pp 1–3
Guo J, Lian S, Zhang Y, Zhang Y, Liang D, Yu Y et al (2021) High-temperature measurement of a fiber probe sensor based on the Michelson interferometer. Sensors 22(1):289
Ho HP, Wu SY, Law WC, Lin C (2004) Highly sensitive surface plasmon resonance optical biosensor based on the use of differential phase Mach-Zehnder interferometer. In: Conference on lasers and electro-optics, 2004 (CLEO), vol 2. IEEE, p 2-pp
Homola J (2008) Surface plasmon resonance sensors for detection of chemical and biological species. Chem Rev 108(2):462–493
Huang Y, Wei T, Zhou Z, Zhang Y, Chen G, Xiao H (2010) An extrinsic Fabry-Perot interferometer-based large strain sensor with high resolution. Meas Sci Technol 21(10):105308
Islam MR, Ali MM, Lai MH, Lim KS, Ahmad H (2014) Chronology of Fabry-Perot interferometer fiber-optic sensors and their applications: a review. Sensors 14(4):7451–7488
Jang TS, Lee SS, Kim YG (2004) Surface-bonded fiber optic Sagnac sensors for ultrasound detection. Ultrasonics 42(1–9):837–841
Jassam GM, Ahmed SS (2023) Tapered PCF Mach–Zehnder interferometer based on surface plasmon resonance (SPR) for estimating concentration toxic metal ions (lead). J Opt 1–6
Jiang L, Yang J, Wang S, Li B, Wang M (2011) Fiber Mach-Zehnder interferometer based on microcavities for high-temperature sensing with high sensitivity. Opt Lett 36(19):3753–3755
Kashif M, Bakar AAA, Hashim FH (2016) Analysing surface plasmon resonance phase sensor based on Mach-Zehnder interferometer technique using glycerin. Opt Commun 380:419–424

Kumar S, Singh R (2021) Recent optical sensing technologies for the detection of various biomolecules: review. Opt Laser Technol 134:106620
Kumar S, Wang Y, Li M, Wang Q, Malathi S, Marques C, Singh R, Zhang B (2022a) Plasmon-based tapered-in-tapered fiber structure for p-cresol detection: from human healthcare to aquaculture application. IEEE Sens J 22:18493–18500
Kumar S, Singh R, Wang Z, Li M, Liu X, Zhang W, Zhang B, Li G (2023) (Invited) Advances in 2D nanomaterials-assisted plasmonics optical fiber sensors for biomolecules detection. Results Opt 10:100342
Kumar S, Agrawal N, Saha C, Jha R (2022) Optical fiber-based plasmonic biosensors: trends, techniques, and applications. CRC Press
Kumari A, Vyas V, Kumar S (2022) Advances in electrochemical and optical sensing techniques for vitamins detection: a review. ISSS J Micro Smart Syst 11:329–341
Levy R, Peled A, Ruschin S (2006) Waveguided SPR sensor using a Mach-Zehnder interferometer with variable power splitting ratio. Sens Actuators B Chem 119(1):20–26
Li YCM, Huang CM, Chang WC, Lin KJ, Chang SC, Yao SC (2011) Novel design of Sagnac interferometry assisted with surface plasmon resonance based sensor technique. Opt Commun 284(13):3369–3377
Li L, Xia L, Xie Z, Liu D (2012) All-fiber Mach-Zehnder interferometers for sensing applications. Opt Express 20(10):11109–11120
Li P, Yan H, Zhang H (2018) Highly sensitive liquid level sensor based on an optical fiber Michelson interferometer with core-offset structure. Optik 171:781–785
Li L, Jia C, Xu T, Ma Q, Sun J (2022) High stability microfiber Sagnac loop liquid refractive index sensor with FBG mode selecting. J Instrum 17(11):P11029
Li M, Singh R, Zhang B, Kumar S (2021) Gold/zinc oxide nanoparticles functionalized tapered SMF-MMF-SMF-based sensor probe for uric acid detection. SPIE
Liang YJ (2007) Detection of acoustic emission based on a fiber optic fizeau interferometer. Acta Photonica Sinica 36(4):681
Liang Y, Qu D, Deng H (2013) Based on optical fiber Michelson interferometer for acoustic emission detection experimental research. In: International symposium on photoelectronic detection and imaging 2013: fiber optic sensors and optical coherence tomography, vol 8914. SPIE, pp 219–226
Liu L, Ren Y, Wang M, Zou M (2020) Optical fiber fabry-perot interferometer sensor fabricated by femtosecond laser-induced water breakdown. Integr Ferroelectr 208(1):55–59
Liu J, Luo C, Yang H, Yi Z, Liu B, He X, Wu Q (2021) Mach-Zehnder interferometer for high temperature (1000° C) sensing based on a few-mode fiber. Photonic Sens 11:341–349
Liu Z, Zhao B, Zhang Y, Zhang Y, Sha C, Yang J, Yuan L (2022) Optical fiber temperature sensor based on Fabry-Perot interferometer with photopolymer material. Sens Actuators A 347:113894
Liu C, Chen H, Chen Q, Gao Z, Wu B, Fan X, Ma M (2023) Sagnac interferometer-based optical fiber strain sensor with exceeding free spectral measurement range and high sensitivity. Opt Laser Technol 159:108935
Liu F, Zhang W, Lang X, Liu X, Singh R, Li G, Xie Y, Zhang B, Kumar S (2023) Development of taper-in-taper-based optical fiber sensors for chemical and biological sensing. Photonics
Lou W, Wang Y, Shen C, Shentu F (2016) An optical fiber humidity sensor based on Mach-Zehnder interferometer coated with a composite film of graphene oxide and polyvinyl alcohol. In: 2016 15th international conference on optical communications and networks (ICOCN). IEEE, pp 1–3
Lu P, Chen Q (2008) Fiber Bragg grating sensor for simultaneous measurement of flow rate and direction. Meas Sci Technol 19(12):125302
Lu G, Cheng B, Shen H, Zhou Y, Chen Z, Yang G et al (2006) Fabry-Perot type sensor with surface plasmon resonance. Appl Phys Lett 89(22):223904
Lu P, Men L, Sooley K, Chen Q (2009) Tapered fiber Mach-Zehnder interferometer for simultaneous measurement of refractive index and temperature. Appl Phys Lett 94(13):131110
Luo Y, Lei X, Shi F, Peng B (2018) A novel optical fiber magnetic field sensor based on Mach-Zehnder interferometer integrated with magnetic fluid. Optik 174:252–258

Luo J, Liu S, Chen P, Lu S, Zhang Q, Chen Y et al (2021) Fiber optic hydrogen sensor based on a Fabry-Perot interferometer with a fiber Bragg grating and a nanofilm. Lab Chip 21(9):1752–1758

Nemova G, Kabashin AV, Kashyap R (2008) Surface plasmon-polariton Mach-Zehnder refractive index sensor. JOSA B 25(10):1673–1677

Omar MA, Cholan NA, Mohd A, Azhan MNM, Talib R, Ngajikin NH (2018) Optical temperature sensor based on Sagnac interferometer. Int J Eng Technol 7(4.30):126–130

Othonos A (1997) Fiber Bragg gratings. Rev Sci Instrum 68(12):4309–4341. https://doi.org/10.1063/1.1148392

Pawar D, Kanawade R, Kumar A, Rao CN, Cao P, Gaware S et al (2020) High-performance dual cavity-interferometric volatile gas sensor utilizing graphene/PMMA nanocomposite. Sens Actuators B Chem 312:127921

Piccotti D, Trevisani M, Pirruccio G, Kalinic B, Cesca T, Mattei G (2023) Polarized coherent emission outside high-symmetry points of dye-coupled plasmonic lattices. Phys Chem Chem Phys 25(12):8641–8650

Raghuwanshi SK, Kumar S, Singh Y (2021) 2D materials for surface plasmon resonance-based sensors. CRC Press

Ravikumar R, Chen LH, Jayaraman P, Poh CL, Chan CC (2018) Chitosan-nickel film based interferometric optical fiber sensor for label-free detection of histidine tagged proteins. Biosens Bioelectron 99:578–585

Razak HA, Sulaiman NH, Haroon H, Mohd Zain AS (2018) A fiber optic sensor based on Mach-Zehnder interferometer structure for food composition detection. Microw Opt Technol Lett 60(4):920–925. https://doi.org/10.1002/MOP.31072

Reyes-Vera E, Cordeiro CM, Torres P (2017) Highly sensitive temperature sensor using a Sagnac loop interferometer based on a side-hole photonic crystal fiber filled with metal. Appl Opt 56(2):156–162

Reyes-Vera E, Gómez-Cardona ND, Chesini G, Cordeiro C, Torres P (2014) Temperature sensibility of the birefringence properties in side-hole photonic crystal fiber filled with Indium. Appl Phys Lett 105(20)

Roldán-Varona P, Pérez-Herrera RA, Rodríguez-Cobo L, Reyes-González L, López-Amo M, López-Higuera JM (2021) Liquid level sensor based on dynamic Fabry-Perot interferometers in processed capillary fiber. Sci Rep 11(1):3039

Sahu S, Kozadaev KV, Singh G (2016) Michelson interferometer based refractive index biosensor. In: International conference on fibre optics and photonics. Optica Publishing Group, pp Th3A-60

Salunkhe TT, Lee DJ, Lee HK, Choi HW, Park SJ, Kim IT (2020) Enhancing temperature sensitivity of the Fabry-Perot interferometer sensor with optimization of the coating thickness of polystyrene. Sensors 20(3):794

Seraji FE, Pazooki F (2017) Performance of Sagnac interferometer with loops made of PCF and PANDA fibers. MOJ Poly Sci 1(6):00034

Shao LY, Luo Y, Zhang Z, Zou X, Luo B, Pan W, Yan L (2015) Sensitivity-enhanced temperature sensor with cascaded fiber optic Sagnac interferometers based on Vernier-effect. Opt Commun 336:73–76

Shao LY, Zhang X, He H, Zhang Z, Zou X, Luo B et al (2016) Optical fiber temperature and torsion sensor based on Lyot-Sagnac interferometer. Sensors 16(10):1774

Shen M, Wang M, Du L, Pan TT (2017) Plasmonic Mach-Zehnder interferometric sensor based on a metal–insulator-metal nanostructure. Chin Phys B 26(7):077301

Singh L, Singh R, Zhang B, Cheng S, Kaushik BK, Kumar S (2019) LSPR based uric acid sensor using graphene oxide and gold nanoparticles functionalized tapered fiber. Opt Fiber Technol 53:102043

Singh R, Wang Z, Marques C, Min R, Zhang B, Kumar S (2023) Alanine aminotransferase detection using TIT assisted four tapered fiber structure-based LSPR sensor: from healthcare to marine life. Biosens Bioelectron 236:115424

Singh R, Kumar S (2020) Chapter nine—Applications of graphene-based sensors for biomedical industries. In: Hussain CM (ed) Comprehensive analytical chemistry. Elsevier

Sun M, Song Y, Wu H, Wang Q (2023) Design and simulation of Au/SiO2 nanospheres based on SPR refractive index sensor. Sensors 23(6):3163. https://doi.org/10.3390/s23063163

Swart PL (2004) Long-period grating Michelson refractometric sensor. Meas Sci Technol 15(8):1576

Tang S, Zou M, Zhao C, Jiang Y, Chen R, Xu Z et al (2022) Fabry-Perot interferometer based on a fiber-tip fixed-supported bridge for fast glucose concentration measurement. Biosensors 12(6):391

Tian K, Yu J, Wang X, Zhao H, Liu D, Lewis E et al (2020) Miniature Fabry-Perot interferometer based on a movable microsphere reflector. Opt Lett 45(3):787–790

Uddin N, Shrestha M, Zheng B, Yoon HJ, Wang X, Fan QH (2017) Liquid sensors based on enhanced Fabry-Perot etalons. IEEE Sens J 17(22):7348–7354

Villatoro J, Minkovich VP, Monzon-Hernandez D (2006) Compact modal interferometer built with tapered microstructured optical fiber. IEEE Photonics Technol Lett 18(11):1258–1260

Wang A, Xiao H, Wang J, Wang Z, Zhao W, May RG (2001) Self-calibrated interferometric-intensity-based optical fiber sensors. J Lightwave Technol 19(10):1495

Wang S, Yang Y, Niu P, Wu S, Liu S, Jin RB et al (2021a) Fiber tip Michelson interferometer for temperature sensing based on polymer-filled suspended core fiber. Opt Laser Technol 141:107147

Wang Y, Zhu G, Li M, Singh R, Marques C, Min R, Kaushik BK, Zhang B, Jha R, Kumar S (2021b) Water pollutants p-cresol detection based on Au-ZnO nanoparticles modified tapered optical fiber. IEEE Trans Nanobiosci 20:377–384

Wang Z, Singh R, Marques C, Jha R, Zhang B, Kumar S (2021c) Taper-in-taper fiber structure-based LSPR sensor for alanine aminotransferase detection. Opt Express 29:43793–43810

Wang Z, Zhang W, Liu X, Li M, Lang X, Singh R, Marques C, Zhang B, Kumar S (2022) Novel optical fiber-based structures for plasmonics sensors. Biosensors 12:1016

Weimin C, Xiaohua L, Wei Z, Xianming L, Changrong L (2018) Recent progress of optical fiber Fabry-Perot sensors. Acta Optica Sinica 38(3):0328010

Wu CM, Jian ZC, Joe SF, Chang LB (2003) High-sensitivity sensor based on surface plasmon resonance and heterodyne interferometry. Sens Actuators B Chem 92(1–2):133–136

Wu C, Tse MLV, Liu Z, Guan BO, Zhang AP, Lu C, Tam HY (2014) In-line microfluidic integration of photonic crystal fibres as a highly sensitive refractometer. Analyst 139(21):5422–5429

Xiao F, Li G, Alameh K, Xu A (2012) Fabry–Pérot-based surface plasmon resonance sensors. Opt Lett 37(22):4582–4584

Yang Y, Zhang X, Yang L, Yu Y, Wang Z, Wang T (2021) Ultrahigh-sensitivity displacement sensing enabled by the Vernier effect with inhibited antiresonance. Opt Lett 46(5):1053–1056

Yu F, Xue P, Zhao X, Zheng J (2019) Investigation of an in-line fiber Mach-Zehnder interferometer based on peanut-shape structure for refractive index sensing. Opt Commun 435:173–177

Yuan L, Zhou L, Jin W (2004) Long-gauge length embedded fiber optic ultrasonic sensor for large-scale concrete structures. Opt Laser Technol 36(1):11–17

Zain HA, Batumalay M, Harith Z, Rahim HRA, Harun SW (2022) Surface plasmon resonance sensor for food safety. In: iopscience.iop.org, vol 2411, p 12023. https://doi.org/10.1088/1742-6596/2411/1/012023

Zhang AP, Shao LY, Ding JF, He S (2005) Sandwiched long-period gratings for simultaneous measurement of refractive index and temperature. IEEE Photonics Technol Lett 17(11):2397–2399

Zhang W, Lang X, Liu X, Li G, Singh R, Zhang B, Kumar S (2023) Advances in tapered optical fiber sensor structures: from conventional to novel and emerging. Biosensors

Zhao Y, Gandhi MSA, Li Q, Liu Z, Fu HY (2021) Vernier effect assisted sucrose sensor based on a cascaded Sagnac interferometer with no-core fiber. Biomed Opt Express 12(12):7338–7347

Zhao H, Wang F, Han Z, Cheng P, Ding Z (2023) Research advances on fiber-optic SPR sensors with temperature self-compensation. Sensors 23(2):644

Zhirnov AA, Choban TV, Stepanov KV, Koshelev KI, Chernutsky AO, Pnev AB, Karasik VE (2022) Distributed acoustic sensor using a double sagnac interferometer based on wavelength division multiplexing. Sensors 22(7):2772

Zhou X, Yu Q (2010) Wide-range displacement sensor based on fiber-optic Fabry-Perot interferometer for subnanometer measurement. IEEE Sens J 11(7):1602–1606

Zhou J, Wang Y, Liao C, Sun B, He J, Yin G et al (2015) Intensity modulated refractive index sensor based on optical fiber Michelson interferometer. Sens Actuators B Chem 208:315–319

Zhou J, Wang Y, Liao C, Yin G, He J, Sun B (2014) Temperature-independent refractive index sensor based on optical fiber Michelson interferometer. In: Asia communications and photonics conference. Optica Publishing Group, pp AF3F-5

Zhu C, Chen Y, Zhuang Y, Du Y, Gerald RE, Tang Y, Huang J (2017) An optical interferometric triaxial displacement sensor for structural health monitoring: characterization of sliding and debonding for a delamination process. Sensors 17(11):2696

Zhu G, Wang Y, Wang Z, Singh R, Marques C, Wu Q, Kaushik BK, Jha R, Zhang B, Kumar S (2022) Localized plasmon-based multicore fiber biosensor for acetylcholine detection. IEEE Trans Instrum Meas 71:1–9

Chapter 8
Application of Geometric-Based SPR Sensors

8.1 Introduction

In the modern age, the importance of optical fiber sensors has increased in the sensing domains owing to real-time monitoring in the following areas: chemical, biological, liquid, gas, drugs, antigens, and antibodies; concentration measurements; and continuous chemical reaction rate monitoring. Chemical and biological analytes can be quantified or analyzed using surface plasmon resonance (SPR) technology, which is affected by the refractive index (RI) of substances on a metal surface (Habib et al. 2019; Lai et al. 2013). Because of their diverse structural designs, SPR sensors based on photonic crystal fiber (PCF) have demonstrated a variety of significant advantages for detecting structures (Kaur and Singh 2017; Senthil et al. 2019; Paul et al. 2018; Luan and Yao 2017; Liu et al. 2017). Metal coatings are applied to the fiber pores in order to detect analytes in PCF-based SPR sensors. For specific wavelengths, incident light is channeled down the fiber core and coupled with surface plasmon polaritons (SPPs) to produce supermodes, which propagate at the contact between the analyte and the metal as a large region of detection. Hollow-core photonic crystal fiber sensors with a silver nanowire in the cladding holes and an analyte channel in the core hole are used to create an SPR sensor to be used in medical applications. Within the 1.33–1.5 refractive index region, the nanowire surface was energized by the resonance characteristics of the core and SPP modes (Wang et al. 2020). At certain wavelengths, resonance peaks can occur between the core mode and higher-order SPP modes, resulting in resonance peak shifts as RI increases (Wang et al. 2020). The broad RI range of the analyte can be measured by the sensor, which may be greater or less than that of the fiber material. This is accomplished by observing that the most sensitive peak is produced by the resonance combination of the x-polarized core mode and the second-order SPP mode. The SPR sensor is equipped with a gold grating-based high-resolution D-shaped photonic crystal fiber. The resonance wavelength can be adjusted with a gold grating to increase the sensitivity of the RI. The PCF and gold grating structure characteristics are evaluated using the

S. K. Raghuwanshi et al., *Geometric Feature-Based Fiber Optic Surface Plasmon Resonance Sensors*, Springer Tracts in Electrical and Electronics Engineering,
https://doi.org/10.1007/978-981-99-7297-5_8

finite element method (FEM) to enhance the performance of the SPR sensor. The results of simulations indicate that refractive index sensitivity is not affected by the pitch of the air holes, diameter of the air holes, gold thickness, or grating constant (Lu et al. 2018). To improve RI sensing resolution, a two-feature (2F) interrogation method has been adopted, which combines wavelength and amplitude interrogation. The theoretical maximum resolution of the SPR sensor in the RI 1.36–1.38 range is 5.98×10^{-6} RIU, and the wavelength sensitivity is 3340 nm/RIU, which is significantly greater than interrogating methods based on wavelength and intensity (Lu et al. 2018). With the creation of an SPR-RI sensor with responsiveness, high sensitivity in real time, and quick reaction time, SPR sensors could be used in a variety of ways (Lu et al. 2018). High-RI analytes can be detected and analyzed in the SPR sensor employing hollow fiber (HF). A waveguide is created by injecting the liquid analyte into the fiber's central hole. To detect high-RI analytes, the incident light can travel via analyte-filled waveguides filled with analytes. Analytes with a lower RI than the fiber material, on the other hand, do not meet the criteria for full reflection, which prevents sensors from working properly. The SPR sensors are in direct contact with the metal. Analytes that can react chemically with the metal cannot be found using these methods. On the other hand, because these fiber holes are only a few microns across, they will be difficult to coat with metal films in real life. A metallic nanowire with SPP modes on its surface has been demonstrated. as an alternative to covering metal films (Wu et al. 2014, 2017; Kaur and Singh 2020). By altering the distance between the core and the nanowire, the coupling mode intensities can be altered (Senthil et al. 2019). Nanowire-filled PCFs can be used with the right design in a wide range of applications. The metal (Au or Ag) used has a lower melting point than the fiber material (silica). Molten metal can be pushed into the air holes of the fiber under high pressure to integrate the metallic nanowire into the PCF (Zhao et al. 2019). The use of nanowires instead of thin metal films allows for simple operation and precise control, unlike the inner coating made using the chemical vapor deposition (CVD) approach (Hao et al. 2013; Otupiri et al. 2014).

Photonic plasmonic sensors have emerged as a promising option in the sensing field because of their selectivity, high sensitivity, design versatility, and fabrication simplicity. Some of the most popular applications for photonic and biochemical biosensors include medicine checking, environmental monitoring, food safety, medical diagnostics, biochemical sensing, gas and liquid detection, and water testing. A prism-based SPR sensor with an EM wave incident at the metal-dielectric contact that is trans magnetic and p-polarized and one side of the prism covered with plasmonic material was described in the literature (Akowuah et al. 2012). In order to promote the movement of electrons along the metal-dielectric contact, it encourages the formation of free electrons on the metal surface. When the incident wave, direction, and frequency match those of free electron oscillations, SPR occurs. SPR results in a high absorption loss due to the surface plasmon wave absorbing all of the incident photonic energy. Traditional SPR sensors have a number of disadvantages, including a narrow range of response and cumbersome size because of the existence of considerable mechanical components. Unique properties such as compact size, high flexibility, regulated birefringence, and single-mode propagation have been created to address these issues. PCF-SPR sensors improve their performance by putting a layer of metal inside or outside of the air hole (Yasli and Ademgil 2018; Sakib et al. 2019). This activates free electrons, which improves the sensor's performance. The PCF-SPR sensors that have been described thus far include internal and external metal coatings, microfluidic channels, and D-shaped sensors. In the instance of PCF-SPR sensors with metal coatings on the inside, there are two major issues to overcome. It's nearly impossible to create a uniformly thin metal sheet inside an air gap. Second, it takes a long time to get liquid into microscopic air holes. This makes the technology less useful.

8.2 SPR Based Sensor with a Different Geometrical Structure of PCF

With their mechanical and optical requirements, Kretschmann-Raether prism geometry sensors have a large footprint. SPR sensors based on optical fibers may be able to overcome these drawbacks. PCFs are ideal for sensing applications because of their capacity to manipulate evanescent fields as well as their unique propagation characteristics, design freedom, and tiny size. Furthermore, structural characteristics such as hole-to-hole spacing, hole diameter, and cladding shape are taken into consideration and can be optimized to improve PCF-based sensor performance (Asaduzzaman and Ahmed 2018; Hossain et al. 2019). For more than a decade, researchers have been analyzing the PCF's structure. The different forms of PCF structures have square, octagonal, and decagonal air-hole arrangements. Cladding holes with square and elliptical shapes have been found to have interesting propagation characteristics in PCF structures (Liu et al. 2016a). Inside silica, PCF-based SPR sensors with circular air holes organized into hexagons coated with gold metallic coatings have been proposed. On the other hand, a PCF-based SPR pressure sensor in the cladding zone with silver-coated air holes was proposed by Yasli and Ademgil (2018). For the octagonal PCF-based SPR temperature sensor, Yang et al. examined the impact of various-shaped silver nanowires (elliptical, square, and circular) on the sensor. Elliptical models are found to have greater peak shifts and greater sensitivity than other geometries (Hasan et al. 2017). When designing a PCF-based SPR sensor, geometrical parameters such as the number, size, location, and shape of the air holes are considered as per the needs. The diameter of the analyte channel and the thickness of its metallic layer have a direct impact on sensor performance (Yan et al. 2018a). The most common plasmonic materials are gold and silver, which have a significant impact on the sensor's performance. In an aquatic environment, gold is more stable because it has a wider resonance peak, which may have an impact on the performance of plasmonic sensors. The oxidation sensitivity of silver is greater than that of gold, despite the latter's sharper resonance peak. As a result, between the analyte and the silver layer, a graphene layer can be inserted to prevent oxidation (Liu et al. 2020a). However, this technique is more difficult to fabricate from the point of view of fabrication than a single gold layer. When it comes to core-guided leaky modes, it is important to note that these modes are strongly influenced by the pattern of the holes (square, hexagonal, or octagonal) as well as the geometrical shape of the cladding holes. In contrast, the metallic layer (gold, silver, or graphene) affects interaction properties under resonance conditions (Yan et al. 2018a; Liu et al. 2020a). Aside from these considerations, the impact of PCF-SPR-based sensor performance against analytes (1.33, 1.34, and 1.35) is evaluated using various geometrical air holes (elliptical, round, square, etc.).

8.3 Photonic Crystal Fiber of Various Geometries

8.3.1 Rectangular Hollow-Core PCF

In the hollow core of the rectangular PCF with a rectangular hollow core, the analyzer was installed. There are circular holes in the cladding and air holes in the middle. The light-directing features are created by meeting boundary conditions. The FEM is used to estimate the high-sensitivity output. With a rise in birefringence, a wide range of chemicals are all affected by numerical aperture and dispersion loss. Terahertz applications make use of this sensor (Kaur and Singh 2019a). A PCF with a hollow core that is used to detect chemical analytes may be the best option. Using a combination of circular and rectangular air holes, a hybrid style of cladding is offered in the design. According to the numerical results, HC-PCF offers a great sensitivity of 89% and minimal confinement loss of 1.15×10^{-9} dB/cm at the 1.7 THz frequency. In addition, this chemical sensor has a 0.007 birefringence, a 0.42 numerical aperture, and minimal dispersion variation as shown in Fig. 8.1. Between 1.1 and 2.4 THz, a value of 0.62 ± 0.275 ps/THz/cm was measured (Kaur and Singh 2019a). Fibers of this kind are being proposed and can be simply realized and manufactured using current technologies. There are numerous applications for chemical sensing.

With analytes such as benzoyl, ethanol, and water injected into the rectangular core air hole, several characteristics such as relative sensitivity, birefringence, and confinement loss will be examined, as well as dispersion, the effective area, and the numerical aperture. Ethanol (1.354) and water (1.33) have refractive indices of 1.33 and 1.66, respectively (Lai et al. 2013). According to a trial-and-error method, when the core length reaches 300 nm, it provides the best guiding qualities. As a result, the suggested sensor core length is set at 300 nm.

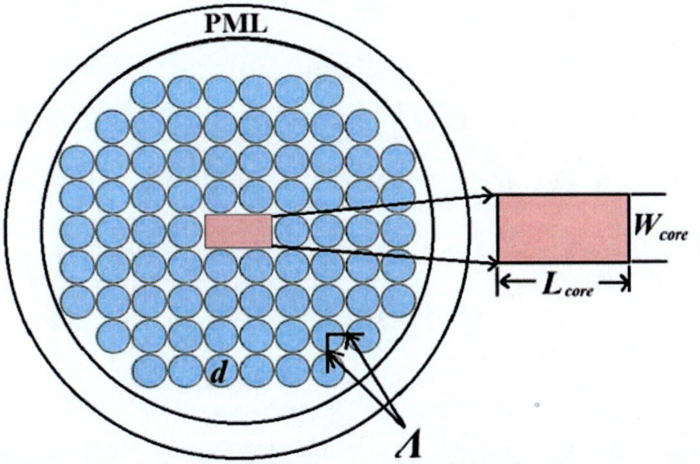

Fig. 8.1 Photonic crystal fiber with a rectangular hollow core. Reprinted with permission from Optical Fiber Technology. Copyright, 2019, Elsevier (Habib et al. 2019)

8.3.2 Multiple Sensing Ring Photonic Crystal Fiber

Multiple sensing ring PCF with spiral cladding is depicted in the cross-section and core design in Fig. 8.2. Multiple sensing rings support maximum relative sensitivity in the PCF sensor design, which also has a spiral pattern of circular air holes to achieve minimal confinement loss. High refractive index sensing liquid is filled into rings embedded in silica material to support TIR propagation along the PCF core. This PCF has four circular sensing rings, each measuring 0.3 μm in width and 0.05 μm apart.

8.3.3 Circular-Pattern Photonic Crystal Fiber

To represent the circular-pattern PCF work for chemical identification motive, a solid-core photonic crystal fiber (PCF) is preferred. Many PCF attributes can be gleaned from finite element models. A PCF is made up of two parts: a core and a cladding. The circular shaped core and internal cladding are shown in Fig. 8.3 as three and five ring-shaped layers, respectively. The cladding section covers the core area.

In this case, three separate core materials were used in the same design. First, glycerol and ethanol are compared, then toluene and ethanol. The extreme sensitivity of ethanol makes it widely used. Glycerol exhibited less confinement loss than ethanol, which is dependent on PCF structure. When compared to ethanol, glycerol and toluene have lower confinement losses. Glycerol has a 65.16% sensitivity and a 2.81 μm^2 effective area. Toluene has a 64.05% sensitivity and an effective area of

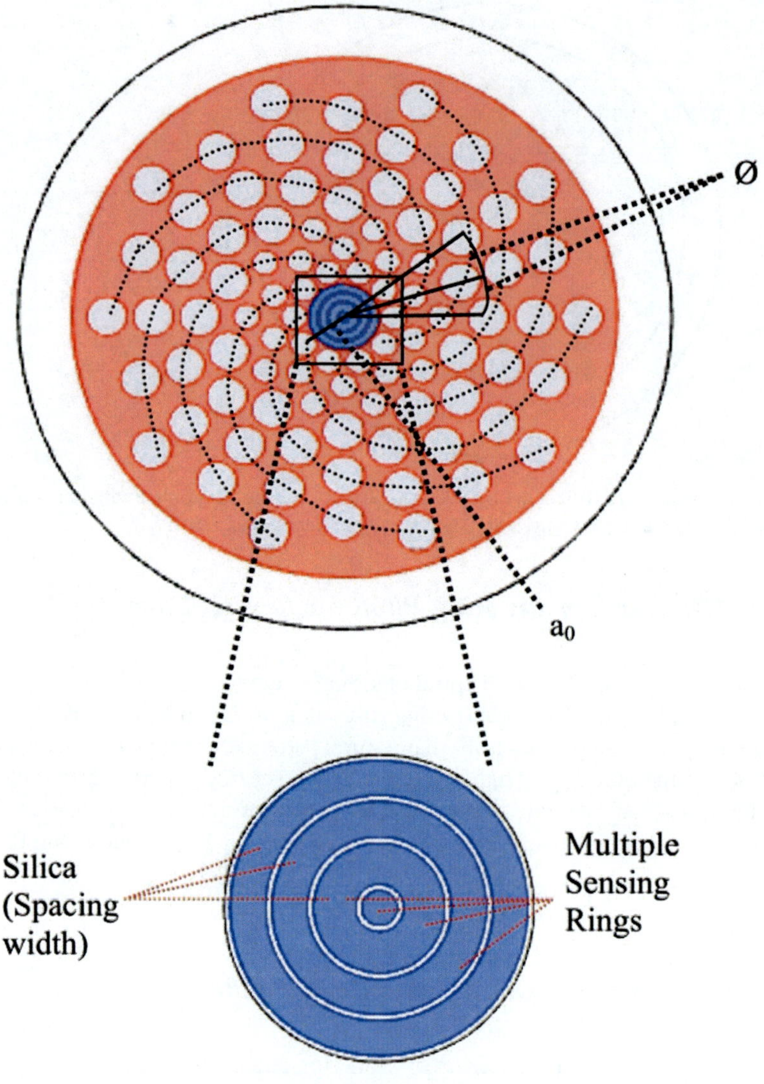

Fig. 8.2 Multiple sensing rings in the core of a proposed circular PCF. Reprinted with permission from Sensing and Bio-Sensing Research. Copyright, 2017, Elsevier (Kaur and Singh 2017)

3.07 μm^2. Due to its large effective area, CP-PCF boosted the sensitivity and effective area (Senthil et al. 2019). The ability to transport more data at a faster rate makes this technology useful in the telecommunications industry. The CP-PCF properties of several liquids, such as toluene and glycerol, were examined in this manner in order to discover potential uses. In addition to biosensing, biomedical protein identification analysis, and telecommunications, this material has a unique sensing property.

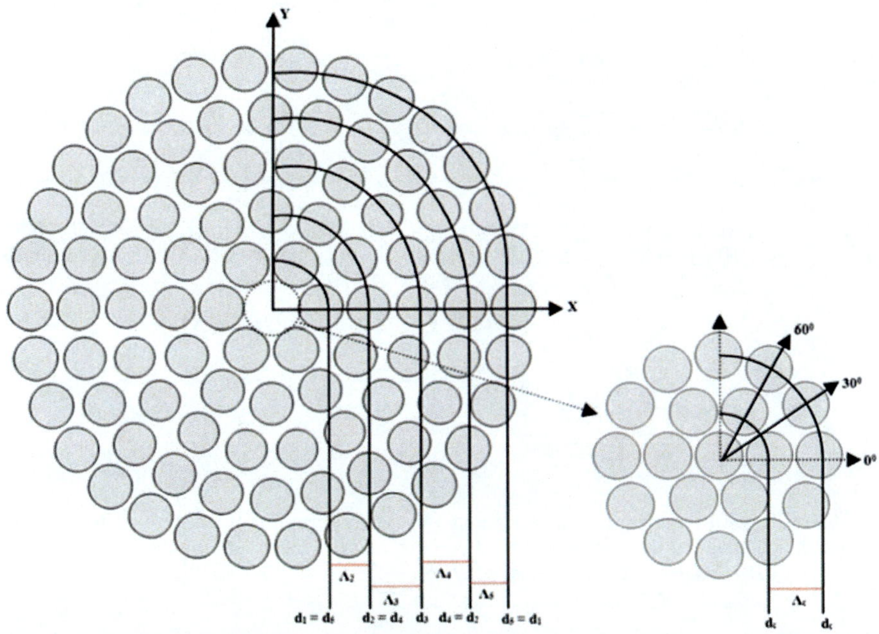

Fig. 8.3 PCF in a circular shape. Reprinted with permission from Sensing and Bio-Sensing Research. Copyright, 2017, Elsevier (Paul et al. 2017)

8.3.4 Quasi Photonic Crystal Fiber

As shown in Fig. 8.4, the Q-PCF can be seen in cross-section. Q-PCF uses a periodic pattern for placing the cladding's air holes. This Q-PCF has a repeating quasi-crystalline pattern filling up all the empty areas in the cladding. Better air filling percent is achieved by using $D_{cl} = 265$ μm for the air holes in the innermost layer, and $D_{cl} = 304$ μm for the air holes in the rest of the cladding with a pitch of $\Lambda_{cl} = 320$ μm. The core has seven evenly spaced air holes of the same width, the length of which is proportional to the diameter of the core. The seven openings range in width from 0.42 to 0.6 D_{core}, 0.89 to 0.97 D_{core}, 0.6 to 0.42 D_{core}, and 0.97 to 0.6 D_{core}, when reading from left to right or right to left. With a value of 0.13 D_{core}, the core pitch (c) is decided upon as the horizontal center-to-center distance between two neighboring slots. To reduce the effect of the surrounding environment on the PCF and to slow the spread of light at the border, a perfectly matched layer (PML) is utilized.

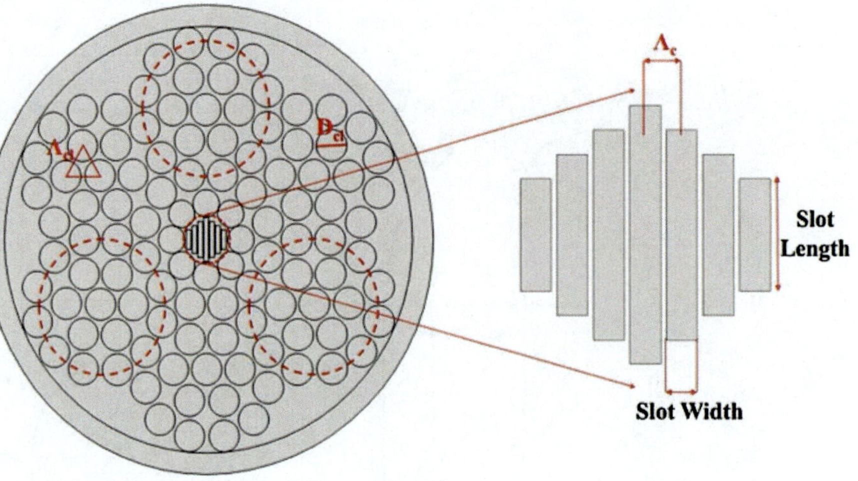

Fig. 8.4 The quasi-photonic crystal fiber's transverse cross-section shows **a** the cladding area and **b** a magnified image of the fiber's innermost core region. Reprinted with permission from Optik, Copyright, 2019, Elsevier (Kanmani et al. 2019)

8.4 Photonic Crystal Fiber of Different Geometries Based on SPR Sensors

8.4.1 Hollow-Core PCF-Based SPR Sensor

A surface plasmon resonance (SPR) refractive index sensor based on a simple hollow-core circular lattice photonic crystal fiber (PCF) has been suggested by Momota and Hasan (2018) as shown in Fig. 8.5. Using the finite element method (FEM), an investigation of the sensing performance is carried out. The plasmonic material for this design is silver, and it is positioned on the exterior of the PCF to make the fabrication process easier. The sensor that has been suggested demonstrates a maximum wavelength sensitivity of 4200 nm/RIU and a sensor resolution of 2.38×10^{-5} RIU. In addition to this, it has been reported that an analyte with a refractive index of 1.37 has a maximum amplitude sensitivity of 300 RIU^{-1} and a resolution of 3.33 10^{-5} RIU.

8.4.2 Square Array PCF-Based SPR Sensor

For the purpose of detecting the annular analyte, a high-sensitivity surface plasmon resonance (SPR) sensor that is based on PCF has been devised and modeled for use in a wide variety of applications for measuring the refractive index (RI) (Jain et al. 2022). PCF has a gold (Au) coating on its outside, and a small layer of titanium dioxide

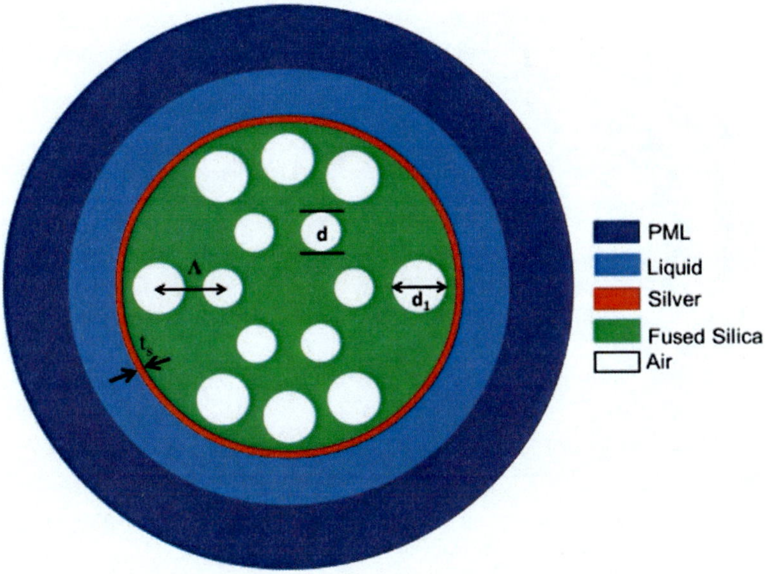

Fig. 8.5 SPR sensor schematic based on hollow-core PCF. Reprinted with permission from Optical Materials. Copyright, 2018, Elsevier (Momota and Hasan 2018)

(TiO_2) acts as an adhesive between the gold (Au) and the silica. The proposed sensor detects the RI of the analyte by utilizing the SPR interaction that occurs between the core mode and the Surface Plasmon Polariton mode.

As shown in Fig. 8.6, the air hole lattice pattern that is employed in the sensor is in the shape of a square and has a consistent pitch (Λ). The pitch is defined as the center-to-center distance between two succeeding air holes. Two square layers are filled with air holes that have a diameter of D (μm), and then four of the air holes in the inner square layer are shrunk down to d (d is the diameter of small air holes). These smaller air holes create a pathway for the light to travel from the center of the fiber to the surface of the fiber. As a result, it offers a solid connection between the core mode and the SPP mode.

8.4.3 D-shape PCF-SPR Sensor

For the purpose of sensing low refractive indices, a novel surface plasmon resonance sensor (SPR) based on D-shaped photonic crystal fiber (PCF) has been developed, and the performance of this sensor has been studied numerically using the finite element approach (FEM) (Liu et al. 2020b). Indium tin oxide, also known as ITO, has been decided upon as the plasmonic material of choice in order to extend the PCF-SPR sensor's operational wavelength range into the infrared area. The wavelength range

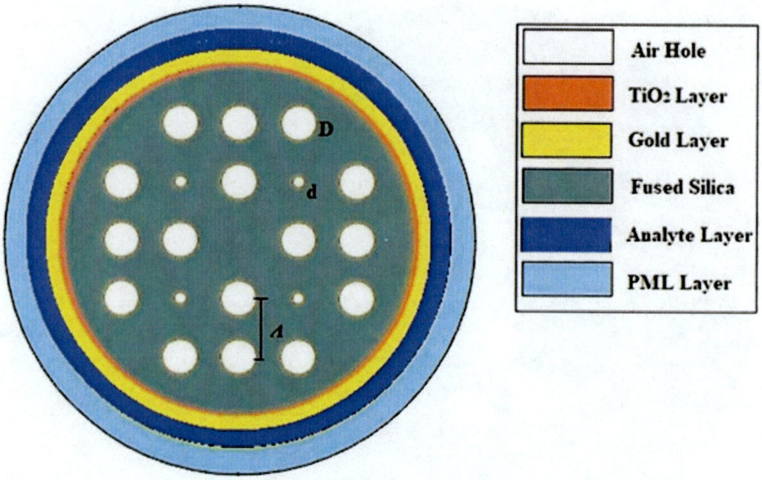

Fig. 8.6 Square array PCF-based SPR sensor. Reprinted with permission from Optical Fiber Technology. Copyright, 2022, Elsevier (Jain et al. 2022)

of 1200–2250 nm was used to validate the spectral sensitivities of 2000–15,000 nm/RIU for refractive indices of 1.22–1.33. The greatest amplitude sensitivity up to 442.47 RIU^{-1} was obtained at 2,010 nm with a high resolution of 6.67×10^{-6} RIU.

The schematic diagram of the PCF-SPR sensor made of fused silica is shown in Fig. 8.7a. In brief, the PCF is polished on one side and covered with indium tin oxide (ITO) to increase its detecting range into the near-infrared. To begin, the PCF sensor can be made using the stack-and-draw and die-cast techniques, with the stacked configuration shown in Fig. 8.7b. The PCF is then prepared for ITO film coating by being subjected to a wheel polishing procedure, which results in a perfectly flat surface. In the middle of the fiber is a regular hexagon of air holes with a diameter of d_1. The larger air holes have a diameter of d_2, and the space between air holes, or "pitch," is Λ. The dimensions of the ITO sheet are 6 μm across and 80 nm in thickness. Specific parameters are ($d_1 = 1.2$ μm, $d_2 = 1.6$ μm, pitch $\Lambda = 3.5$ μm, and $D = 2.8$ μm), and a schematic representation of the sensor and 3D structure is shown in Fig. 8.7c.

8.4.4 D-shaped PCF Polished on One Side SPR-Based Sensor

The graphical representation of PCF is uncomplicated and easy to understand. Using as its foundation, the commercial PCF ESM-12 (manufactured by NKT Inc.), Fig. 8.8 depicts the basic geometry of the PCF-SPR sensor (Luo et al. 2021). The cladding is made with a hexagonal pattern of air holes over all six layers. Polishing depth h, measured from the fiber core to the polished surface, is used to smooth the top of the

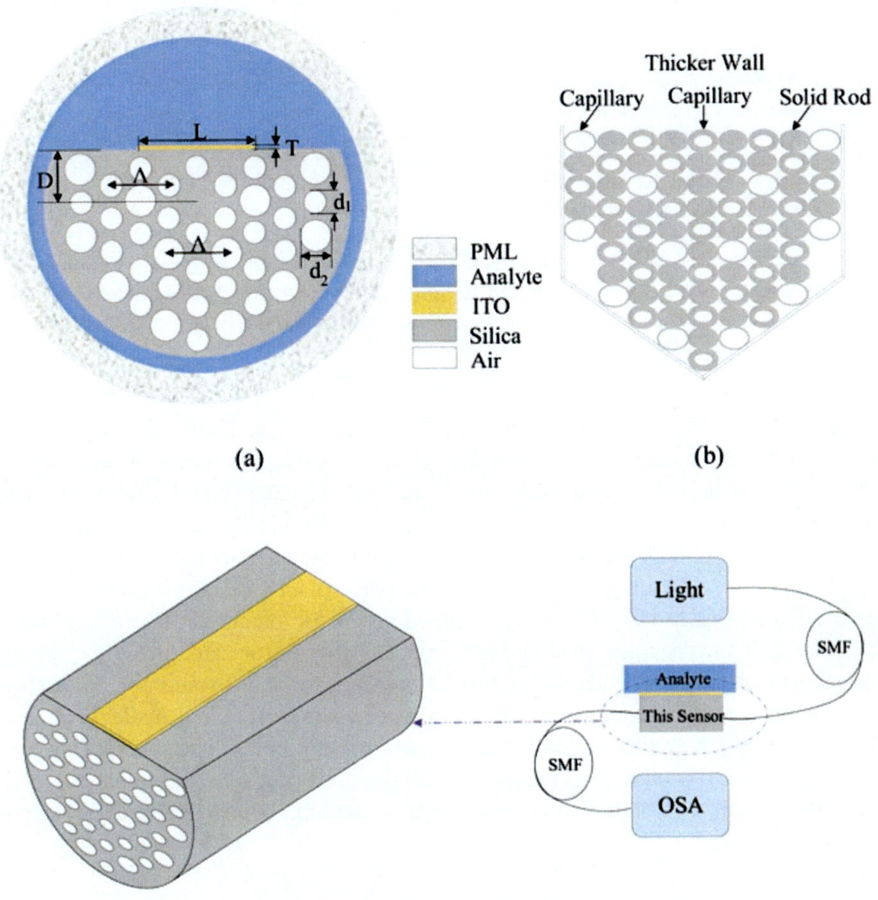

Fig. 8.7 **a** Cross-section schematic of the sensor, **b** stacked structure of the sensor, and **c** 3D illustration of the sensor. Reprinted with permission from Optics Communications. Copyright, 2020, Elsevier (Liu et al. 2020b)

PCF. The dimensions of the lattice are 7.9 μm for the pitch (Λ) and 3.9 μm for the diameter of the air holes. After the core has been exposed by polishing the sides, a gold coating one nanometer thick is coated on the smoothed surface.

8.4.5 *Surface Plasmon Resonance Sensor with D-shape Dual-Core PCF*

The D-shape dual-core PCF surface plasmon resonance (DD-PCF-SPR) sensor has an exterior coating of Au that serves as the plasmonic material, making it more preferable from the standpoint of modern technology. For external bio-target refractive index

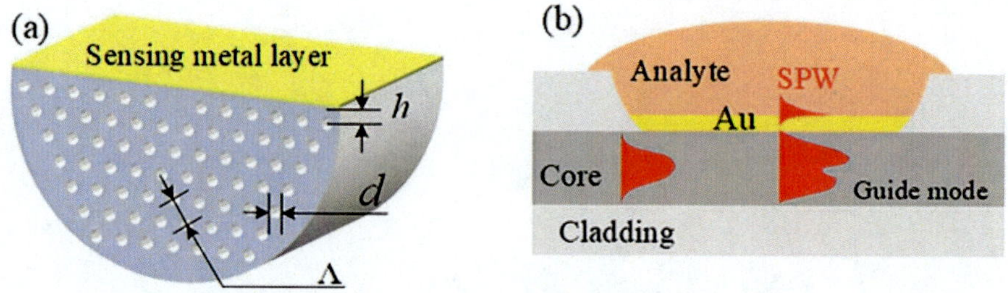

Fig. 8.8 **a** D-shaped PCF-SPR sensor's cross-section, based on ESM-12. **b** Mode coupling between the SPW and core-guided transmission modes: a schematic diagram. Reprinted with permission from Measurement. Copyright, 2021, Elsevier (Luo et al. 2021)

detection, it has higher sensitivity than most. The maximum amplitude sensitivity of the DD-PCF-SPR sensor was 700 RIU^{-1}, and the sensor resolution was 1.7857×10^{-5} RIU. Furthermore, it has the same wavelength sensitivity and resolution as the RIU with a 1.25×10^{-5} RIU resolution, which is 8000 nm/RIU.

The gold layer thickness and pitch parameters are used to alter the sensor's geometrical parameters, resulting in improved sensor performance. According to 138 RIU^{-1} as the Figure of Merit (FoM), the optimized geometric parameter (Kaur and Singh 2020). While the proposed structure is simple and straightforward, the gold (Au) and sensing layer on the outside of the circular air holes make it stand out. Figure 8.9 depicts the biosensor's graphical presentation. The lattice of PCF is made up of three rings of air holes, but the second ring has two missing holes. The second ring's air holes are used to produce birefringence, and the third ring's two air holes are made relatively small to produce an evanescent field, making it possible to easily excite surface plasmons.

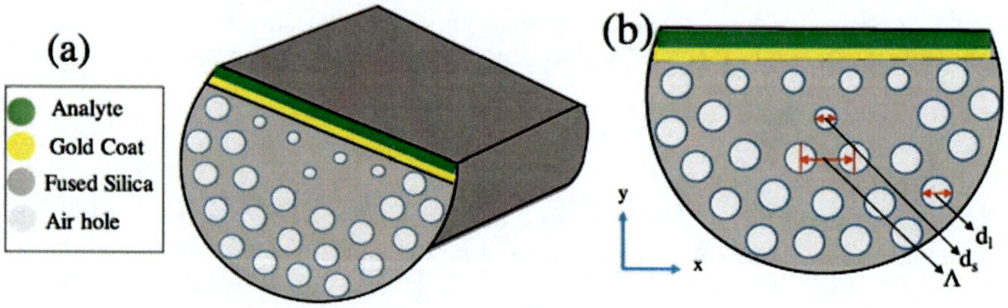

Fig. 8.9 **a** A three-dimensional image of the proposed DD-PCF-SPR **b** the suggested Au-coated D-shaped PCF sensor cross-section with $\Lambda = 1.9$ μm, $d_s = 0.25\ \Lambda$, $d_1 = 0.5\ \Lambda$, and $t_g = 30$ nm. Reprinted with permission from Results in Physics. Copyright, 2019, Elsevier (Sakib et al. 2019)

8.4.6 PCF with a Dual Symmetrical Eccentric-Core Configuration-Based SPR Sensor

The COMSOL Multiphysics software is used to simulate and design the series-wound construction using the FEM approach. The sensor schematic is shown in Fig. 8.10, and it includes PCFs with a dual symmetrical eccentric-core configuration that are wound in series. Simple fabrication makes this structure more readily attainable for construction. It is also important to package the sensor ends and detecting area so that the platform can be supported and fixed, allowing it to reliably identify analytes in real-world sensing situations. There are regular hexagonal air holes embedded in pure silica that serve as the cladding material. Light loss can be minimized, and basic mode transmission is ensured, with such a cladding construction.

Total internal reflection allows the incident light to pass through the PCF and support an evanescent field that can trigger SPR (Zhao et al. 2019). With this sensor and a single-mode fiber (SMF), an optical coupler can be explored for transmitting light. The coupling effect can be achieved more effectively with this twin eccentric-core arrangement. In the outer layer, there are six smaller air holes arranged in two groups of varying diameters, and d_2 refers to the air holes' lower diameter; the larger d_1 refers to all the other air holes. A 1.5 μm gap separates the air openings. Because of its greater chemical stability, gold is preferred over silver as a surface plasmon material for initiating surface plasmon polaritons (SPPs). Furthermore, the fabrication method is simplified by designing a symmetric gold nanowires configuration and

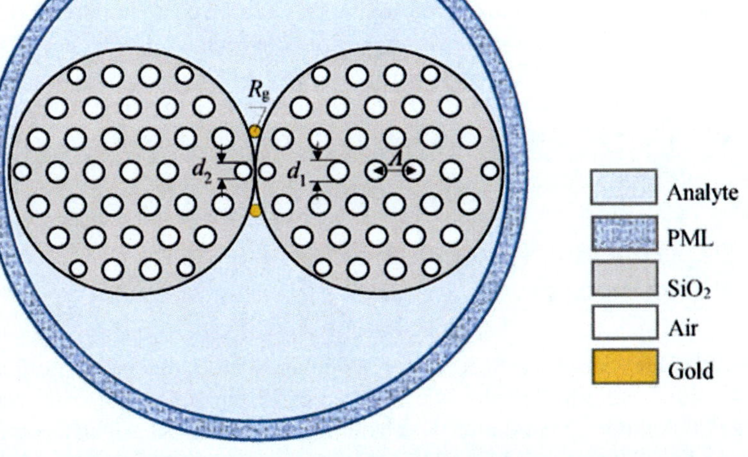

Fig. 8.10 The PCF-SPR sensor's schematic diagram. Reprinted with permission from Results in Physics. Copyright, 2020, Elsevier (Wang et al. 2020)

mounting them between two neighboring silica core fibers rather than generating a gold coating on the air holes' inner surface.

8.4.7 PCF-Based SPR Sensors Designed with Circular, Square, and Elliptical Air Holes

These PCF-based sensors feature seven various-sized air openings, as shown in Fig. 8.11. Which are placed hexagonally on the silica background, with gold (Au) layers and analyte channels surrounding them. Dimensions d_1, d_2, and d_3 refer to the diameters of the air holes, which vary from structure to structure. All structures have a 1.2 μm hole-to-hole spacing. For purposes of fair comparison, all structures have the same number and location of air holes. Analytes (n_a) have refractive index values of 1.33, 1.34, and 1.35, respectively. With an initial 40 nm gold layer thickness, Johnson and Christy's data is used to calculate the gold permittivity (Hao et al. 2013).

To model silica as a substrate material, the Sellmeier equation is employed (Otupiri et al. 2014). Initially, the SPR sensor was built with round air holes in the cladding region, as shown in Fig. 8.11a. Because birefringence is a desired effect, the diameter of the side holes is slightly more than that of the cladding holes. If you use this method, the refractive indexes of the modes will be different. Multianalyte/multichannel sensing can benefit from this phenomenon (Akowuah et al. 2012). Circular air holes have diameters of $d_1 = 0.359$ μm, $d_2 = 0.565$ μm, and $d_3 = 0.424$ μm, respectively, and areas of 0.10 $(\mu m)^2$, 0.25 $(\mu m)^2$, and 0.14 $(\mu m)^2$, respectively (Yasli and Ademgil 2018). The design parameters for elliptical and square-holed models are similar to those for a circular air-holed model. All models' air hole areas have been standardized so that they can be compared fairly. The ellipticity constant is set to 0.5 for the elliptical structure because it is defined as the ratio *db*/*da* (Yasli and Ademgil 2018). Table 8.1 shows the PCF comparisons with various design characteristics (Sakib et al. 2019; Asaduzzaman and Ahmed 2018; Hossain et al. 2019; Liu et al. 2016a; Hasan et al. 2017; Yan et al. 2018a).

8.5 Experimental Set-up of Hollow Dual-Core PCF-SPR-Sensor

The hollow dual-core PCF-SPR sensor is shown, with double-sided gold coatings on both sides and silica in the two fiber cores in the schematic cross-section in Fig. 8.12a. Adding a medium to the cladding reduces the RI slightly compared to using just fibers. An optical fiber with a step-type structure reduces transmission loss and ensures fundamental mode transmission by limiting the amount of light that can enter its cores. An externally reflected light beam is used to excite SPR in the PCF

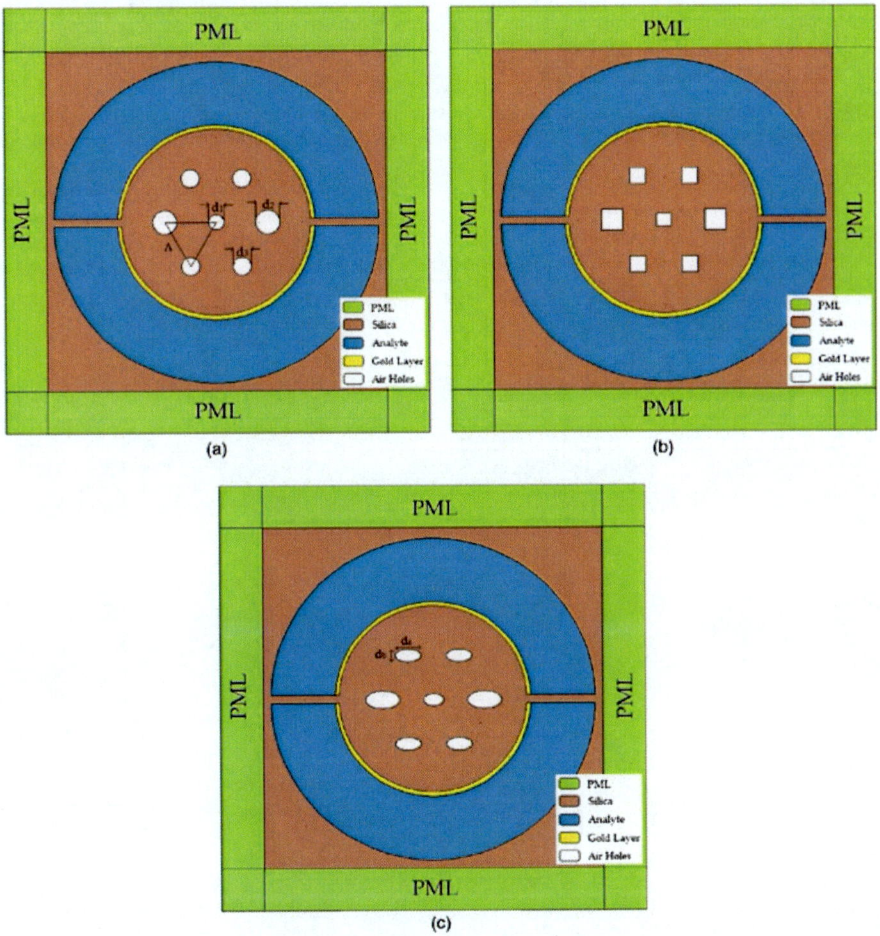

Fig. 8.11 **a**, **b**, and **c** Show schematics for PCF-based SPR sensors that have been designed with circular, square, and elliptical air holes, respectively. Adapted from Yasli and Ademgil (2018)

using total internal reflection is a light-guiding mechanism. As an added benefit, PCF is a great SPR platform that can be made using the stack-and-draw technique by Liu et al. (2020a), Kaur and Singh (2019a). The SPP mode uses a gold film because it has better chemical stability than silver. It is possible to apply gold films to the thin cladding's inner and outer surfaces by using high-pressure chemical vapor deposition (CVD), thermal evaporation, sputtering, electro less plating, or wet chemistry by Liu et al. (2020a). Hole diameter (d_L), radius of the silica core (r_c), and thickness of gold layers on both sides (t_{Au}) are all optimized for this particular design, which has an aperture width (D) of 6.01 μm. The experimental setup is shown in Fig. 8.12c, and spectral response can be observed using an optical spectrum analyzer (OSA).

Table 8.1 Photonic crystal fiber (PCF) comparisons with various design characteristics (Sakib et al. 2019; Asaduzzaman and Ahmed 2018; Hossain et al. 2019; Liu et al. 2016a; Hasan et al. 2017; Yan et al. 2018a)

Base structure type	Pattern recognition	The refractive index range of the sensing layer	Thickness of the sensing layer	Amplitude sensitivity	Spectral sensitivity	Amplitude interrogation for sensor resolution	Wavelength interrogation for sensor interrogation
SPR-PCF gold coated	Coating externally	1.33–1.37	Gold = 30 nm	478	4000	2.1×10^{-1}	2.5×10^{-5}
SPR-PCF hollow Core	Coating externally	1.33–1.37	Silver = 30 nm	300	4200	3.33×10^{-5}	2.38×10^{-5}
SPR-PCF with a square array form	Coating externally	1.38–1.42	Gold = 0.03 um	478	4000	2.1×10^{-1}	2.5×10^{-5}
D-shaped PCF polished on one side	Coating externally	1.33–1.34	Gold = 45 nm	–	21,700	–	–
SPR-PCF duplex core	Coating externally	1.39–1.40	Gold = 35 nm	1770	10,700		9.34×10^{-6}
Two ring, hexagonal PCF		1.34–1.36	–		8500		
Three ring, hexagonal PCF	Coating externally	1.34–1.37	Gold = 40 nm	265	4000	3.7×10^{-5}	$\times 10^{-5}$

The Sellmeier equation is used to calculate the dispersion relationship between the cladding layer and fused silica.

As can be seen in Fig. 8.12b, a more refined grid division model is used. We can quickly locate the most appropriate mode to the grid 257,196 domain units and 15,068 boundary units. The core mode confinement loss, which is crucial for most PCF-SPR sensors, is used to characterize SPR excitation. The mode effective index's imaginary portion is used to calculate the loss in confinement by Liu et al. (2019a).

The center-to-center distance of an air hole is denoted by pitch (Λ), the diameter of a small air hole is denoted by d_s, and the diameter of a big air hole is denoted by d_1. In this case, $\Lambda = 1.9$ μm, $d_s = 0.25\ \Lambda$, and $d_1 = 0.5\ \Lambda$. The PCF has a diameter of 9.5 μm using the chemical vapor deposition (CVD) approach, a variable thickness (t_g) gold layer is coated at the external region of the PCF, acting as the noble and

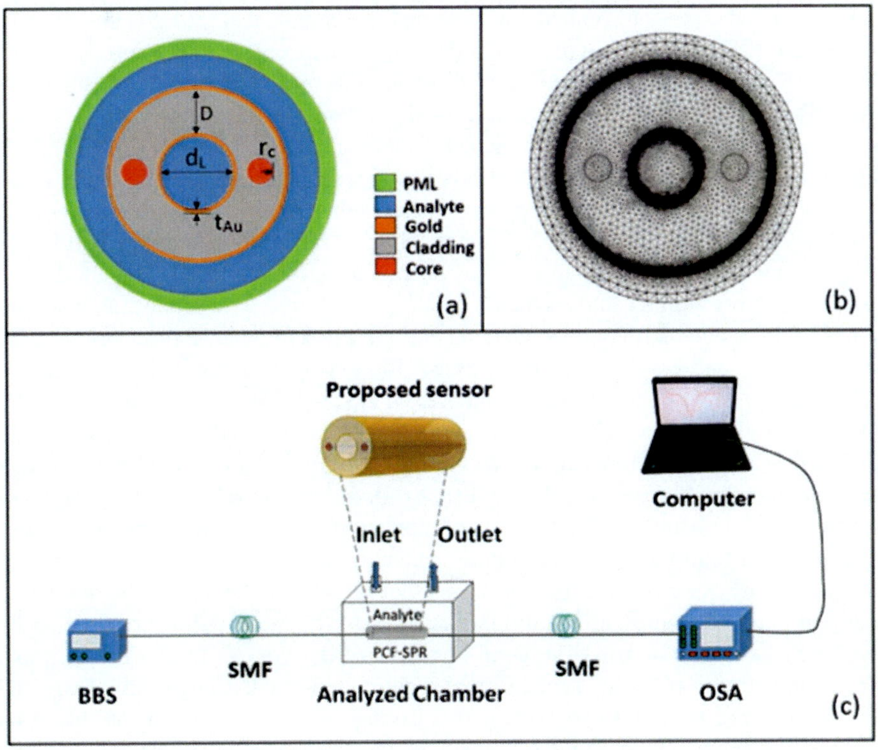

Fig. 8.12 **a** Sensor cross-section **b** model grid division **c** experimental setup schematic. Reprinted with permission from Results in Optics. Copyright, 2020, Elsevier (Liu et al. 2020c)

active plasmonic metals for the proposed device. The fiber circumstantial material is fused silica. Characterize the fused silica's complicated RI.

8.6 Numerical Investigation of D-shaped PCF-SPR Sensors with Various Material Coatings

By etching a portion of the PCF and then polishing it, a PCF-SPR sensor with a D shape is created. The D-shaped hollow-core sensor with gold metal coating, 2900 nm/RIU wavelength, and 120 RIU^{-1} magnitude sensitivity was proposed by Luan et al. (2015). To make D-shaped PCF-SPR sensors, a part of the PCF is etched and polished; Dash and Jha (2015). In the refractive index range of 1.330–1.370, a D-shaped graphene-coated biosensor with a wavelength sensitivity of 3700 nm/RIU has been developed (Dash and Jha 2015). A gold coating inside the core of a D-shape design having a wavelength sensitivity of 6430 nm/RIU at its maximum was accomplished by Wang and colleagues (2016). With a 5700 nm/RIU wavelength sensitivity and layers of ITO and graphene on top, Patnaik et al. (2015) constructed a D-shaped PCF-SPR. Tian et al. (2012) described silver-coated silica core D-shaped PCF-SPR sensors with a sensitivity of 7300 nm/RIU and an F1 value of 216 RIU^{-1}. According to Peng et al. (2015), a PCF-SPR sensor in the shape of a D with a rectangular pattern of air holes achieved wavelength sensitivities of 7481 nm/RIU and 478.3 RIU^{-1}. In addition to the ITO coating, graphene was used to increase the wavelength sensitivity of these sensors, which resulted in a wavelength sensitivity of 10,693 nm/RIU (Hasan et al. 2017; Kaur and Singh 2019a; Singh and Prajapati 2019). A refractive index greater than 1.33 has been analyzed using the D-shaped design. Analyte sensors with RIs of less than 1.33 are still required. The proposed sensor has a sensitivity range of 1.30–1.38. Water (1.330), plasma (1.340), Cancer tissue (skin, breast, lungs, and cervical), diabetic tissue, food-borne infections, toxic methanol (RI = 1.314), and human intestinal mucosa (RI 1.328–1.334) are all found in WBCs (1.360), Hb (1.360), and RBCs (1.40) have distinct refractive indices (Kaur and Singh 2020). The most commonly used tests in human illness diagnosis are the COMBS test, blood count, biopsy, and blood glucose test. In order to take advantage of these techniques. Micro and Nano integrated photonic sensors are easier to use in terms of cost and time than their predecessors. For manufacturing and infiltration purposes, the D-shaped sensor is more flexible since a flat surface can be equally textured with conventional techniques, and liquid infiltration into the metal surface may be easily accomplished with the use of a pump or a flow of the liquid. Based on a PCF and ITO conductor layer for the near-infrared region, with a sensitivity of 2000 nm/RIU for RIs ranging from 1.33 to 1.35. It was discovered that the sensor produced by Liu and colleagues had a range of 1.33–1.38 V, an average sensitivity/RIU ratio of 5990, and a resolution ranging from 1.68 to 10.5 RIU. Singh and Prajapati placed a gold-graphene layer on the surface to build a D-shaped PCF-SPR sensor (Singh and Prajapati 2019). The sensitivity was 33,500 nm/RIU and the effective RI resolution was 2.98 105 RIU for analyte RIs between 1.32 and 1.40. In the sensing range of 1.33–1.36, Hasan et al. demonstrated gold-coated PCF-SPR sensors with maximal sensitivities of 2200 nm/RIU and resolutions of 3.75×10^{-5} RIU (Hasan et al. 2017). Metal deposition on PCF or the inner surface of a small PCF structure's air holes is problematic, despite approaches such as stack-and-stack (Kaur and Singh 2019a) and side polishing (Dash and Jha 2016). Recently, PCF-SPR sensors with refractive indexes of at least 1.3 have been widely used. Sevoflurane, liquid CO_2, and fluorine-containing organics are all examples of analytes having RIs less than 1.30 can be detected via low refractive index detection (Liu et al. 2018a). Despite their rarity, these applications can be utilized. Due to the difficulty of detecting species with low refractive indices, it is vital that the correct structure be determined for PCF-SPR sensors in order to increase their application in environmental monitoring, biochemical analysis, and food safety. The new PCF-SPR sensor built on a series-wound gold nanowire device. As an oxidation-resistant plasmonic material, gold is frequently chosen over other metals. It is easier to build sensors using this PCF-SPR nanowire structure than using a gold coating on the inside or outside of the photonic crystal fiber, for example. The structural features of size, pitch, and species of air holes, as well as gold nanowires can be adjusted with a resolution of 5.71×10^{-6}

RIU, identify analytes with low refractive indices from 1.13 to 1.35 at 17,500 nm/RIU. The sensor's near-infrared wavelength range of 865–1675 nm can be used for many different things in environmental engineering, biology, and healthcare.

8.7 Chemical Detection and Concentration Measurements PCF-Based SPR Sensors

In the last few years, a great deal of work has been done on PCF for sensing applications. There has been a slew of new geometries proposed for a variety of sensing applications for chemical sensors. Strong sensitivity, low losses, high birefringence, low dispersion, and other unique guiding and sensing properties, of PCF in the THz regime have made it popular for chemical sensing applications. In addition, its small size, light weight, and adaptability, along with its durability and toughness, allow it to be used in a variety of sensing applications, including those involving hazardous materials. To achieve these goals, proper geometric design is the main challenge. Because of this, PCF's sensing properties must be maximized. Improving geometrical properties like lattice size and placement, hollowed-out areas in both the core and outer layer. SPR chemical sensor with a microstructured optical fiber may sense the visible to near-infrared refractive index (RI). In SPR, unbound electrons in metals absorb the energy of light to form collective oscillatory patterns (Esfahani Monfared 2020). Sophisticated detection of analytes' refractive indexes is possible because it is extremely sensitive to changes in the refractive indexes (RIs) of the analytes in its environment, resulting in wavelength shifts. Indeed, Because of its advantages, such as fast reaction, real-time dynamic monitoring, high sensitivity, label-free detection, and no sample degradation, SPR sensing technology has been used for chemical safety monitoring and environmental engineering (Li et al. 2020). An optical fiber known as a microstructured optical fiber (MOF) is a recent innovation. Design flexibility, tunable dispersion, and single-mode operation indefinitely (Liu et al. 2018b, 2020a), have led to a significant amount of effort to create MOF for a variety of devices such as amplifiers (Liu et al. 2019b), filters (Yan et al. 2018b), and sensors that can be applied in various fields such as strain, temperature (Liu et al. 2016b), and refractive index (Rifat et al. 2017). After the rapid advancement of SPR sensing technology, interest in MOF-based SPR sensors has also been piqued. MOF-SPR sensors have been developed and studied, although the range of analyte RIs that can be detected is still rather limited. Only a few sensors are capable of detecting RIs between 1.00 and 1.10, which means that most detectable analytes have RIs greater than or equal to 1.33. Furthermore, the analyte's gas–liquid form is rarely mixed, and only one sensor can typically detect it. Sensors made of MOF-SPR with a larger RI range are very desirable for the detection of gaseous and liquid analytes at the same time in order to broaden their use. Asaduzzaman et al. developed a hybrid PCF for chemical sensing applications, achieving a sensitivity of 49.29% for benzene detection with a confinement loss of 3.13×10^{-10} dB/m, a sensitivity of 49.17% for ethanol detection with a confinement loss of 2.75×10^{-10} dB/m, and a sensitivity of 48.85% for water detection with a confinement loss of 2.75×10^{-9} dB/m (Asaduzzaman et al. 2016). For liquid chemical detection, Ademgil et al. developed a PCF-based sensor with high birefringence and negligible condensing loss. However, they only achieved a 26% sensitivity with their proposed PCF geometry, which includes air holes in the core region that are elliptical in shape and a V- or H-shaped cladding structure (Ademgil and Haxha 2015). Islam et al. have proposed a chemical detector with a complex kagome structure (Islam et al. 2020). According to the numerical output, the most sensitive point is at 86% relative to the baseline. For liquid samples, Islam et al. developed a hollow-core THz sensor on the PCF platform and achieved a maximum sensitivity of 96% at 1.4 THz (Islam et al. 2018). Because benzene, methanol, and ethanol are particularly dangerous to human health, a low-loss PCF geometry was created to be highly sensitive in chemical detection while requiring little power. For precise chemical detection, symmetrical hexagonal holes in the cladding and a hollow core are considered here. At a frequency of 3 THz, the PCF relative sensitivity is nearly 99%, other PCF performance metrics, such as effective material loss and fine-tuning loss, are also analyzed. Effective area (EA), numerical aperture (NA), effective refractive index (ERI), dispersion, V parameters, and spot size are also included in the calculations. Because benzene, methanol, and ethanol are particularly dangerous to human health, a low-loss PCF geometry was created to be highly sensitive in chemical detection while requiring little power. These have been greatly improved and are now very close to the existing chemical sensors (Asaduzzaman et al. 2016; Ademgil and Haxha 2015; Islam et al. 2018, 2020). The full-vector finite element approach is used to build and analyze an SPR sensor based on MOF with an extremely large RI. For practical purposes, the sensor has fewer air holes placed in the outer cladding to generate the upper and lower hexagonal symmetrical construction. In addition, the exterior surface of the entire MOF structure is coated with a gold film, which eliminates the need to fill the air holes with analyte and makes uniform coating much easier than with a D-shaped structure (Li et al. 2019). The SPR effect is generated for efficient sensing in this structure by the evanescent wave penetrating the gold coating. Analytes can be detected in a wide range of states (gas and liquid) using the MOF-SPR sensor. There is no requirement for a separate light source for low RIs detection because the working wavelength ranges from 420 to 640 nm in the visible spectrum. A high RI detection from 1.44 to 1.45 is also suitable for this method. Results show that the wavelength sensitivity (WS) is 15,000 nm/RIU and the magnitude sensitivity (AS) is 1,603.37 RIU^{-1}. The proposed sensor and method have huge practical potential and also represent a breakthrough and innovation in the design and functions of SPR sensors due to their simple structure, flexible design, ease of manufacture, and many capabilities.

8.7.1 Design and Analysis of Chemical Detection Sensor

The proposed sensor has a huge practical potential and also represents a breakthrough and innovation in the design and functions of SPR sensors due to their simple structure, flexible design, ease of manufacture, and many capabilities. Figure 8.13a shows the MOF-SPR sensor's cross-section with the substrate shown as fused quartz. In order to confine the bulk of incident light to the fiber core during propagation, the MOF structure's air pores lower the cladding's effective refractive index. The full-vector FEM is used to analyze the model, which is run through the COMSOL MULTIPHYSICS software. To reduce unwanted electromagnetic reflection, a circular perfectly matched layer (C-PML) is also used as an absorbing boundary condition in two-dimensional modeling (Liu et al. 2020d). Convergence tests are carried out during the simulation process by adjusting the mesh size and C-PML thickness. A triangular ultra-refined grid divides the computing space into 31,962 domain units and 2,190 border units. This exact and accurate segmentation makes it easy to find viable forms of transportation. The first, second, and third air holes in the cladding have dimensions of 0.7 m, 0.8 m, and 1.6 m, respectively. Three of the six first air holes form a group, whereas the other two groups are symmetrically organized. A total of eight air holes are positioned in a hexagon around the central second air hole. The strong oxidation resistance in aqueous solutions and the broad resonance wavelength shifts of gold make it an ideal plasmonic medium. A chemical vapor deposition (CVD) process is used to ensure a uniform thickness of the gold layer on the surface of the MOF (Tong et al. 2018). A lot of advancements have been made in MOF manufacturing technologies in recent years. Figure 8.13b shows a stack-and-draw approach for fabricating MOFs. A fiber is created by stacking and inserting numerous small capillaries into a glass tube with an exterior cladding that matches the capillaries on the outside. An additional 125 μm of cladding is then added to the MOF to complete the process. We may regulate the diameter of the air holes as well as the pressure applied to keep them open during the fiber drawing process (Liu et al. 2018c).

8.7.2 Experimental Set-up for Chemical Substance Detection and Testing

An experimental platform for photochemical sensing is depicted in Fig. 8.14. Figure 8.14a depicts the evanescent wave's propagation process in the fiber, as well as the excitation principle that results in resonance. Air gaps in the MOF can be filled to reduce the cladding's effective refractive index. The optical fiber used in the MOF's design is a step type, which limits incident light transmission in the fiber core and so reduces transmission loss. Evaporative photorespiration (SPR) can be stimulated by using total internal reflection (TIR) to guide light into the MOF. Surface plasmons are excited at the metal–dielectric medium (analyte) interface by

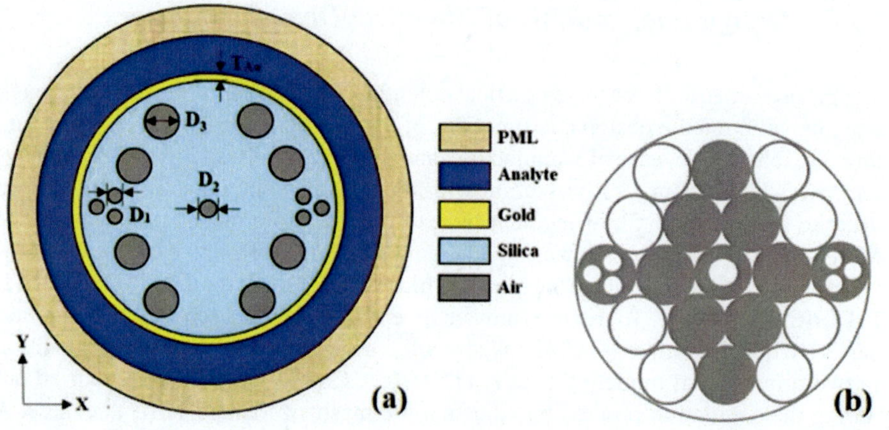

Fig. 8.13 **a** MOF cross-section and **b** stacked configuration in the model. Reprinted with permission from Optics Express. Copyright, 2021, Optica (Liu et al. 2021)

the guided/propagating rays. The wavelength, fiber characteristics, probe geometry, metal layer qualities, and sensing medium (analyte) all have a role in the evanescent wave's surface plasmon coupling (Lu et al. 2018; Ermatov et al. 2020).

Mode coupling is strongly influenced by the RIs of the external medium (analyte). An evanescent wave that is transmitted to the metal's surface can induce collective oscillations of the internal free electron population, resulting in the SPR effect. Experimentation for photochemical sensing is depicted in Fig. 8.14b. Transmitting light into the single-mode fiber (SMF), which then strikes the MOF's interior, comes from a wide-bandwidth light source (BBS). A gold-coated RI sensor is inserted into the chamber to ensure proper contact with the analyte. The optical spectrum analyzer (OSA) detects and analyzes shifts in the resonance wavelength (blue or red) in samples with various RIs on the gold film surface. Detection of gas differs from that of liquid since it necessitates the use of an enclosed space. A gas generator is required to produce and deliver gas to the experiment gas chamber. The gas flow rate may be carefully controlled with the help of a flow meter. A hygrometer is used to monitor and control humidity, and a desiccant is placed in the gas chamber to dry the gas. The aforementioned photochemical sensor experiment equipment can detect the gas in real time as it enters the gas chamber. Figure 8.14c depicts a succession of sealed tubes, such as those containing liquid pollutants and hazardous gases, which were used to collect samples. In addition, a drying oven and ultrasonic bath for cleaning and drying the containers, as well as a refractometer for measuring analyte RIs, are used. As a result, dangerous compounds in the environment can be more efficiently monitored and regulated, enhancing measurement security and quality.

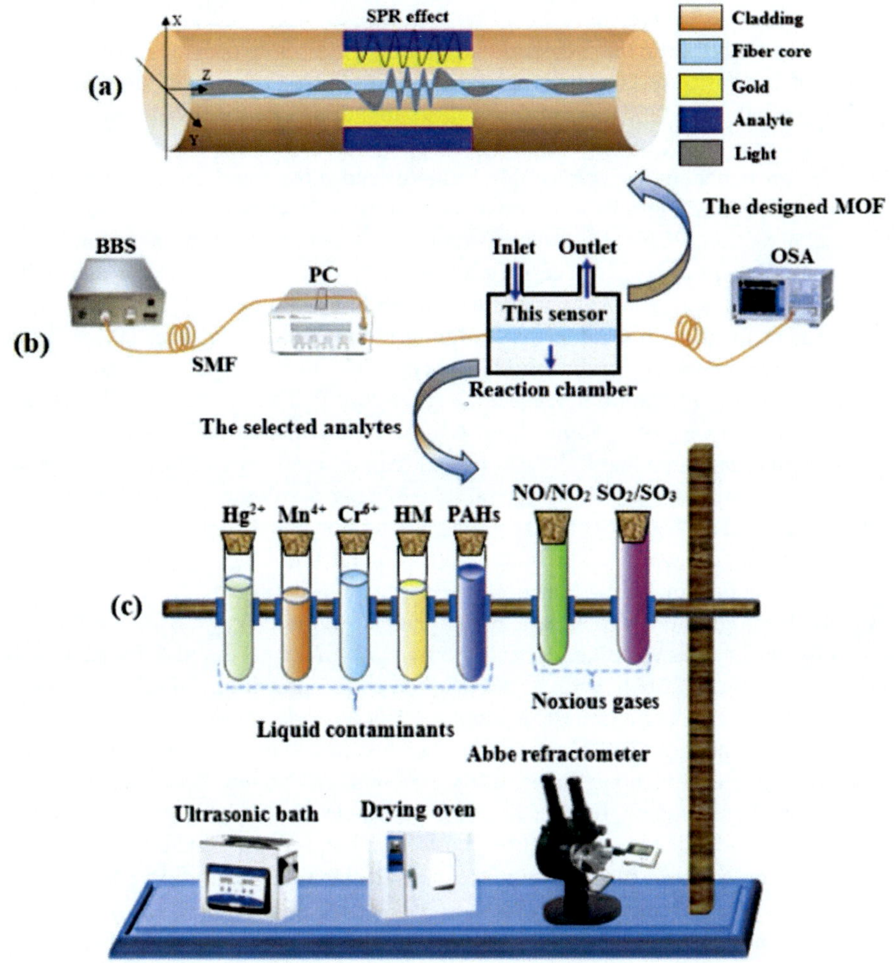

Fig. 8.14 **a** MOF light-directing mechanism and SPR resonance concept; **b** photochemical sensing experimental setup; **c** chemical substance detection and testing setup. Reprinted with permission from Optics Express. Copyright, 2021, Optica (Liu et al. 2021)

8.8 Biological Detection/Biomedical Diagnosis Geometric-Based SPR Sensors

Biosensors are being developed to meet the growing demand for real-time diagnostics at the point of patient care. Optical label-free sensors have attracted a lot of attention from the biomedical industry because of their extremely high sensitivity to changes in the ambient refractive index (RI), good selectivity, low cost of production, and design flexibility. Photonic crystal fibers have emerged as a prospective research focus in the optical sensing sector in the previous decade (Lee et al. 2009). For the manufacturing of PCF, silica is employed, and the air holes can be arranged in a variety of ways. In addition, the infiltration of a detecting liquid into the air hole provides an additional degree of flexibility for sensing applications. So far, PCF manufacturing has used a variety of advanced methods of fabrication, including chemical vapor deposition, sol–gel casting, slurry casting, capillary stacking, drawing, and extrusion procedures. PCF design is becoming increasingly miniaturized throughout the world as many sectors and research groups look for ways to improve its performance. Aside from medical detection, PCF sensors have been in great demand for biosensing and drug detection applications. It is necessary to precisely organize air holes in PCF with a hollow core because of its narrow transmission spectrum, high sensitivity, and requirement for this. In spite of this, solid-core photonic crystal fiber (SC-PCF) is able to overcome the constraints of hollow-core PCF, although at a reduced level of sensitivity. In the literature, index-guided solid-core PCFs have been used as smart biosensors for a variety of applications (Xu et al. 2018; Kaur and Singh 2019b). It has been demonstrated that hollow-core PCF can be used for cancer cell detection, with a detection limit of 100 pg for samples having a volume of 10 nL or less. A photonic crystal waveguide (PCW) for illness detection has been proposed by Chopra et al. (Chopra et al. 2016). The body's most crucial fluid, blood, carries out all of its essential duties, including transporting oxygen and supplying nutrients and waste to cells. Hemoglobin and platelets, white and red blood cells (WBCs and RBCs), minerals, water, plasma, proteins, and vitamins all make up the blood's many cell types (Chopra et al. 2016). Ambiguities in blood cell composition might lead to major health problems. Detecting these discrepancies at an early stage has the potential to save a great number of lives. Malaria, anemia, leukemia, lymphoma, hemophilia, and hepatitis B and C are among the most frequent blood disorders. For the detection of these anomalies, photonic biosensors are the ultimate device, and these apparatuses are more dependable, more compatible, cost-effective, accurate, and tiny in size with the human body. Human abnormalities, such as diabetic tissue, cancer tissue, and blood samples, need to be detected by a PCF sensor.

By measuring changes in transmission, hollow-core microstructured optical fibers (HC-MOFs) may detect liquid analytes. HC-MOFs have the benefit of allowing a huge volume for analyzing light-analyte interactions, which enhances the RIS over cavity-based alternatives (Cubillas et al. 2013). Resonant shifts in the scattering spectra, reflection, and transmission associated with changes in the analyte's RI and concentration are generally exploited by optical sensors (Farka et al. 2017). Antibodies, aptamers, and other analyte binders can be added to sensor template nanostructures to enhance their specificity to target biomolecules (Shandilya et al. 2019). In addition to being inefficient and costly, this approach ignores RI optical dispersion, which can serve as a basic fingerprint for liquid biosample samples and allow for real-time monitoring of compositional changes. Some disorders can be detected by looking at the optical dispersion of blood serum, which is closely linked to changes in blood components such as albumin content and the emergence of albumin-conjugated forms (Wang et al. 2011).

PCF-based SPR sensors have been utilized in a number of different applications, including medical diagnostics, biosensing, bioimaging, environmental virus detection, and other biomedical applications (Lu et al. 2018; Liu et al. 2020b; Haider et al. 2020; Sandeep et al. 2016). When the imposing light travels through the core, electromagnetic scattering occurs, creating an evanescent field. The free electrons on the metal surface are excited by this evanescent field, which strikes the plasmonic layer (Wang et al. 2019). At the metal–dielectric interface, there are free electrons in contact that collectively oscillate. Establishing the existence of the SPR phenomena requires careful consideration of the plasmonic material used. Materials used in SPR sensors now include aluminum, silver, gold, copper, and a few oxides (Indium tin oxide, Titanium dioxide).

For optoelectronic sensing purposes, bio-instruments based on surface plasmonic resonance (SPR) are highly suggested. In pharmacological research, bioimaging, food eminence measurement, antigen–antibody collaboration, environmental observation, and so on, the plasmonic sensor has been widely used (Rifat et al. 2016). The prism connecting to metallic film was the SPR design used in older SPR approaches, such as the Kretschmann configuration (Mishra and Mishra 2016). Conventional SPR instruments have not been accepted because of their enormous size, unjustifiable expense of optical and mechanical components, remote sensing incapacity, and smallness irregularity. A PCF-SPR-based bio-instrumentation system was first published by Jorgenson in 1993. A characteristic of PCF-SPR biosensor operation is the interaction of the SPP electron and momentary field with each other (Mishra and Mishra 2016). The wavelength of the fundamental core mode and SPP mode in real time is equivalent to the real part of the effective refractive index (n_{eff}) at a given wavelength, called the resonance point. The release of SPP electrons from the metal/dielectric interface is due to the impact of the core-clad momentary field. As a result, SPWs have been created. Au is preferred as the most striking plasmonic material for use in SPR sensing, although its sensing range is limited (Lu et al. 2018). As a result, in order to increase sensitivity, the plasmonic metal silver (Ag) can be utilized in place of gold in Rifat et al. (2015). A number of studies have been carried out in order to gain a better understanding of their sensitivity, detection accuracy, quality factors, and detection ranges. The majority of PCF-SPR biosensors fall into two categories. Starting with an internal detection method based on PCF-SPR sensors, which coat the inside PCF with a thin layer of plasmonic material before allowing real-time liquid analyte to pass through. Another option is to use PCF-SPR sensors that detect plasmonic material on the outside of the PCF. These sensors have recently been explored and proposed (Liu et al. 2018c; Asaduzzaman and Ahmed 2018). PCF-SPR sensitivity is in the range of 1.33–1.39 RIU, with a maximum wavelength sensitivity of 6000 nm RIU^{-1} in the sensing range (An et al. 2017). PCF-SPR biosensors based on open ring channels have been reported and sold that can detect analytes with modest refractive indices (RI) from 1.23 to 1.29 RIU in real time. The amplitude sensitivity is 333.8 RIU^{-1} and the wavelength sensitivity is 5500 nm RIU^{-1}.

A bio-inspired butterfly PCF-based externally gold-coated plasmonic sensor is proposed to detect unknown analytes with a wide range of RIs from 1.33 to 1.42. Control of light propagation and detection range, cost-effectiveness, and fabrication feasibility are all improved using the bio-inspired plasmonic structure (Mashrafi et al. 2021). The manufacturing of sensors is made possible by an exterior metal–dielectric layer and a simple air-hole configuration. In addition, the suggested sensor operates in the visible to near-infrared spectrum, where commercially accessible light sources are both affordable and readily available for development.

SPR biosensor with an exterior coating of silver, which is more suitable from the perspective of updating technology, is statistically evaluated for its superior sensitivity to a hollow-core photonic crystal fiber (PCF). It has a maximum wavelength and amplitude of 21,000 nm/RIU and 2456 RIU^{-1} for external bio-target refractive index detection applications. and matching resolution of 4.76×10^{-6} RIU and 4.07×10^{-6} RIU. Geometrical characteristics like silver layer thickness and plasmonic metal type can be changed to improve sensor performance. Pitch and analyte fluctuation can also improve sensor performance.

8.8.1 Design and Modeling of Biological Detection of Different Structures

8.8.1.1 Hollow-Core Microstructured Optical Fibers for Liquid Biosample Detection

Static and real-time in-fiber multispectral optical sensing (IMOS) of liquid biological samples occurs when a liquid bioanalyte flows through specifically engineered liquid chambers in a hollow-core microstructured optical fiber. The sensing concept depends on detecting spectral changes of maxima and minima in the transmission spectrum (Fig. 8.15a). Bio analytical RIs are clearly linked to the core capillary wall Fabry–Perot resonances that are associated with these resonant characteristics. Measurement of the RI over a range of wavelengths up to 3000 nm/RIU is possible with only a single fiber, and the figure of merit (FOM) can be as high as 99 RIU^{-1} in the visible and near-infrared spectrums. Using a layer-by-layer (LbL) assembly of oppositely charged polyelectrolytes (PEs) that is highly controlled and reproducible, we were able to expand the number of acquisition wavelengths by coating the capillaries of numerous HC-MOFs with significantly shifted transmission windows. Figure 8.15a shows the configuration that we have used to test the optical transmission of liquid-filled HC-MOFs. Inserting the tips of HC-MOFs into miniature liquid cells (LCs) equipped with tubing interfaces and optically transparent windows results in the liquid filling of fiber capillaries. Coating the fiber capillaries is illustrated in Fig. 8.15b. Pre-determined volumes of PE solutions are fed into the fibers by the pump, which controls and maintains the flow rate at a predetermined volume. As a result of this technique, the PE layers are uniformly deposited inside the fiber capillaries as a result. It was decided to employ PAH and PSS (polyanionic polystyrene sulfonate) as polycationic poly (allylamine hydrochloride) as PEs (Zhang et al. 2018).

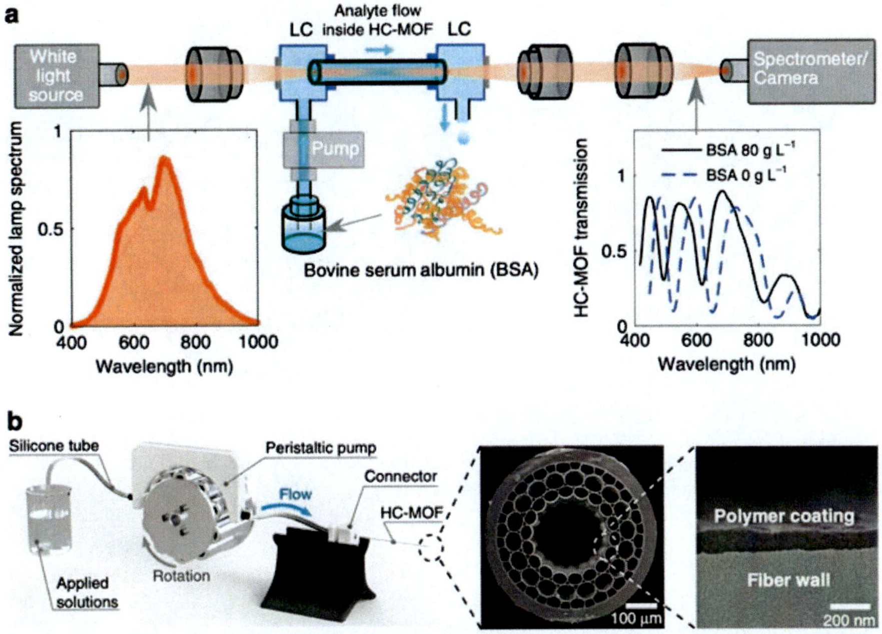

Fig. 8.15 Illustrates the principle of dynamic multispectral sensing for liquid samples. Adapted from Ermatov et al. (2020)

Figure 8.15a shows the setup used to characterize HC-MOF transmission properties. Fluids can be pumped via fiber capillaries while the transmission spectrum is being measured because the fiber facets are attached to liquid cells (LCs), which are optically accessible through thin glass windows. Using the HC-MOF and a CCD (charge-coupled device) camera, we were able to record the output mode profile by following the red rays from the broadband halogen lamp. Input and output spectra of BSA dissolved in PBS at various concentrations are shown in the insets of Fig. 8.15b. The HC-MOF LbL functionalization setup is shown. pumping of polyelectrolyte water solution and pure water through the length of the fiber by a peristaltic pump, results in the creation of a polymer coating on the core capillary's inside wall. Bilayers are created by repeating the method using inversely charged polyelectrolytes. Photos of the capillary wall and fiber end face taken with a scanning electron microscope (SEM) are shown in the insets.

8.8.1.2 Hollow-Core Circular Shaped PCF-SPR Biosensor

The sensor pattern is made up of there is one circular air hole in each of the five hexagon shaped circles. Figure 8.16 Schematic and visual representation of the hollow-core circular PCF-SPR biosensor offered, as shown in cross-section. To make a hollow core, one air hole is removed from the fourth circle. The four air holes, two

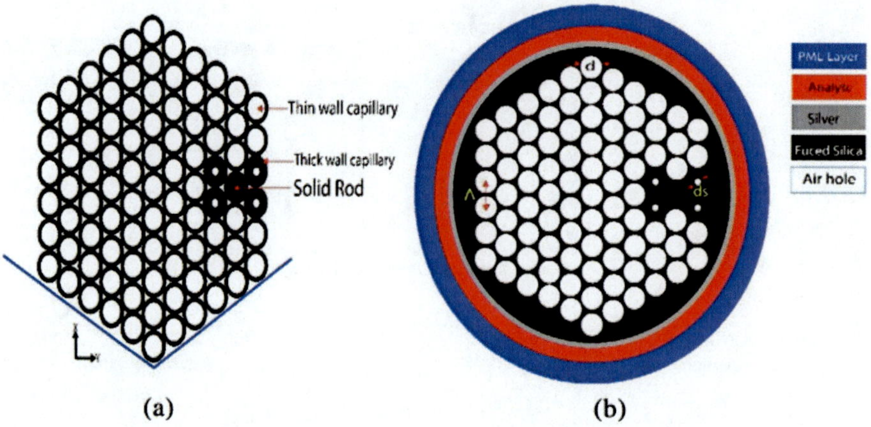

Fig. 8.16 **a** The provided design is stacked in 2D along the *X–Y* plane. **b** A 2D cross-section of the hollow-core circular shaped PCF-SPR biosensor with $d_s = 0.35\Lambda$, $d = 0.82\Lambda$, and $t_s = 30$ nm. Reprinted with permission from Results in Physics. Copyright, 2020, Elsevier (Hossain et al. 2020)

from the third circle and two from the fifth, are positioned below where the distance between the dielectric and the core should be reduced. As a result, SPs can easily engage with metallic surfaces. The side half of the hexagonally photonic crystal air holes were removed to make the circular ring. The center-to-center distance of neighboring air holes is defined by pitch and expressed by, where the diameter of a tiny air hole (s) can be denoted as d_s, and the diameter of a large air hole (d) can be indicated by d. First, these variables are given the values of 3 m, $d_s = 0.35$, and $d = 0.82$. The plasmonic material is a silver coating with a thickness of $t_s = 30$ nm that is deposited on the device's planar surface using the chemical vapor deposition (CVD) process. As a background material, we used fused silica.

The suggested bio-inspired plasmonic sensor is shown in cross-section in Fig. 8.17a. In a hexagonal layout, the sensor has four circular air-hole rings that resemble butterflies. Following a modified-total internal path, the light travels through the suggested PCF. A moment of contemplation (M-TIR). In order to keep the light in the fiber core, the first air hole ring is eliminated from the core. The second and third (each with two air holes) are missing a few of their air holes (four air holes) For ease of light penetration and the generation of evanescent fields, there are four (four air-hole) rings. The free electrons on the metal surface are excited when objects are struck against the plasmonic surface. Also, with its covering, the core's light is kept from leaking out into the surrounding open space. An air hole in the fiber's center is reduced in size in order to provide a more evanescent field. In order to the y-axis is protected from direct connection by two scaled-down air holes.

Figure 8.17b shows the stacked preform of the proposed sensor. A thin-walled capillary is used to denote regular air holes, while a thicker-walled capillary is used to denote microscopic air holes. An example of an experimental setup is depicted schematically in Fig. 8.17c. Using a broadband light source, the proposed PCF core

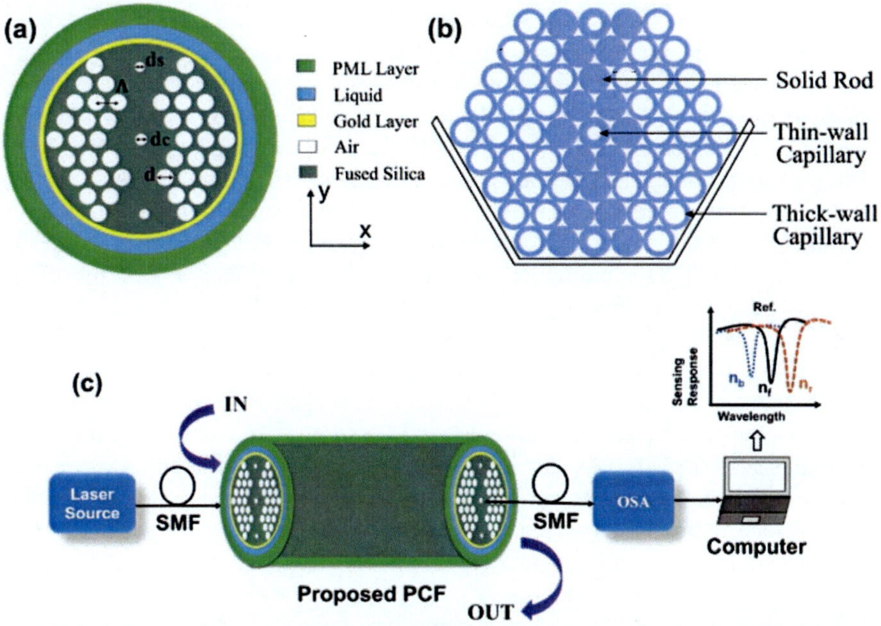

Fig. 8.17 **a** Cross-section view **b** a butterfly-shaped plasmonic RI sensor preform stack inspired by nature. **c** Schematic diagram of a proposed experimental setup. Reprinted with permission from OSA Continuum. Copyright, 2021, Optica (Mashrafi et al. 2021)

that excites the analyte channel can be illuminated. By using micro-capillary needles with tapered heads, a syringe pump can be used to infiltrate liquid into the sensing channel. Analyzing the loss spectrum can be done using an optical spectrum analyzer (OSA). Loss spectra will exhibit either blue or red shifts as a result of an analyte RI change.

8.9 Liquid and Gas Detection of Photonic Crystal-Based SPR Sensors

The development of PCFs, which have overcome the limitations of conventional fibers. Many applications, including communication and sensors, have been developed around PCF in recent decades, its light-guiding properties and potential interactions with gases or liquids. Photonic crystal fibers with an air-holed micro-core have been proposed for liquid sensing applications. PCF circular pattern with a large effective area and high sensitivity for various liquid sensing applications (Senthil et al. 2019). For describes a carbon dioxide (CO_2) gas sensor based on HC-PCF technology. The HC-PCF-based sensor is more sensitive than solid-core PCF and porous-core PCF sensors for monitoring a wide range of liquids and gases (Sardar et al. 2021). Air pollution is a comparatively recent concern, and it is becoming more and more polluted. Humans are poisoned by airborne poisons. Human lungs and tissue are particularly vulnerable to the effects of ultrafine airborne particles. Gas sensors are employed to check the quality of the air that is circulated. Several sensors have been developed to help identify dangerous toxic gases. The small size, high sensing accuracy, and ease of installation make PCF-based sensors a popular choice. GeO_2 doped silica-based defective core PCF was introduced in 2015 by M. Morshed et al. for gas sensing applications (Morshed et al. 2015a). The absorption line for methane (CH_4) and hydrogen fluoride (HF) gases at $k = 1.33$ lm offered a 27.58% absorption efficiency (Morshed et al. 2015b). The same year, the same author introduced a pure silica-based modified hexagonal form PCF for gas sensing applications. In the case of CH_4 and HF gases, it was discovered that operating at a wavelength of 1.33 μm yielded the maximum sensitivity response is 42.27%. Asaduzzaman et al. presented a silica-based PCF for the detection of flammable and poisonous gases in 2016 (Asaduzzaman and Ahmed 2016). At an input wavelength of $k = 1.33$ l m, this gas sensor has a sensing response of 53.02%. Single-mode spiral-shaped PCF-based gas sensor with improved gas detection sensitivity was proposed by M. I. Islam et al. in 2017. At a wavelength of 1.33 lm, the sensor's sensing response was 57.61%.

The development of SPR sensors has received increasing attention due to a wide range of potential future applications (Sharma et al. 2007; Chen and Ming 2012; Wong and Olivo 2014; Hutter and Fendler 2004). Numerous sensor types have been developed and extensively evaluated. Ammonia, hydrogen sulphide, chlorine, hydrogen, and nitrogen can all be detected using a SPR-based fiber optic gas sensor, according to Mishra et al. (2014). Fiber SPR sensing probes for liquid concentration measurement were developed by Y. Zhao et al. using a new chemical method based on the silver mirror reaction (Zhao et al. 2014). Liquid refractive index detection is becoming more popular with the use of optical sensors based on photonic crystal fiber structures. In the realm of optics, SPR is a common occurrence. Light incident on a metal surface is resonant because the evanescent light wave matches the plasma wave vector of the metal surface. It is at this point in time that surface plasma waves absorb most of the light's energy, reducing the reflected light's energy. It also serves as the foundation for SPR applied to optical sensors. Light, on the other hand, has a smaller wave vector than metal plasma. As a result, the two waves arriving at the wave vector should be matched using a special structure of photonic crystal fiber (Dash and Jha 2016). SiO_2 is the material of choice for optical fibers made of conventional materials. These materials can't be used to detect liquids with lower refractive values because their refractive indices are too high. PCF sensors, a new fiber optic device, can accomplish (Malka et al. 2017). Due to the unique periodic air-hole cladding structure, they are able to lower the core's refractive index, making detection possible. Apply a metal membrane to the air holes of optical fibers, and use selective filling of liquid to determine how much liquid is in each air hole. Because of their limited mobility and detection speed, sensors built using this method are often highly sensitive and capable (Liu et al. 2015; Fan et al. 2015). The second method is to coat the outside of the optical fiber with a layer of metal. All of the body sensor's components are placed in the liquid to be detected during actual measurements (Hasan et al. 2017; Tan et al. 2014). Real-time detection is performed. Making a sensor with no internal air holes means that the manufacturing process can be streamlined and is relatively easy to implement. The uneven thickness of the metal film produced by the two methods, on the other hand, is a problem that must be addressed. SPR sensor based on PCF structure with high sensitivity.

There are numerous advantages to using gold nanowires coated with two gold nanowires, including a reduced impact on the original gold film thickness and an improved resonance effect and sensor sensitivity (Lee et al. 2008; Lu et al. 2012). Axially aligning the two gold nanowires and the PCF will result in errors, given the reality. Consequently, we'll examine the effects of sensor alignment and misalignment in subsequent chapters. Sensor sensitivity is unaffected by axial misalignment within the specified tolerances. In addition, the sensor's high sensitivity allowed for more accurate detection and determination of liquid refractive indexes and concentrations.

8.9.1 Design and Analysis of Liquid and Gas Sensor

A silica-based material with six large air holes and ten small air holes was used in this design. Figure 8.18. depicts the proposed sensor; the innermost layer was made of plastic. The inner lining's large air holes A hexagonal arrangement is used for the layers. In addition, the outer layer's small air holes surround the same angle as the center of the circle. The metal surface can be illuminated by a suitable optical path. The sensor is axisymmetric in all aspects. The analyte channel was coated with two gold layers on the outer layer nanowires. Internal air-hole distances, distance from outer air hole to center, Analyte channel width, r_d, and r_c were all set to 4 μm and 2 μm, respectively, 1 dBa = 1 m, and so on. Inner and outer air hole radii were calculated as $r_a = 0.5$–0.6 μm and $r_b = 0.2$–0.4 μm, respectively. The nanowires have a radius of $r_g = 0.2$–0.32 μm. The analyte refractive index ranged from 1.33 to 1.36. As depicted in Fig. 8.19a, it is this asymmetric regular hexagonal air hole arrangement with a lattice constant that creates the cladding layer. The air holes' diameters are d_1 and d_2, and the refractive index is $n_{air} = 1$, respectively. These air holes reduce confinement loss and somewhat raise birefringence by increasing confinement loss.

To position the liquid to be detected, an elliptical sensor channel coated in gold is introduced on the left side of the core. The elliptical channel hole wall has a Tox gold layer covering to evaluate the sensor system's capabilities. The ideal air hole values are = 2 μm, $d_1 = 0.8$ and $d_2 = 0.6$ after careful analysis. We have retained the thickness of gold at 40 nm when using gold as the plasmonic material. This channel's initial length is 0.8, its elliptically is equal to *a/b*, and the initial value for this channel is $e^0 = 0.7$. The liquid's refractive index is Na, which is used to establish the system's basic structural parameters. Figure 8.19b depicts the proposed sensor's experimental setup. The detection channel is filled with the liquid to be

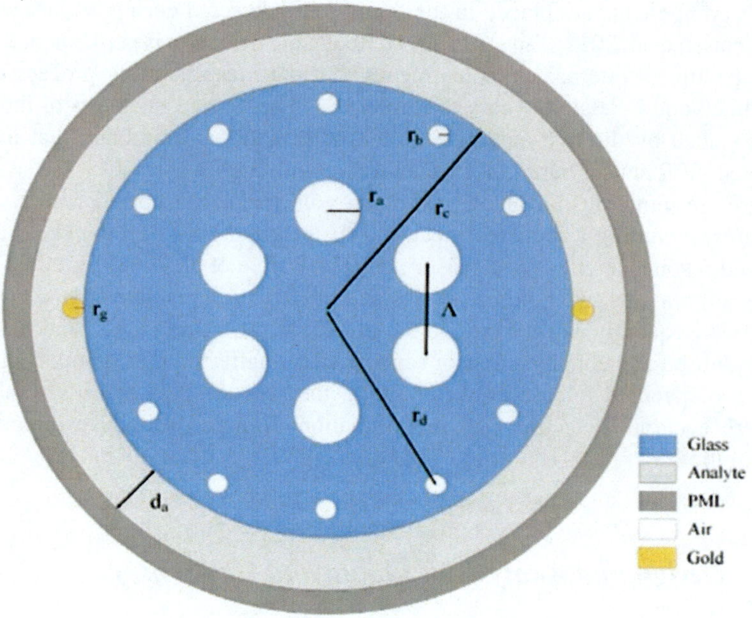

Fig. 8.18 The liquid sensor construction is depicted schematically. Adapted from Yan et al. (2018a)

detected, and as the surrounding environment changes, the liquid refractive index changes. The PCF's single-mode fiber (SMF) is joined at both ends by a fusion process. The contained loss is 3–4 dB/cm, so we recommend using 0.5 m SMF and 1 cm PCF. After passing through the PCF, the light source emits light. If it is too long, the unnecessary loss will increase and results will be affected. You can use the Hogen Tungsten Light Sourcevis-Nir (360–2000 nm) to emit light in the 360–2000 nm range, which is ideal for us. The optical spectrum analyzer then receives and analyzes this data (OSA). To ensure that the incident light is parallel to PCF, the OSA directly analyzes emitted light intensity. The emitted light's intensity drops during OSA analysis. The trough will move if the liquid changes significantly. In order to find the resonance wavelength at which the SPR effect occurs, we make use of this phenomenon. The wave trough in our design shifts to the long wave direction as the refractive index rises, which is in line with the shift in the paper's loss peak. Sensor based on the SPR effect with an elliptical detecting channel that uses PCF refractive index cladding. Detection ranges from 1.43 to 1.49 nm/RIU and a sensitivity of up to 12,719.97 nm/RIU for liquids with high refractive indexes can be achieved with this technology.

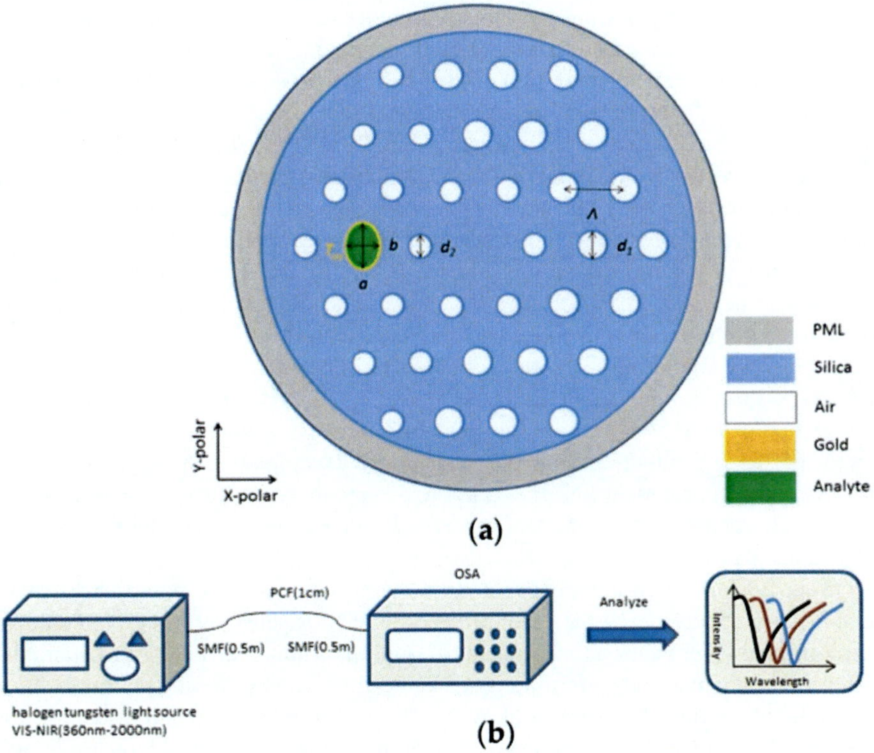

Fig. 8.19 **a** Surface plasmon resonance photonic crystal fiber (SPR-PCF) refractive index sensor's cross-sectional schematic structure **b** illustration of the intended sensor experimental setup. Adapted from Yan et al. (2021)

8.10 Future Prospects

The chapter on the application of geometric-based SPR sensors presents a comprehensive overview of the advancements and potential applications of photonic crystal fiber (PCF)-based SPR sensors (Chaudhary et al. 2021a, b, 2022a). The future prospects of this chapter lie in further exploring and expanding the capabilities of PCF-SPR sensors in various fields (Chaudhary et al. 2022b, 2023). Such as (1) Sensor Integration: Integration of PCF-SPR sensors with other technologies and systems holds promise for developing advanced sensing platforms. This integration can enable real-time, multiplexed detection of multiple analytes, facilitating more comprehensive and efficient analysis in diverse applications. (2) Biomedical and Clinical Applications: The use of PCF-SPR sensors in biomedical and clinical settings is an area with significant potential (Jain et al. 2022; Chaudhary et al. 2023; Kaur et al. 2022). Further research can focus on developing specific sensor designs and surface functionalization techniques for targeted biomolecular interaction analysis, disease diagnostics, and therapeutic monitoring. (3) Environmental Monitoring:

PCF-SPR sensors have shown promise in environmental monitoring applications such as water quality assessment, pollutant detection, and gas monitoring. Future research can explore the integration of PCF-SPR sensors with portable and wearable devices for on-site monitoring of environmental parameters. (4) Nanomaterial Integration: Incorporating novel nanomaterials into PCF-SPR sensors can enhance their sensitivity, selectivity, and detection range. Research efforts can focus on exploring new nanomaterials, such as plasmonic nanoparticles, graphene, or quantum dots, for improved sensor performance and expanded sensing capabilities (Kaur et al. 2023). (5) Sensor Optimization: Ongoing efforts can be directed toward optimizing the design parameters of PCF-SPR sensors, such as air hole form, size, and placement, to achieve even higher sensitivity and improved confinement loss. Computational modeling and simulation techniques, such as finite element analysis, can aid in the design optimization process (Kumar et al. 2022). (6) Industrial Applications: The integration of PCF-SPR sensors into industrial processes, such as chemical manufacturing, food production, and pharmaceutical development, holds potential for quality control, process monitoring, and real-time analysis. Research can focus on developing robust and reliable sensing systems that can withstand harsh industrial environments (Kumar et al. 2021).

In conclusion, the future prospects of the chapter on geometric-based SPR sensors revolve around advancing the capabilities of PCF-SPR sensors through integration, exploring new applications in biomedical and environmental fields, incorporating innovative nanomaterials, optimizing sensor designs, and expanding their utilization in industrial settings (Mishra et al. 2022). These prospects will contribute to the continued growth and development of PCF-SPR sensors as powerful tools for optical sensing and analysis (Rachana et al. 2022; Zhang et al. 2023).

8.11 Summary

SPR sensor has great potential application in diverse fields such as chemical sensing, biomolecules, gas and liquid detection, medical diagnostics, chemical reaction rate monitoring, and biomolecular interaction. In recent years, PCF-SPR sensors have become increasingly popular because of their high sensitivity and effectiveness. These sensors are found to have a better sensing response in terms of sensitivity, selectivity, response time, recovery time, and repeatability than uncoated and prism-based optical sensors. In this article, a number of PCF-SPR sensors are reviewed, whose different geometries, such as hollow core, circular pattern, multiple sensing ring, quasi-periodic pattern, square array, D shape, D-shape dual core, and designs, could be used in a wide variety of chemical, biological, gas, and liquid detection applications. The PCF-SPR sensor can be utilized to detect drugs, analytes, antigens, and antibodies in the medical and chemical fields. To detect changes in the refractive index of analytes, PCF-SPR sensors with different structures have been proposed. These structural characteristics, such as the shape, number, location, and size of the air holes, can be altered to improve sensitivity and range.

Due to its high sensitivity, the SPR technique is commonly employed in sensing. Medical diagnostics, antibody–antigen interaction, and a variety of biological and chemical analyte detection are possible uses for SPR-based sensors. Due to its label-free sensing, quick reaction, and high sensitivity, SPR is an effective and strong optical detection technology for analyzing label-free biomolecular interactions in real time in a variety of different biomedical applications. This is a widely used technique that is extremely sensitive to changes in molecule-induced refractive indices. There are a number of exciting new advancements in fiber optics, including PCF. Because of their design flexibility, SPR-based PCF sensors have lately emerged as a viable use of PCF and the simplicity of modifying optical characteristics. Sensing performance can be improved by altering the analyte refractive index. A chemical sensor with microstructured SPR. The refractive index (RI) sensing optical fiber was developed for visible to near-infrared wavelengths as well as gas–liquid pollutant detection to realize coupling and manufacturing ease, inert gold with plasmonic characteristics, and analyte. The MOF exterior surface is coated with a sensing layer. The sensor is examined by the computer. The wavelength and amplitude interrogation methods, as well as the full-vector FEM, are used to determine the sensing properties. FEM is used to construct and analyze a new MOF-RIs sensor based on the SPR effect. Gold is a precious metal. Because of its chemical stability in aqueous conditions, it was chosen as a plasmonic material. Because of its flexible design, it was chosen as a plasmonic material. The MOF-SPR sensor provides remarkable sensing properties and industrial compatibility. Carbon dioxide concentration detection, ozone monitoring, and other possible applications. In the environmental field, water quality testing is done. The refractive index range is quite large, the sensitivity is likewise high, the linear correlation is closer to one, and the linear correlation is better. Thus, it has a certain significance in the future for biological and chemical sensing. Using the FEM, a PCF sensor based on gold nanowires is capable of sensing changes in SPR. The analyte of the sensor was placed outside of the optical fiber in order to permit real-time detection. The sensor-detecting qualities were examined in relation to the diameters of the air holes and gold wires. For liquids with a refractive index between 1.33 and 1.36, the sensor was created. Refractive index measurements can be made using a SPR sensor. The sensor structure and performance were investigated numerically using the finite element approach. The system layers and boundary scattering features were properly matched to absorb the emitted energy. Air-hole radius, gold line radius, and analyte refractive index impacts on confinement loss were all studied by optimizing structural characteristics for greater sensitivity.

References

Ademgil H, Haxha S (2015) PCF based sensor with high sensitivity, high birefringence and low confinement losses for liquid analyte sensing applications. Sensors 15(12):31833–31842

Akowuah EK, Gorman T, Ademgil H, Haxha S, Robinson GK, Oliver JV (2012) Numerical analysis of a photonic crystal fiber for biosensing applications. IEEE J Quantum Electron 48(11):1403–1410

An G, Hao X, Li S, Yan X, Zhang X (2017) D-shaped photonic crystal fiber refractive index sensor based on surface plasmon resonance. Appl Opt 56(24):6988–6992

Asaduzzaman S, Ahmed K (2016) Proposal of a gas sensor with high sensitivity, birefringence and nonlinearity for air pollution monitoring. Sens Bio-Sens Res 10:20–26

Asaduzzaman S, Ahmed K (2018) Investigation of ultra-low loss surface plasmon resonance-based PCF for biosensing application. Results Phys 11:358–361

Asaduzzaman S, Ahmed K, Bhuiyan T, Farah T (2016) Hybrid photonic crystal fiber in chemical sensing. Springerplus 5:1–11

Chaudhary VS, Kumar D, Kumar S (2021a) Gold-immobilized photonic crystal fiber-based SPR biosensor for detection of malaria disease in human body. IEEE Sens J 21(16):17800–17807

Chaudhary VS, Kumar D, Kumar S (2021b) SPR-assisted photonic crystal fiber-based dual-wavelength single polarizing filter with improved performance. IEEE Trans Plasma Sci 49(12):3803–3810

Chaudhary VS, Kumar D, Kumar S (2022a) Au-TiO_2 coated photonic crystal fiber based SPR refractometric sensor for detection of cancerous cells. IEEE Trans NanoBiosci 1–1

Chaudhary VS, Kumar D, Mishra GP, Sharma S, Kumar S (2022b) Plasmonic biosensor with gold and titanium dioxide immobilized on photonic crystal fiber for blood composition detection. IEEE Sens J 22(9):8474–8481

Chaudhary VS, Kumar D, Pandey BP, Kumar S (2023) Advances in photonic crystal fiber-based sensor for detection of physical and biochemical parameters—a review. IEEE Sens J 23(2):1012–1023

Chen Y, Ming H (2012) Review of surface plasmon resonance and localized surface plasmon resonance sensor. Photonic Sensors 2:37–49

Chopra H, Kaler RS, Painam B (2016) Photonic crystal waveguide-based biosensor for detection of diseases. J Nanophotonics 10(3):036011–036011

Cubillas AM, Unterkofler S, Euser TG, Etzold BJ, Jones AC, Sadler PJ, Russell PSJ et al (2013) Photonic crystal fibres for chemical sensing and photochemistry. Chem Soc Rev 42(22):8629–8648

Dash JN, Jha R (2015) On the performance of graphene-based D-shaped photonic crystal fibre biosensor using surface plasmon resonance. Plasmonics 10:1123–1131

Dash JN, Jha R (2016) Highly sensitive D shaped PCF sensor based on SPR for near IR. Opt Quant Electron 48:1–7

Ermatov T, Noskov RE, Machnev AA, Gnusov I, Atkin V, Lazareva EN, Gorin DA et al (2020) Multispectral sensing of biological liquids with hollow-core microstructured optical fibres. Light Sci Appl 9(1):173

Esfahani Monfared Y (2020) Refractive index sensor based on surface plasmon resonance excitation in a D-shaped photonic crystal fiber coated by titanium nitride. Plasmonics 15(2):535–542

Fan Z, Li S, Liu Q, An G, Chen H, Li J, Tian W et al (2015) High sensitivity of refractive index sensor based on analyte-filled photonic crystal fiber with surface plasmon resonance. IEEE Photonics J 7(3):1–9

Farka Z, Jurik T, Kovar D, Trnkova L, Skládal P (2017) Nanoparticle-based immunochemical biosensors and assays: recent advances and challenges. Chem Rev 117(15):9973–10042

Habib MA, Anower MS, Abdulrazak LF, Reza MS (2019) Hollow core photonic crystal fiber for chemical identification in terahertz regime. Opt Fiber Technol 52:101933

Haider F, Aoni RA, Ahmed R, Chew WJ, Mahdiraji GA (2020) Alphabetic-core assisted microstructure fiber based plasmonic biosensor. Plasmonics 15:1949–1958

Hao CJ, Lu Y, Wang MT, Wu BQ, Duan LC, Yao JQ (2013) Surface plasmon resonance refractive index sensor based on active photonic crystal fiber. IEEE Photonics J 5(6):4801108–4801108

Hasan MR, Akter S, Rifat AA, Rana S, Ali S (2017) A highly sensitive gold-coated photonic crystal fiber biosensor based on surface plasmon resonance. Photonics 4(1):18. MDPI

Hossain MB et al (2019) Numerical analysis and design of photonic crystal fiber based surface plasmon resonance biosensor. J Sens Technol 09(02):27–34. https://doi.org/10.4236/JST.2019.92003

Hossain MB, Islam SR, Hossain KT, Abdulrazak LF, Sakib MN, Amiri IS (2020) High sensitivity hollow core circular shaped PCF surface plasmonic biosensor employing silver coat: a numerical design and analysis with external sensing approach. Results Phys 16:102909

Hutter E, Fendler JH (2004) Exploitation of localized surface plasmon resonance. Adv Mater 16(19):1685–1706

Islam MS, Sultana J, Rifat AA, Dinovitser A, Ng BWH, Abbott D (2018) Terahertz sensing in a hollow core photonic crystal fiber. IEEE Sens J 18(10):4073–4080

Islam MR, Mou FA, Rahman MM, Hassan Bhuiyan MI (2020) Hollow core photonic crystal fiber for chemicals sensing in liquid analytes: design and analysis. Int J Mod Phys B 34(28):2050259

Jain S, Choudhary K, Kumar S (2022) Photonic crystal fiber-based SPR sensor for broad range of refractive index sensing applications. Opt Fiber Technol 73:103030

Kanmani R, Ahmed K, Roy S, Ahmed F, Paul BK, Rajan MM (2019) The performance of hosting and core materials for slotted core Q-PCF in terahertz spectrum. Optik 194:163084

Kaur V, Singh S (2017) Extremely sensitive multiple sensing ring PCF sensor for lower indexed chemical detection. Sens Bio-Sens Res 15:12–16

Kaur V, Singh S (2019a) Design of titanium nitride coated PCF-SPR sensor for liquid sensing applications. Opt Fiber Technol 48:159–164

Kaur V, Singh S (2019b) A dual-channel surface plasmon resonance biosensor based on a photonic crystal fiber for multianalyte sensing. J Comput Electron 18:319–328

Kaur V, Singh S (2020) Design of D-shaped PCF-SPR sensor with dual coating of ITO and ZnO conducting metal oxide. Optik 220:165135

Kaur B, Kumar S, Kaushik BK (2022) Recent advancements in optical biosensors for cancer detection. Biosens Bioelectron 197:113805

Kaur B, Kumar S, Kaushik BK (2023) Novel wearable optical sensors for vital health monitoring systems—a review. Biosensors 13(2):181

Kumar S, Guo Z, Singh R, Wang Q, Zhang B, Cheng S, Liu FZ, Marques C, Kaushik BK, Jha R (2021) MoS_2 functionalized multicore fiber probes for selective detection of shigella bacteria based on localized plasmon. J Lightwave Technol 39(12):4069–4081

Kumar S, Agrawal N, Saha C, Jha R (2022) Optical fiber-based plasmonic biosensors: trends, techniques, and applications. CRC Press

Lai WC, Chakravarty S, Zou Y, Chen RT (2013) Multiplexed detection of xylene and trichloroethylene in water by photonic crystal absorption spectroscopy. Opt Lett 38(19):3799–3802

Lee HW, Schmidt MA, Tyagi HK, Sempere LP, Russell PSJ (2008) Polarization-dependent coupling to plasmon modes on submicron gold wire in photonic crystal fiber. Appl Phys Lett 93(11):111102

Lee B, Roh S, Park J (2009) Current status of micro-and nano-structured optical fiber sensors. Opt Fiber Technol 15(3):209–221

Li C, Yan B, Liu J (2019) Refractive index sensing characteristics in a D-shaped photonic quasi-crystal fiber sensor based on surface plasmon resonance. JOSA A 36(10):1663–1668

Li T, Zhu L, Yang X, Lou X, Yu L (2020) A refractive index sensor based on H-shaped photonic crystal fibers coated with Ag-graphene layers. Sensors 20(3):741

Liu C, Wang F, Lv J, Sun T, Liu Q, Mu H, Chu PK (2015) Design and theoretical analysis of a photonic crystal fiber based on surface plasmon resonance sensing. J Nanophotonics 9(1):093050–093050

Liu C et al (2016a) Numerical analysis of a photonic crystal fiber based on a surface plasmon resonance sensor with an annular analyte channel. Opt Commun 382:162–166. https://doi.org/10.1016/j.optcom.2016.07.031

Liu C, Wang F, Lv J, Sun T, Liu Q, Fu C, Chu PK et al (2016b) A highly temperature-sensitive photonic crystal fiber based on surface plasmon resonance. Opt Commun 359:378–382

Liu M, Yang X, Zhao B, Hou J, Shum P (2017) Square array photonic crystal fiber-based surface plasmon resonance refractive index sensor. Mod Phys Lett B 31(36):1750352

Liu C, Yang L, Liu Q, Wang F, Sun Z, Sun T, Chu PK et al (2018a) Analysis of a surface plasmon resonance probe based on photonic crystal fibers for low refractive index detection. Plasmonics 13:779–784

Liu C, Su W, Liu Q, Lu X, Wang F, Sun T, Chu PK (2018b) Symmetrical dual D-shape photonic crystal fibers for surface plasmon resonance sensing. Opt Express 26(7):9039–9049

Liu M, Yang X, Shum P, Yuan H (2018c) High-sensitivity birefringent and single-layer coating photonic crystal fiber biosensor based on surface plasmon resonance. Appl Opt 57(8):1883–1886

Liu C, Fu G, Wang F, Yi Z, Xu C, Yang L, Chu PK et al (2019a) Ex-centric core photonic crystal fiber sensor with gold nanowires based on surface plasmon resonance. Optik 196:163173

Liu E, Tan W, Yan B, Xie J, Ge R, Liu J (2019b) Robust transmission of orbital angular momentum mode based on a dual-cladding photonic quasi-crystal fiber. J Phys D Appl Phys 52(32):325110

Liu Q, Sun J, Sun Y, Ren Z, Liu C, Lv J, Chu PK et al (2020a) Surface plasmon resonance sensor based on photonic crystal fiber with indium tin oxide film. Opt Mater 102:109800

Liu C, Wang J, Wang F, Su W, Yang L, Lv J, Chu PK et al (2020b) Surface plasmon resonance (SPR) infrared sensor based on D-shape photonic crystal fibers with ITO coatings. Optics Commun 464:125496

Liu W, Wang F, Liu C, Yang L, Liu Q, Su W, Chu PK et al (2020c) A hollow dual-core PCF-SPR sensor with gold layers on the inner and outer surfaces of the thin cladding. Results Opt 1:100004

Liu C, Wang J, Jin X, Wang F, Yang L, Lv J, Chu PK et al (2020d) Near-infrared surface plasmon resonance sensor based on photonic crystal fiber with big open rings. Optik 207:164466

Liu W, Shi Y, Yi Z, Liu C, Wang F, Li X, Chu PK et al (2021) Surface plasmon resonance chemical sensor composed of a microstructured optical fiber for the detection of an ultra-wide refractive index range and gas-liquid pollutants. Opt Express 29(25):40734–40747

Lu Y, Hao CJ, Wu BQ, Huang XH, Wen WQ, Fu XY, Yao JQ (2012) Grapefruit fiber filled with silver nanowires surface plasmon resonance sensor in aqueous environments. Sensors 12(9):12016–12025

Lu J, Li Y, Han Y, Liu Y, Gao J (2018) D-shaped photonic crystal fiber plasmonic refractive index sensor based on gold grating. Appl Opt 57(19):5268–5272

Luan N, Yao J (2017) A hollow-core photonic crystal fiber-based SPR sensor with large detection range. IEEE Photonics J 9(3):1–7

Luan N, Wang R, Lv W, Yao J (2015) Surface plasmon resonance sensor based on D-shaped microstructured optical fiber with hollow core. Opt Express 23(7):8576–8582

Luo W, Meng J, Li X, Xie Q, Yi D, Wang Y, Hong X (2021) Temperature effects on surface plasmon resonance sensor based on side-polished D-shaped photonic crystal fiber. Measurement 181:109504

Malka D, Cohen E, Zalevsky Z (2017) Design of 4 × 1 power beam combiner based on multiCore photonic crystal fiber. Appl Sci 7(7):695

Mashrafi M, Kamrunnahar QM, Haider F, Haider R, Aoni RA, Ahmed R (2021) Bio-inspired butterfly core-shaped photonic crystal fiber-based refractive index sensor. OSA Continuum 4(4):1179–1190

Mishra AK, Mishra SK (2016) Gas sensing in Kretschmann configuration utilizing bi-metallic layer of rhodium-silver in visible region. Sens Actuators B Chem 237:969–973

Mishra SK, Tripathi SN, Choudhary V, Gupta BD (2014) SPR based fibre optic ammonia gas sensor utilizing nanocomposite film of PMMA/reduced graphene oxide prepared by in situ polymerization. Sens Actuators B Chem 199:190–200

Mishra GP, Kumar D, Chaudhary VS, Kumar S (2022) Design and sensitivity improvement of microstructured-core photonic crystal fiber based sensor for methane and hydrogen fluoride detection. IEEE Sens J 22(2):1265–1272
Momota MR, Hasan MR (2018) Hollow-core silver coated photonic crystal fiber plasmonic sensor. Opt Mater 76:287–294
Morshed M, Hasan MI, Razzak SA (2015a) Enhancement of the sensitivity of gas sensor based on microstructure optical fiber. Photonic Sens 5:312–320
Morshed M, Hassan MI, Roy TK, Uddin MS, Razzak SA (2015b) Microstructure core photonic crystal fiber for gas sensing applications. Appl Opt 54(29):8637–8643
Otupiri R, Akowuah EK, Haxha S, Ademgil H, AbdelMalek F, Aggoun A (2014) A novel birefrigent photonic crystal fiber surface plasmon resonance biosensor. IEEE Photonics J 6(4):1–11
Patnaik A, Senthilnathan K, Jha R (2015) Graphene-based conducting metal oxide coated D-shaped optical fiber SPR sensor. IEEE Photonics Technol Lett 27(23):2437–2440
Paul BK, Ahmed K, Asaduzzaman S, Islam MS (2017) Folded cladding porous shaped photonic crystal fiber with high sensitivity in optical sensing applications: design and analysis. Sens Bio-Sens Res 12:36–42
Paul BK, Ahmed K, Vigneswaran D, Ahmed F, Roy S, Abbott D (2018) Quasi-photonic crystal fiber-based spectroscopic chemical sensor in the terahertz spectrum: design and analysis. IEEE Sens J 18(24):9948–9954
Peng L, Shi F, Zhou G, Ge S, Hou Z, Xia C (2015) A surface plasmon biosensor based on a D-shaped microstructured optical fiber with rectangular lattice. IEEE Photonics J 7(5):1–9
Rachana M, Charles I, Swarnakar S, Krishna SV, Kumar S (2022) Recent advances in photonic crystal fiber-based sensors for biomedical applications. Opt Fiber Technol 74:103085
Rifat AA, Mahdiraji GA, Chow DM, Shee YG, Ahmed R, Adikan FRM (2015) Photonic crystal fiber-based surface plasmon resonance sensor with selective analyte channels and graphene-silver deposited core. Sensors 15(5):11499–11510
Rifat AA, Mahdiraji GA, Shee YG, Shawon MJ, Adikan FM (2016) A novel photonic crystal fiber biosensor using surface plasmon resonance. Procedia Eng 140:1–7
Rifat AA, Ahmed R, Yetisen AK, Butt H, Sabouri A, Mahdiraji GA, Adikan FM et al (2017) Photonic crystal fiber based plasmonic sensors. Sens Actuators B Chem 243:311–325
Sakib MN, Hossain MB, Al-tabatabaie KF, Mehedi IM, Hasan MT, Hossain MA, Amiri IS (2019) High performance dual core D-shape PCF-SPR sensor modeling employing gold coat. Results Phys 15:102788
Sandeep V, Takasumi T, Ghanshyam S, Manish T (2016) Ultraflat broadband supercontinuum in highly nonlinear Ge11. 5As24Se64. 5 photonic crystal fibres. Ukrainian J Phys Opt 17(3):132–139
Sardar MR, Faisal M, Ahmed K (2021) Simple hollow core photonic crystal fiber for monitoring carbon dioxide gas with very high accuracy. Sens Bio-Sens Res 31:100401
Senthil R, Soni A, Bir K, Senthil R, Krishnan P (2019) Circular-pattern photonic crystal fiber for different liquids with high effective area and sensitivity. Plasmonics 14:1783–1787
Shandilya R, Bhargava A, Bunkar N, Tiwari R, Goryacheva IY, Mishra PK (2019) Nanobiosensors: point-of-care approaches for cancer diagnostics. Biosens Bioelectron 130:147–165
Sharma AK, Jha R, Gupta BD (2007) Fiber-optic sensors based on surface plasmon resonance: a comprehensive review. IEEE Sens J 7(8):1118–1129
Singh S, Prajapati YK (2019) Highly sensitive refractive index sensor based on D-shaped PCF with gold-graphene layers on the polished surface. Appl Phys A 125:1–7
Tan Z, Li X, Chen Y, Fan P (2014) Improving the sensitivity of fiber surface plasmon resonance sensor by filling liquid in a hollow core photonic crystal fiber. Plasmonics 9:167–173
Tian M, Lu P, Chen L, Lv C, Liu D (2012) All-solid D-shaped photonic fiber sensor based on surface plasmon resonance. Opt Commun 285(6):1550–1554
Tong K, Wang F, Wang M, Dang P, Wang Y (2018) Three-core photonic crystal fiber surface plasmon resonance sensor. Opt Fiber Technol 46:306–310

Wang Z, Tangella K, Balla A, Popescu G (2011) Tissue refractive index as marker of disease. J Biomed Opt 16(11):116017–116017

Wang G, Li S, An G, Wang X, Zhao Y, Zhang W, Chen H (2016) Highly sensitive D-shaped photonic crystal fiber biological sensors based on surface plasmon resonance. Opt Quant Electron 48:1–9

Wang Y, Li S, Li J, Zhang Z, Zhang S, Wu J (2019) High-sensitivity photonic crystal fiber refractive index sensor based on directional coupler. Opt Fiber Technol 49:16–21

Wang J, Liu C, Wang F, Su W, Yang L, Lv J, Chu PK et al (2020) Surface plasmon resonance sensor based on coupling effects of dual photonic crystal fibers for low refractive indexes detection. Results Phys 18:103240

Wong CL, Olivo M (2014) Surface plasmon resonance imaging sensors: a review. Plasmonics 9:809–824

Wu T, Liu Y, Yu Z, Peng Y, Shu C, Ye H (2014) The sensing characteristics of plasmonic waveguide with a ring resonator. Opt Express 22(7):7669–7677

Wu T, Shao Y, Wang Y, Cao S, Cao W, Zhang F, Wang Y et al (2017) Surface plasmon resonance biosensor based on gold-coated side-polished hexagonal structure photonic crystal fiber. Opt Express 25(17):20313–20322

Xu K, Huang L, Zhang Z, Zhao J, Zhang Z, Snyman LW, Swart JW (2018) Light emission from a poly-silicon device with carrier injection engineering. Mater Sci Eng B 231:28–31

Yan X, Li B, Cheng T, Li S (2018a) Analysis of high sensitivity photonic crystal fiber sensor based on surface plasmon resonance of refractive indexes of liquids. Sensors 18(9):2922

Yan B, Wang A, Liu E, Tan W, Xie J, Ge R, Liu J (2018b) Polarization filtering in the visible wavelength range using surface plasmon resonance and a sunflower-type photonic quasi-crystal fiber. J Phys D Appl Phys 51(15):155105

Yan X, Wang Y, Cheng T, Li S (2021) Photonic crystal fiber SPR liquid sensor based on elliptical detective channel. Micromachines 12(4):408

Yasli A, Ademgil H (2018) Geometrical comparison of photonic crystal fiber-based surface plasmon resonance sensors. Opt Eng 57(3):030801–030801

Zhang L, Guo H, Sun J (2018) Salt effects on the structural tailoring of layer-by-layer assembled polyelectrolyte complexes and salt-containing polyelectrolyte films. Thin Solid Films 653:258–266

Zhang W, Lang X, Liu X, Li G, Singh R, Zhang B, Kumar S (2023) Advances in tapered optical fiber sensor structures: from conventional to novel and emerging. Biosensors 13.https://doi.org/10.3390/bios13060644

Zhao Y, Deng ZQ, Wang Q (2014) Fiber optic SPR sensor for liquid concentration measurement. Sens Actuators B Chem 192:229–233

Zhao Y, Tong RJ, Xia F, Peng Y (2019) Current status of optical fiber biosensor based on surface plasmon resonance. Biosens Bioelectron 142:111505